ALPINE OPERATIONS

by

Mark V. Lonsdale

Additional Books by
Mark Lonsdale

SNIPER/COUNTER SNIPER
SNIPER II
SNIPER DATA BOOK
RAIDS
CQB (Close Quarter Battle)
ADVANCED WEAPONS TRAINING
SRT DIVER
BODYGUARD
ALPINE OPERATIONS

www.sttu.com

ALPINE OPERATIONS

by

Mark V. Lonsdale

DISCLAIMER

The author, STTU and those who contributed to this book take no responsibility for the use or misuse of the material herein, nor the loss damage, injury or death resulting from information contained in, or omitted from, this guide. ALPINE OPERATIONS should not be used as a how-to training manual nor a substitute for professional instruction, hands-on training and experience in a realistic training environment. Any individual or team venturing into the mountains or participating in climbing, mountaineering, rescue or tactical activities must be prepared to take full responsibility for their own actions and safety.

Alpine operations and mountaineering, as with tactical training, are potentially dangerous activities that could lead to serious injury or death, if not properly organized and supervised - and even when everything is done correctly, there is still a significant level of risk involved.

The training methods depicted in this book should be undertaken only by select personnel under the supervision of qualified instructors, team leaders and mountain guides. Each and every individual involved in alpine operations should act as a Safety Officer, and be constantly alert to any potential safety violations.

For economy, I have frequently used the generic "he", "him" and "his" throughout this book. I realize that there are many female climbers, mountaineers, rescue team members, law enforcement officers and special agents involved in various aspects of special operations, so where ever the reader sees reference to the male of the species, please read in "she" and "her" with equal weight.

Even though many official groups are discussed and depicted in this book, ALPINE OPERATIONS is not a publication or training manual of any government agency or military unit - nor should the material here in be taken as the policy or procedure of any group mentioned.

ALPINE OPERATIONS

First Printing January, 2000

ISBN 0-939235-07-2

Library of Congress Catalog Card Number: 99-97322

PRINTED IN THE UNITED STATES OF AMERICA

DEDICATION

To
Sir Edmund Hillary
A man who inspires so many

To the memory of
Rob Ngaira
NZSAS Mountain Troop

To the WWII veterans of the
10th Mountain Division
They laid the foundation for
modern mountain warfare

ACKNOWLEDGEMENTS

For the two years that I have been working on this book, many teams, agencies, companies and individuals have contributed their time, products and invaluable expertise to this work in progress. I would like to thank each and every one individually, however there are in fact too many to remember - but as I sit here doing the final edit, the following come to mind:

The staff and instructors of the USMC MWTC, and in particular Colonel O'Toole, Major Sawicki, Captain Tontini, Captain Yost, Sergeant Major Bressler, and Sergeant Clearman. Also special thanks to Colour Sergeant Molly Macpherson of the Royal Marines, and Sergeant Major Duke Speed.

Lt.Colonel Paro, Captain Nate Lord, SFC Robert Gordon and the staff instructors of the Army National Guard Mountain Warfare School

Los Angeles Sheriff Lee Baca, SEB Captain Cathy Taylor and the Deputies of the LASD Emergency Services Detail – Sgt Connolly, Sgt Thompson, Hitchcock, Wilkerson, Mitchell, Despues, Ascherin, Harrell, Hernandez, Airhart, Edwards, Hoffman, Jennings, McCabe, Rathburn, Sambolich and Vilander. The LASD Aero Bureau Air 5 pilots, particularly the steady hands of Dave Martin and Larry Stewart.

Captain Rick Homan and the Montrose Search & Rescue team

Robert SP Parker of the Sierra Mountain Center for his valuable input

James Frank of CMC Rescue for his contributions to the mountain rescue community

Mike Goldsmith of the National Wilderness Training Center for Law Enforcement (NWTC) for his hours of proof reading

Team Aconcagua: Major Mark Burger (USMC), Lt. Col Rich Sheppard (US Army), Lt. Col Dan Curfiss (USMC), Major Mike Morris (USMC), Captain Matt Marcenelle (USMC), and Colonel Eduardo Garate of the Chilean Army.

Gendarmeria Nacional de Argentina – Grupo Rescate E/27 and our liaison officer Lt Pablo Badie

The American Alpine Club for the use of their moutaineering accident reports

Doug Fesler and Jill Fredston of the Alaska Mountain Safety Center

The American Mountain Guides Association

The Mountain Rescue Association

Thomas Dietz, MD and the High Altitude Medicine Guide

Dr Glen Egstrom for his expertise in kinesiology and human performance

Ang Chuldim Sherpa of Explore Himalaya

Gordon Janow, Barbara Winkler and Vern Tejas at Alpine Ascents International

Assistance with equipment for evaluation and photography from Robbie Barrkman at Robar, Michael Slack at Leupold, Bill Wilson, Tim Davis at OR, Gary French at Gregory, Nick Yardley and Helmut Lenes at Climb High, Nick Allen at WL Gore, Malcolm Daly at Trango, Mike McDonald at Atlas, Kenny Glasscock and Geoff Lane at BD, Mike Grimm at Misty, Peter Gibbs, Bill Griggers at Omega, Dave Ashburn at PMI, Margaret Brown at SMC, John Carver at Eagle, Gary Richter at Garmont, Rick Halling at Kolflach, Charlie Oliver at La Sportiva, Paula at Jagged Edge, Georg Reichart at Silvretta, Willy Crear at Sterling, Jud Barr and Amy at Moonstone, Andy Marker at TNF, Scott Whipps at SD, Susan at Tecnica, Jim Foley at Leatherman, Justin at Petzl, Jennifer Gombas at MSR, Bill and Donna at CM, Josh at Pika, Richard Nelson at Dermatone, Doug and Maureen at SOS, Mike at Blackhawk, Brooke at Metolius, Steve Parola and Howard Wright at New England, Mike Hatrupp at K2, Debbie at Leki, Steve Hardesty at TUA, Jon Cox at Brunton, Bill Pegler at TEA, Michael Naughton at Kestrel, Benjamin at A16, Debbie Lazenby at Thorlo, James Markley at Smartwool, Lauren Zimmerman at Suunto, Rod Allen at AlpineAire, David and Renee at Danner, Hobbs Allison at Surefoot, and Theresa Neddermyer at Nonin Medical

Jeff Lowe and Teri at Arctic Wolf and Mike O'Donnell at San Juan Mountain Guides

Brant Didden at Mountain Adventures

Scott and Patricia Helvenston, Dan O'Shea and Billy Trolan

Photographic materials from Sgt Brent Clearman/USMC, James Frank/CMC, Rick Lipki/Conterra, Steve Dundorf, John Claussen/Atlas, SFC Gordon/USA, Sgt Alan Edwards/LASD, Charlie Shimanski/MRA, Chinook Medical, Lt Com Alderson and Patricia O'Connor/NSW, and the Royal Marines.

Photo shoot assistance from Jens Rubschlager, Lt. Bill Angione, Jimmy Romano, Tony Pennello, Dave Kilde and Sabrina, and computer graphics by Darlene Williams

Finally the crew at CP Graphics and in particular Kent Chesney, Allan Farson and Tracy Dragoo

My apologies to any that I have overlooked, and to all of you, my deepest thanks.

AUTHOR'S NOTE

I am often asked how long it takes to write a book, but it is a difficult question to answer since it is hard to put a specific date on when I actually begin the writing. To be an advisor and instructor one must also be a researcher, aggressively seeking out new methods and equipment to be evaluated. As such, it seems that I am constantly collecting information, developing data, learning from others and gaining experience that I try to keep track of in some form or another.

Over the years these small pieces of information are integrated into more formal files, papers, articles and lecture notes to be used in training but with no plan of writing a book on a specific subject. Then one day, driven by a need, the thought of writing a book rears its ugly head.

Knowing that my work had taken me to some of the more remote corners of the earth, in 1997 I was asked for my recommendation for an existing text that would help prepare special operations personnel for work in cold, mountainous terrain. But, after some research I found that there was no one book that would provide an operator or team leader all the information required. Granted, there are several excellent books available on various aspects of mountain travel, mountaineering and mountain rescue but none specifically written for military and law enforcement personnel who are, or could be, required to operate in an alpine environment.

About this same time, it was finalized that Salt Lake City, Utah would be the host of the 2002 Winter Olympic Games. I was again asked what information we had at STTU that would assist local and government law enforcement agencies to begin planning, equipping and training for an event of this nature.

In looking through my notes and files I found that indeed we did have a lot of information that would be helpful to both law enforcement and military personnel operating in mountainous or snow covered terrain. So at the end of 1997 I began officially working on Alpine Operations as a source book for agencies involved in or anticipating alpine or cold weather operations.

Like other STTU manuals, Alpine Operations is not intended as a stand-alone textbook, nor a step-by-step "how to" manual on climbing or mountaineering. It is designed for use before and after the actual training. When read before training it will give an individual or agency an overview of the equipment and training required for alpine operations and insight into the time and budgetary commitment necessary. After training it is a valuable resource that can be used to review subjects covered in training and to refresh the memory before actual operations. As an actual training manual it is designed for use in conjunction with a formal mountaineering text such as "Mountaineering—The Freedom of the Hills."

Alpine Operations is a compilation of my own personal experience and the experience of several mountain guides and mountain warfare instructors with whom I have had the distinct pleasure to work and train. I have also gleaned valuable pieces of information from other civilian and military manuals that are listed in the appendix.

Anyone required to work in the mountains will find this book useful. Even though several of the government and military agencies that this book is targeted at have distinctly different missions, the equipment and basic training required to operate safely and efficiently in the mountains is very similar. The basic principles of cold weather survival, mountain travel and alpine operations remain the same, as do the essential equipment requirements.

Lastly, Alpine Operations is formatted similar to other STTU texts in that the material is presented in the order that one would approach training or operations. Preparation, planning, equipment, training, and finally operations—all geared towards preparing the reader to operate more effectively in the unforgiving environment found above the snowline.

"Train with a Purpose" —Mark Lonsdale

CONTENTS

PART II—BASIC ALPINE TRAINING

PART III—ALPINE RESCUE

PART IV
MILITARY & LAW ENFORCEMENT OPERATIONS

APPENDIX:

INTRODUCTION

The mountains have come to symbolize many things to many people. For the average family looking for a weekend outing they represent peace, clean air and a place for a picnic; for the hiker or backpacker an opportunity to commune with nature; and for the fisherman it is the tranquility of clear unpolluted streams. Snow skiers look to the snow-capped mountains for a winter playground; snow-boarders for the thrills and spills of deep powder; while climbers and mountaineers see an endless chain of challenges that need to be met.

But mountains are more than just recreational wonderlands. For several countries they have historically represented isolation and protection, and yet for those with disputed borders they represent generations of ugly war and constant conflict. For the conventional army mountains are perceived as an obstacle and a danger to troops, but for the guerilla fighter a refuge - a place to vanish into the mist, to rest and regroup.

In some parts of the world, a trip though the mountains represents hope and freedom for the brave souls struggling to escape poverty or an oppressive regime. However, they are also a land of greed and opportunity for bandits, thieves and smugglers. Criminals, moonshiners, drug cultivators and poachers also see the mountains as a place to hide and to conceal their illegal activities.

Mountains also symbolize mystery. For the geologist, archeologist or botanist they are a place to be explored, probed and studied with the promise of secrets to be unlocked; and for the environmentalist, a sensitive eco-system to be defended from encroaching civilization.

For most the mountains are a place of pristine beauty and a celebration of the human spirit, but for others who have lost loved ones they cast a dark shadow of sadness and death. For all their breath-taking majesty, the mountains can be as cruel as they are beautiful—a deathtrap for the careless, the foolhardy and the ill prepared. But millions continue to flock to the mountains each year to escape the over-crowded gray world of the concrete jungles.

So with dramatic increases in wilderness use, comes an equally dramatic increase in the number of rescues and problems for law enforcement. Add to this the numerous military actions, border conflicts and guerrilla activity found in the mountain zones of the world, and it becomes inevitable that mountain-trained professionals will always be needed to venture into this unforgiving world of rock and ice to find the lost, assist the weary, rescue the injured, recover the bodies, enforce the laws, patrol the borders and keep the peace.

(Lonsdale)

PART I
PREPARATION

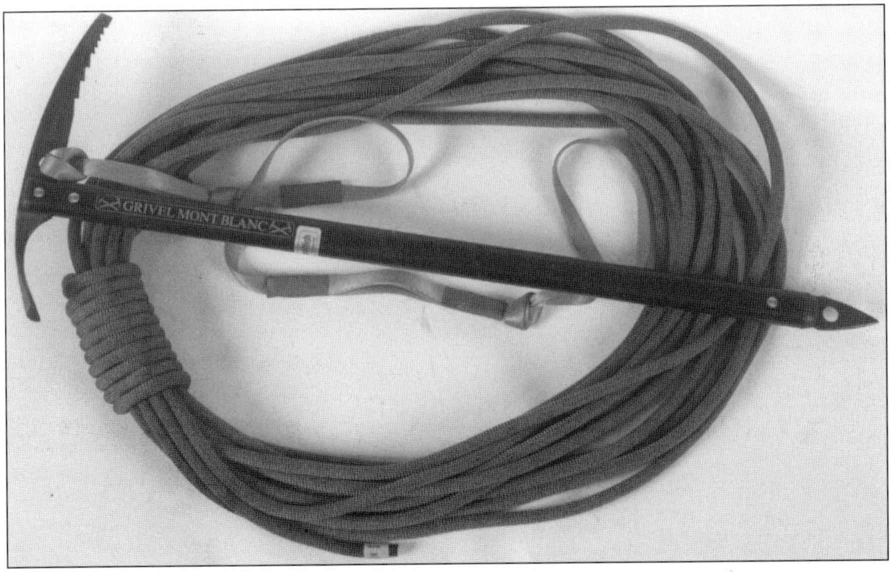

The barren terrain above 15,000 feet in the Himalayas

Summer snowline in the Northern Cascades at about 6,000 feet. In winter it can drop down as low as 2,000 feet.

1

THE ALPINE ENVIRONMENT

Picture a desolate, rocky, wind-swept place where any exposed skin will freeze in seconds; the wind is so strong it can blow a man off his feet; the air so thin that it barely supports life; getting dressed and putting your boots on in the morning takes over an hour and leaves you exhausted; the snow is so deep that movement is impossible; and in every direction, a three thousand foot icy drop into a dark abyss. Now imagine being trapped in a small tent, winds gusting to over 80mph threaten to tear it from the small ledge carved in the snow; it is thirty degrees below freezing inside and the tent walls are covered in ice. You have a splitting headache, a hacking cough, your throat is raw, you haven't slept well for several days, and to make things worse, your hands and feet are numb with cold. You have no energy, no appetite, your lips are cracked and you stink—but that's the least of your problems. You ate your last energy bar yesterday, you have no water, no fuel for the stove to melt snow, the batteries in your radio are dead, your tent mate is incapacitated with altitude sickness and the only thing blocking the wind outside is the frozen body of the third member of your climbing team. This is the high alpine zone that many climbers have come to know all too well.

The alpine environment can add considerable complexity to otherwise simple operations and is very unforgiving of errors. Poorly trained or ill-equipped individuals working at higher elevations can easily become chilled, fatigued, dehydrated, energy depleted and demoralized, and if not carefully monitored and corrected, can incapacitate a team causing an entire operation to falter and stall. So before a team ventures into the high mountains, they should first understand exactly what makes an alpine region such a beautiful, uniquely challenging and sometimes unmerciful place to work. Only through understanding the environment will they be able to equip and train accordingly.

Over forty percent of the earth's land mass is classified as mountainous, and topping those mountains we find the alpine regions, also known as high mountain zones. Some define these zones as the area above the snowline but the snowline can change radically from sea level in the arctic and sub-arctic regions to over 19,000´ (5,800 meters) in others. The actual altitude, or more precisely the elevation, where we find conditions associated with the alpine zone, usually above the tree line, can vary depending on geographic location and season.

Countries that are further from the equator, north or south, are also further from the sun. As such, they have a higher propensity for the cold conditions associated with snow-capped mountains. Similarly, regions that may enjoy a relatively high snowline in summer, for example the Andes and Himalayas, may be inundated in the winter with deep snow down to the lower elevations.

ALPINE OPERATIONS

In many areas, like the Northern Cascades of Washington, the snowline is generally found down between four and seven thousand feet (1,200—2,100 m), whereas in more temperate climates such as found in the Andes, bordering Argentina and Chile, one may not hit the snowline in summer until above 17,000 feet (5,200 m). This means that mountains that are easily negotiated in summer can become impassible in winter with trails and passes remaining blocked well in to spring and even early summer.

From an operational standpoint, this appreciation of geographic location, season and prevailing weather patterns is critical. For example, during the Falklands Conflict (1982) in the South Atlantic, a British SAS Mountain Troop was deployed directly from a ship, by helicopter, onto the inhospitable Fortuna Glacier, which runs from sea level all the way up to 10,000 feet. Even though well trained and equipped, it became quickly apparent that even at the lower elevations, they would not be able to sustain themselves in the frigid conditions made worse by howling sub-arctic winds.

Apart from human stupidity and poor judgment, there are five factors that make the alpine regions of the world some of the most challenging environments in which to operate.

- Remoteness of the Location
- Ruggedness of the Terrain
- Seasonal & Climatic Extremes
- Changeable Weather Patterns
- Thin Air at Elevation

For the recreational backpacker, trekker or climber, the severity of the conditions found in the mountains is minimized by the fact that they can choose where and when they will enter the mountains. If the weather looks like it is changing for the worse, they can beat a hasty retreat off the mountain. However, the rescue professional or soldier seldom has that luxury and must often work in the very conditions that have trapped others on the mountains. So it is only through better understanding of this pristine, changeable and sometimes harsh world above the snowline that one has any reasonable expectation of working and surviving.

TERRAIN

"Steep, rocky, snowy and icy" are the words that first come to mind to describe the terrain found in the high mountain zone. But in summer, it is also not unusual to find gently sloping valleys, large open areas and even fertile farmland where the snow has melted. It is this diversity that can lull some into a false sense of well being in the mountains.

Created by volcanic or geologic activity millions of years ago, an alpine zone could cover the relatively small area that encompasses a single peak or a massive alpine wilderness like the Cascades with scores of peaks over 8,000´, countless snowfields and over 400 glaciers. It is the size, gradient and elevation of this terrain that will have a direct affect on a team's work efficiency. Climbing steep mountainsides up to elevations of thin air and lower oxygen partial pressures can slow a team's movement to a snail's pace—especially if they are not in good physical condition or had the opportunity to acclimate.

But long before reaching the point where lack of oxygen is a factor, a team will have to move through many miles of rocky mountainous terrain with all their essential provisions and equipment on their backs. In addition, although the measured distance between two points in the mountains may not be significant when viewed on a map, the trail or route between those two points will not be a straight line. The route will more than likely follow terrain features, old animal trails, river beds, ridgelines and mountain passes. In addition to physical fitness, this requires competency in land navigation and terrain appreciation.

Even following the trails marked on a map is no guarantee of an easy route since they may vary from well-worn paths to loose scree, crumbling rock, huge boulder fields, deep snow and even near vertical rock faces. Rain, snow, ice, wind, rockslides and avalanches can change terrain features every season with the steeper, more exposed areas invariably being the most hazardous and a greater impediment to travel.

Deep snow is another perennial problem in the mountains since it can severely slow movement, fatigue the team, increase the incidence of cold weather injuries, contribute to snow blindness, reduce visibility and in blowing white-out conditions prevent any possibility of search, rescue or operational movement.

Mountains with significant snowfall will also have areas susceptible to avalanche conditions, and with approximately a million a year worldwide and over 100,000 in the US, the probability of having to operate in avalanche country is very high.

Fast flowing rivers are common in the mountains as illustrated by this spring run-off in the Himalayas at 10,000 feet

Moving through 16,500 feet where there is a total absence of vegetation

At 17,500 feet in the spring, the upper Khumbu Glacier is a desolate moonscape of rock and ice. In winter this would be covered in deep snow

SEASONS

Most people associate snow with the winter months, but at the higher elevations snow can be a year round occurrence. In fact, most recreational alpine mountaineering takes place in the spring and summer (May-August in the US) because of the low temperatures, harsh weather and depth of snow encountered in winter.

Mountaineering, in any season, exposes the climber to a wide variety of climatic conditions. Even in mid-summer, the high alpine region is usually still snow covered and produces daily temperatures below freezing, especially at night.

There are however, some advantages to operating in the winter if a team has suitable equipment and training. For one, the colder conditions make for firmer, more stable snow throughout the day. In warmer seasons the sun heats the snow making it wet, slushy and unstable to climb on, plus creating conditions conducive to avalanches. But in the Himalayas for example, the climbing seasons are spring and autumn, since the winter snowfall is just too deep to be navigable and summer is the rainy season when torrential rain and snowmelt turn streams into raging rivers making travel and climbing extremely hazardous.

WEATHER

The best descriptive word for the weather found in the mountains is "changeable", especially if affected by moist coastal weather patterns. In the low lands, weather is usually a daily constant, but on the exposed ridgelines of the mountains, weather comes in patterns that are somewhat predicable for the experienced weather watcher but dangerously changeable to the uninitiated.

Hikers and climbers lose their lives every year in the mountains because they fail to pack adequate clothing and supplies to handle inclement conditions. This is particularly true of alpine style climbers traveling light and fast and underestimating the speed or ferocity of an approaching weather front, only to find themselves huddled on the side of a mountain without a shovel to dig a snow cave, a stove to melt snow for drinking water or a bivi sack to keep them warm and dry.

However, being cold and wet are not the only dangers. The higher elevations are closer to the sun where the air is clearer and dryer, the atmosphere is thinner and the heat of the sun can be brutal, especially when reflected off of a snowfield or glacier. This increases the potential for severe sunburn, dehydration, heat stroke, and snow blindness for those caught without glacier glasses.

The weather in the mountains can vary from hot sunny days to freezing nights; from calm conditions to strong 100mph winds; from light rain to thunder storms; with heavy snowfall, dense fog and low clouds all thrown into the mix. Each requires a different level of protection, awareness and training.

On any given day, one can be sweating under a blazing hot sun that is suddenly obscured by low clouds then quickly deteriorates to rain, snow, and whiteout conditions. Then when the sun goes down, temperatures plunge well below freezing. Even in the milder summer conditions of the Pacific Northwest, the fast moving weather patterns can change almost on an hourly basis. It is not unusual to have several rain showers a day requiring climbers to constantly alternate layers between lightweight Poly-pro, fleece and Gore-Tex.

One of the most dangerous situations that this creates occurs when a climbing team leaves high camp for the summit in beautiful clear weather, but a few hours later finds themselves in total whiteout conditions. With an obscured view of their tents, they are left to wander in crevasse filled snow fields. (Unless they were smart enough to shoot compass bearings and lay down marker wands as they left camp earlier that morning).

Wind is another major concern in the mountains since it takes cold conditions and makes things many times worse with severe wind-chill factors, blowing debris, driven snow and reduced visibility. On Aconcagua (22,829´), where the snowline resides at over 17,000 feet in summer, it can be all hot sun and T-shirts during the day, followed by freezing nights with ice forming on the inside of the tents. A clear, windless day will give little warning of the approaching Viento Blanco (White Wind), the 80-100 mph winds that rage across the exposed slopes at night taking with it any tents or equipment not well anchored down.

One of the factors contributing to these changeable weather patterns is that high mountains are natural barriers that form a catcher's mitt for the warm moisture-laden clouds and weather patterns being pushed in off the ocean or rising up from the warmer lowland areas. In many alpine regions one can sit on a high mountain in the afternoon and watch these thick clouds accumulate far below, and then roll up valley until they hit the cooler temperatures at altitude. This moisture is then deposited in the mountains in the form of rain in warmer climates or snow if it is cold enough. These clouds can also bring storms that severely limit mobility and can trap a team for so long that their food runs out and they lose the energy or will to save themselves.

In any mountain region, just as heavy rain can dislodge rocks and cause rivers to rise to dangerous levels, heavy snowfall creates its own hazards in the form of unstable ridgelines, overhanging cornices and avalanche prone slopes. Heavy snow can also obliterate trails, bury essential equipment, limit visibility, impede movement and exhaust those trying to break trail.

Above: Aconcagua on a typical calm and clear afternoon. Below: Within one hour, summit teams are faced with total whiteout conditions, and by sunset there will be 80 mph winds hammering the high camps

ALTITUDE

Another important consideration in alpine operations is altitude (Covered in more detail in Chapter 15). Movement above 9,000´ (2,750 m) must be approached with caution and rapid ascent above 14,000´ (4,250 m) must be taken very seriously in terms of altitude adaptation.

The general rule when moving above 10,000 feet (3,048 m) is to gain no more than 1,000 feet per day and to take an additional rest day every 3,000 feet. But even following this procedure, it must be anticipated that some team members will be slowed or incapacitated by the thinner air and the resulting maladies such as Acute Mountain Sickness (AMS), High Altitude Pulmonary Edema (HAPE) and in extreme conditions, High Altitude Cerebral Edema (HACE).

If this happens, those individuals must be immediately moved to a lower elevation and carefully monitored for worsening symptoms since all can be fatal if ignored or left untreated.

Apart from mountain sickness, dehydration also raises its ugly head again at altitude. Higher altitude equates to drier air and more moisture loss through respiration, accelerated by heavy exertion that burns many calories and causes increased fluid loss through sweat and the body's attempts to maintain thermal equilibrium. The cold also acts as a diuretic adding further to fluid loss and major problems with dehydration, and all in a region where water is harder to find and the only source may be to melt snow, which requires both a reliable stove and an adequate supply of suitable fuel.

Having the knowledge to avoid a survival situation, and yet still being prepared for a worst-case scenario, is the key to surviving in the mountains. This means not just education, but also considerable training, good equipment and time-honored experience. People who have had a bad experience in the snow or dislike the winter can often attributed this to their lack of suitable clothing. As the Scandinavians are heard to say, "There is no such thing as bad weather, just bad clothing." This is somewhat over simplified but it does highlight the fact that with the right gear and training, the mountains can be a great place to work—but without, they are a death trap for the unwary.

"Mountains are not fair or unfair—
they are just dangerous."
—Reinhold Messner

Navy Seals undergoing cold weather training in Kodiak Alaska (NSWG)

2

ALPINE OPERATIONS AN OVERVIEW

Alpine operations can be defined as any professional endeavor that takes a team into the high mountain zone above the snowline. Recreational climbing expeditions and mountain guiding aside, these can be divided in to primarily three categories: Mountain Search & Rescue, Law Enforcement Operations, and Mountain Warfare.

Numerous county and state agencies maintain a mountain rescue team of some form with their standards of training and equipment influenced by the type of mountains within their jurisdiction, the number of call-outs and available budget. Most counties cannot afford the expense of a full-time dedicated alpine rescue team so become dependent on trained volunteer teams supported by a core group of sworn personnel.

One of the few federal agencies to have a professional level of expertise in the area of alpine rescue is the National Park Service with their mountaineering rangers who patrol most of the major mountains such as Denali in Alaska and Mt Rainier in Washington. Both of these parks see significant climbing activity each season and a corresponding number of rescues.

Outdoor wilderness tactical operations, particularly in mountainous, snow-covered terrain, are usually considered beyond the scope of conventional law enforcement, but units like the Royal Canadian Mounted Police have proven their capability in the mountains and frozen northern regions of Canada with a long history of tracking and apprehending fugitives from justice. Similarly in the US, outdoor operations have now become a common task for agencies involved in narcotic interdiction. DEA, DOJ, State Narcotics, and county Sheriffs have all become involved in locating and destroying marijuana patches and agencies like BATF get drawn in when illegal weapons and explosives are involved.

Border Patrol is another group than can be forced to operate in the mountains. Natural barriers such as mountain chains often form the borders between many countries and have been the scene of numerous military conflicts, with the current clashes in the Kashmir Mountains between India and Pakistan being a good example. Argentina and Chile have also had their share of problems in the Andes with bandits and smugglers.

In countries where the military does not handle internal law enforcement, civilian border patrol units often find themselves in an alpine environment trying to locate and apprehend the myriad of bandits, smugglers and illegal aliens trying to take advantage of more remote and inaccessible routes. The danger to these officers is compounded by the fact that they are often working in ones and twos, are miles from backup, and many alien and drug smugglers have shown no hesitation in using deadly force to protect their precious cargos.

ALPINE OPERATIONS

US Fish and Wildlife Service (USFWS) special agents routinely patrol high mountain zones looking for poachers and illegal hunting operations, areas where it is safe to assume that everyone they approach is armed. Some agents have even been shot in the line of duty resulting in massive manhunts for their killers.

National Park Service (NPS) also has rangers that patrol our national parks and wilderness preserves enforcing park regulations, protecting the environment and serving essential rescue functions. As with anyone in the wilderness, they can come into contact with dope growers, moon-shiners, poachers, fugitives, para-military survivalist groups, and a variety of other wackos with guns.

The fact that the United States will be hosting the 2002 Winter Olympics in Salt Lake City, Utah has also created a rush on outdoor clothing, equipment and training by a variety of local, state, and federal agencies that will have security responsibilities before and during the games. An event of this magnitude and public exposure must always be considered a possible target of terrorism, even if it is just the homegrown variety that we have seen in the Oklahoma City bombing. The FBI in particular has already begun equipping and training agents for this event, and the HRT has begun developing a more comprehensive cold weather capability.

When considering the development of a cold weather alpine operations capability for an agency, one must first look at the nature of man's approach to the mountains and mountaineering in general.

Man often finds it necessary to pit himself against adversity as a gauge of his own self worth, but in the mountains it is really not man against the mountain since the mountain just sits there and is not actively trying to stop him. Nor is it man against the elements since the weather patterns are usually a known variable. But when man ventures up into the alpine region lacking in conditioning, equipment and training, gambling that the weather will be perfect and he can return to camp before dark, then nature often finds it necessary to humble him—sometimes with fatal results.

Although equipment can be a contributing factor, more often than not, injury and death in the mountains are a result of poor training, poor judgement, lack of suitable physical conditioning or a combination of all three.

As with any tactical operation, preparing a team to operate in alpine terrain requires a systematic approach, and although the skills and equipment required to operate above the snowline are quite specialized, they also have application in many other theatres of operation that will become obvious later in this text.

Number One on the "Must Do" list is to seek out professional advice and guidance. Operating effectively and efficiently in the mountains is a specialized field of endeavor, but fortunately there are a large number of highly skilled and experienced people available in both the civilian and military arenas, from AMGA certified guides to seasoned mountain warfare instructors.

Number Two on the list is a careful study of the anticipated mission of the team and the proposed areas of operation. For without this focus, training time will be wasted and equipment found to be unsuited to the task.

Number Three—personnel. Define the manning requirements for the team, and then select command staff, training staff and operational personnel.

Number Four—equipment. Identify the needs of the team and put into place a process for the careful evaluation and selection of all necessary clothing and equipment.

Number Five—location. Select a suitable location to headquarter the team and preposition operational assets and equipment. Then identify suitable training areas with progressively more challenging terrain.

Number Six—training. Establish training progressions, standards, skill levels and the training time needed to meet the operational requirements of the mission.

Number Seven—budget. See if the funds are available for all of the above training and equipment.

SELECTION OF PERSONNEL

Once the feasibility study has been done and the funds are on line, the training can begin. To develop a long term, self-propagating program, it will take time to first evaluate and then train the in-house instructional staff in several areas of expertise. This period can also be used to evaluate the equipment and identify any deficiencies in the training program. Part of the initial instructor program should be to have the instructors help develop and formalize the curriculum and training aids for the first batch of students to follow.

In the initial phases of program development, the instructional staff will have to be recruited from the general population with special attention given to those with prior experience and in excellent physical condition. However, once the program is up and running, then future instructor candidates can be selected from the best of the students as they graduate training.

Whether it is mountain rescue or military operations, any individual aspiring to be an alpine operator must be evaluated first for fitness and endurance and then tested for aptitude and suitability for the exposure inherent in mountaineering. The larger the pool that the team has to draw from, the higher the probability that there will be individuals with prior mountain experience and aptitude, or the levels of fitness and commitment suited to mountain training.

Once an individual has been selected for initial training, and after having demonstrated a good grasp of basic individual skills, there must follow a period of group training to develop and perfect operational procedures and hone their ability to operate as a team. This training period is also used to further evaluate the individual's suitability for the team and to weed out the weak, the careless and the potentially dangerous.

The final part of any training program should be formal test and validation exercises based on real-world scenarios and in the anticipated area of operation. At the end of all this, and it may take several weeks or several months depending on the standard and prior experience of the individuals coming into the program, will be the final certification of team members.

EQUIPMENT CONSIDERATIONS

An alpine operator or mountain traveler must be totally self-sufficient when venturing into remote locations devoid of logistical support or re-supply. This makes mountaineering a very equipment dependent activity with many lives lost each year because of lack or loss of essential items.

Everything needed to live and operate in the mountains must be carried in a pack, and since weight has a direct effect on speed, comfort and efficiency, experienced professionals tend to carry only essential items. So if almost everything being carried can be considered important to survival, every single item of clothing and equipment should be selected and purchased as if it were life-support equipment.

Chapters 3 & 4 have lists of the clothing and equipment essential for alpine operations, but each item should be selected for not only suitability to the season and the mission, but also for fit and comfort. Something as basic as inferior or ill-fitting boots or socks can result in range of problems from minor blisters to severe frostbite, either of which can degrade the efficiency of a team in the mountains. Similarly, clothing or sleeping gear inappropriate for the local alpine temperatures can contribute to a number of thermal injuries ranging from heat stroke to hypothermia—both potentially life-threatening conditions. The failure of a tent, stove, or even a lighter could be the difference between life and death in inclement conditions. On a glacier or high angle climb, the lack of or failure of a crampon, a rope, a harness, one piece of hardware or protection could also lead to the untimely demise of an individual team member or an entire team.

However, even with the fittest personnel and the best equipment, the team is doomed to failure if both are not welded together with professional, realistic and frequent training.

Training is without a doubt the single biggest recurring cost within any organization both from the standpoint of dollars and man-hours, so before anticipating involvement in alpine operations, an organization or agency must ensure that they have the resources to support the training for an operation of this scope and magnitude.

(Training is covered in more detail in Part II).

US Military climbing team leaving the Refugio in TNF Kichatna jackets, fleece and Kolflach plastic boots

3

CLOTHING & FOOTWEAR

To survive in any environment, man requires warmth and shelter, but on the rocky side of a mountain, both are in short supply. Without ready access to shelter and warmth, the cold alone can disable and kill. To be caught in the open cold and wet is even worse. Add to this a wind-chill factor and survival expectancy is not good.

This makes clothing the first and most important line of defense for the climber. Therefore, clothing should be selected with the same care that one would select any "life support" equipment. The same is true of boots and socks. In remote locations, where walking is the only practical means of travel, having feet that are severely damaged by blisters or frostbite can render one immobile and incapable of self-rescue.

CLOTHING SYSTEMS

Dressing for success in the mountains is not a simple matter of selecting a single suitable item of clothing, but rather assembling a "clothing system." This system must give the individual the flexibility to adjust to rapidly changing weather conditions, dramatic temperature shifts as well as protection from wind, sun and rain. Weight and durability are also important factors.

It may sound complicated but isn't in actuality. There is no big secret to putting together a technical clothing system since it is a simple process of layering oneself with the correct fabrics and design features. But even though the axiom of "Wick, Warmth and Wind," is well known in the outdoors community, sadly many still venture into the mountains each year with a poor selection of clothing and end up as statistics.

There are several fabrics and clothing designs that may be fine under ideal conditions, but in inclement conditions the problems with these quickly add up. For example, those favorite cotton T-shirts that hold sweat and keep us cool on a sunny day, are slow to dry so will only accelerate heat loss when the sun goes down and the wind picks up. Similarly, lightweight nylon windbreakers that work well to cut the wind don't breathe at all, and with exertion will hold the sweat against the body so that it cannot evaporate, resulting in further chilling and heat loss. So the T-shirt and windbreaker that were fine on a sunny day, not only offer no thermal insulation at night, they hold moisture that accelerates heat loss—a deadly combination in the sub zero temperatures that can be encountered on a mountain.

A good technical clothing system suitable for alpine operations must be able to transport sweat away from the body, dry quickly, supply adequate warmth while allowing for ventilation and still give protection from wind and rain. To achieve this, professionals use a system made up of three primary components: an underwear layer to wick moisture away from the body; a fleece or pile layer for thermal protection; and an outer shell layer that also breathes to allow body moisture to evaporate.

Ice covered climbers on Cotapaxi, Ecuador (SP Parker)

Cotton clothing is not recommended for alpine operations but can be worn on long hot approaches and in summer conditions where sweat will have time to dry and temperatures are not expected to plunge suddenly

UNDERWEAR—THE WICK LAYER

The purpose of the underwear layer is to provide some comfort and thermal protection but primarily to wick moisture away from the surface of the body. To achieve this, most alpinists will carry two sets of lightweight long underwear, consisting of separate tops and bottoms of a synthetic fiber such as polypropylene (aka Poly-Pro), or wool but definitely not cotton.

Lightweight grade underwear is used for its versatility since it can be worn singly in warmer conditions and double layered for colder conditions, but underwear is also available in Medium and heavier Expedition weights for trips where the conditions warrant.

Other desirable features include zippered high neck tops, which allow for more ventilation and long sleeves for protection against the sun. Some professionals also recommend one set of white underwear for intense sunny days on the glacier and one set of dark colored underwear for faster drying. In an effort to travel light, many climbers will make their approach in only their underwear with just a pair of shorts over the top.

Marmot DriClime and Patagonia Capilene both have excellent reputations and North Face Micronamics offer a high performance base layer in Aerobic or Alpine weight. These use a special bi-component construction that allows freedom of movement while fitting closely and retaining body warmth

Like many climbers, AAI guide Vern Tejas negotiates a steep alpine approach in just synthetic underwear

Special attention must be given to protection from both direct and reflected sun light

THE THERMAL LAYER

The key to thermal protection is a loose layer of clothing that traps a layer of air that the body then heats. The fabric should limit the circulation of air but still allow moisture from sweat to be expelled. For many years, wool has been the traditional thermal layer of the alpinist and is still an excellent insulator, but the best value for weight in this category are synthetic pile or fleece jackets and pants.

The pile jacket should be a Mid- to Heavyweight pile such as Polartec 200-300 depending on individual cold tolerance, and should fit fairly loosely to give a full range of motion but without being too bulky. A full-front zippered version is easier to put on and has better ventilation than a pullover and underarm zippers will also aid in ventilation when exerting on a climb.

The pile pants should be Midweight, Polartec 200, with full separating side zippers that allow the pants to be donned without having to remove boots and crampons. This is important for not only ventilation but also for ease of dressing up or down when conditions change in the middle of a climb or operation.

Some fleece and pile products are treated with DWR (durable water repellent) for better wear and weather resistance, while others are utilizing Windstopper technology to make them more suitable as outer wear. Gore Windstopper is a knit or fleece laminate with a wind-stopping membrane that is designed to cut wind and reduce wind-chill when worn as an outer garment. In dryer climates, a pile or Windstopper vest is an excellent addition to the clothing system since it leaves the arms free for a wider range of movement while maintaining body core temperature.

In colder regions and at higher altitudes a down jacket will be another essential item. Even though it must be kept dry, down is still the best lightweight insulation available to climbers and for extreme high altitude climbing a one piece technical climbing suit may even be required.

North Face Windstopper fleece jacket and OR Flex-Tex Granite climbing pants make a very practical combination for work and mountaineering

Moonstone's HP Sweater in BiPolar Polartec 200 with underarm zippers for ventilation

Moonstone's Momentum Tech Parka made from lightweight 3-ply rip-stop Gore-Tex with reinforced elbows and shoulders

SHELL/OUTER-WEAR

For protection against rain and snow, Gore-Tex shell pants and a shell jacket with hood are essential. The jacket should have a long front zipper; a draw-cord hood that will accommodate a climbing helmet; a roomy fit for extra layers of clothing; "arms-up" construction for ease of climbing; and underarm zippers which go well below the armpits. These zippers are an essential part of the ventilation system and allow the climber to cool down without removing the jacket, a major hassle when wearing a pack and harness.

A well-designed hood should fit over a climbing helmet, have a drawstring closure and allow good range of side-to-side movement of the head.

The pants must have full length separating side zippers so that they can be donned and doffed without removing boots or crampons. For mountain rescue work and similar rugged use, North Face, Mountain Hardwear, Sierra Designs and Moonstone all make pants or bibs with extra reinforcing material, such as Cordura, in the knees and seat.

Gore-Tex sets the standard for outerwear with its waterproof yet breathable fabric that is designed to keep water out while allowing body moisture to evaporate and permeate out through the material. Jackets and pants are available in 2 or 3 ply construction and although strong and durable, they can still be cut by the careless use of an ice-axe or crampon spike, and burned if allowed too close to the stove or fire.

Several companies make excellent Gore-Tex jackets in various weights and for various applications. North Face's Mountain Light, Sierra Designs Quantum and Moonstone's Momentum Tech parkas are all widely used and the heavier Kichatna and Mountain Guide are popular with rescue teams. The military issue Extended Cold Weather Clothing System (ECWCS) consists of camouflaged Gore-Tex Best Defense parka and pants and the Gen II models adopted by the Marine Corps have all the desirable features of top-of-the-line technical climbing wear.

Author in TNF Mountain Light Gore-Tex jacket. Alpine Ascents guide Barbara Winkler wears a Patagonia shell jacket.

Underarm vents on military Gen 2 ECWCS Gore-Tex parka

Sierra Designs Quantum Parka made with lightweight Gore-Tex ripstop twill

Down jackets like this North Face Summit model supply superior warmth at altitude

ADDITIONAL CLOTHING

The above items represent the Big Three in alpine clothing systems and the standard for most operations. However, depending on the type of operations, location and climate, there are other items that can be added or substituted, especially when temperatures are predictably warmer or working below the snowline.

As stated earlier, to save on weight, many mountaineers will climb in their long underwear when it is warm and only add the pile and shell when needed. In striving for the minimum weight, no extra clothing is taken to altitude except that which is essential, but for long approaches at lower elevations and when working in the mountains on a daily basis, something more substantial and practical than just underwear is required.

One very versatile piece of clothing is stretch climbing pants, which offer more protection from abrasion than underwear with less bulk and more freedom of movement than pile pants. Outdoor Research offers two excellent products in this area that we use extensively. The first is the OR Flex-Tex Granite Pants made from Spandura with double layer knees, zippered and gussetted ankles and 5 zippered pockets. The second, OR Flex-Tex Rhyolite Tights are also made from Spandura, giving 50% stretch in every direction with the strength of Cordura and the stretch of Lycra. North Face has the El Cap and Big Wall pant in the same category and First Light offer good climbing pants made from electro-static nylon with reinforced knees and ankle zippers.

Moonstone's sleeveless windstopper pullover ideally suited to climbing and work in warmer conditions

Another item that should be considered, especially in warmer regions, is long sleeved shirts. Several outdoor clothing companies now make synthetic, quick drying lightweight shirts that are very practical for work below the snowline, on long approaches and used primarily for protection from the sun.

Fleece vests like this model from SD supply core warmth but leave the arms free for work or climbing

GLOVES & MITTS

Keeping the hands warm and functioning in the mountains is of critical importance and many a mountaineer has lost his life or fingers for no other reason than losing a glove. This happens all too often when a climber removes a glove to tie a rope or open a zipper, only to drop the glove and see it slide away into the abyss—a situation that could have been prevented if the gloves were attached to the jacket with an "idiot cord".

Without gloves the hands quickly become cold and then cease to function, especially when working in cold, wet snow or icy conditions. Without protection and re-warming, the hands eventually become

French Peloton de Gendarmerie de Haute-Montagne climbing team traveling light and fast in fleece jackets and stretch climbing pants

The Jagged Edge down vest is super lightweight and packs up into a very small package

stiff, numb, frozen and succumb to frostbite leaving the climber disabled and unable to handle ropes or move down the mountain.

Depending on the temperature and conditions, the recommendation for all mountaineers is to carry two pairs of synthetic gloves, either bunting or fleece, which will fit comfortably inside mitts or waterproof gloves. The heavier fleece gloves do a better job of keeping the hands warm when wet than lighter polypropylene or Capilene gloves, which are more suited to warmer regions.

For more extreme conditions, the climber may also need a pair of pile mitts and Gore-Tex over-mitts. The inner mitts allow the fingers to hold heat and warm each other while the over-mitts keep the hands and pile mitts dry. Modular glove systems also work well and even though not quite as warm, are much easier to work in.

Outdoor Research makes excellent Modular Mitts made from a 3 layer Cordura Gortex shell with a tough palm that is more abrasion resistant and has a higher melting point than that used on the Expedition Mitts. These also come with a warm liner, a 1" cinch strap, an extra long gauntlet and removable idiot cords. For more dexterity, OR's line of Professional gloves are more suited to mountain work, and the military has opted for the OR Mutant Mitt which has a three finger configuration.

Another glove system that we have found excellent for a broad range of alpine operations is the Black Diamond Ice Glove system that has both a warm pile liner and a lined shell with padded knuckles. This is without a doubt one of the best ice climbing gloves that exist. They do an excellent job of keeping the hands warm in freezing conditions when working hard, even without the liners installed. For spring and summer conditions, the BD Light Weights also give excellent dexterity for climbing and handling hardware.

We have also evaluated several other gloves at STTU for their suitability to mountain operations and have identified those that have useful features. The neoprene Glacier Gloves are good for wet work but must be worn quite loose and with liners to retain heat. Charlet Moser wool Glacier Gloves have proven effective in ice climbing since they will actually adhere to the ice, and TNF Windstoppers are good all around travel gloves when not working in wet snow.

Since the hands and therefore gloves are so important in mountaineering, it is recommended that operators always carry spare gloves and liners and to have outer gloves and mitts connected to the jacket or wrists with an "idiot cord". In an emergency, improvised gloves can be made from a spare pair of socks and a couple of small stuff sacks or plastic bags.

HEAD & EYE WEAR

There are three elements that the head needs to be protected from in the mountains, each requiring different headgear—the cold, the sun and miscellaneous gravity powered missiles (rocks, ice and hardware).

BD Ice gloves with padded knuckles and detachable liners are excellent all around gloves and suited to more than just ice climbing

Since a disproportionate amount of heat can be lost through the head, it is essential that mountain operators have a warm hat of wool or pile, and a balaclava to give more complete protection. For extreme conditions where little or no skin should be exposed, the OR Gorilla Balaclava made with Gore Windstopper fleece is totally windproof and comes with a removable facemask with nose protector and a mesh exhaust port.

For sun protection, most climbers will use a ball cap in conjunction with a bandanna to add protection to the neck and ears. Even better are the broader brimmed floppy hats found in many outdoor stores that offer more protection, are lightweight and quick drying.

When moving in a vertical environment of rock or ice, a lightweight climbing helmet is another essential. The helmet should be adjustable so that is can accommodate the balaclava with the Petzl Ecrin Roc being one of the most popular in this category. The helmet is also used in conjunction with a headlamp like the Petzl Zoom or Princeton Tec for night movements.

Glacier Glasses are another essential for protection from the blinding glare of the sun reflecting off the surface of the snow. They should be 100% UV, high quality optical lenses designed for mountain use, with side covers and no more than 6% light transmission. Glasses like the CEBE Ultimate 40 are an excellent choice with 100% UV protection and filtering of 100% IR and absorption of 92%-

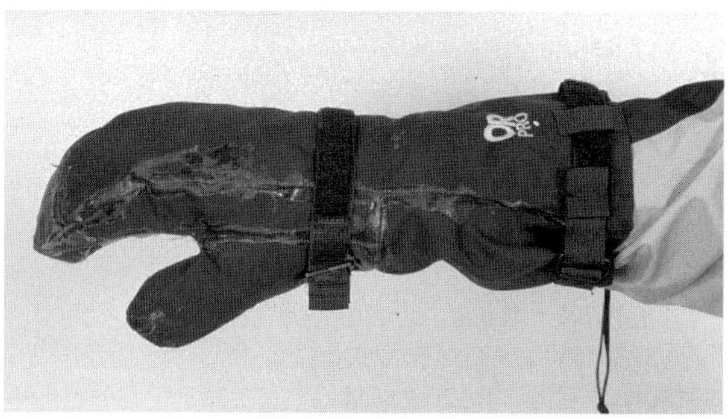

OR Pro series modular mitts

97% visible light. Bolle, RayBan and Julbo are other options and for climbers who wear glasses it is recommend that they have prescription glacier glasses made up in gray or green and always carry a back-up pair.

OR Gorilla mask for really harsh conditions on the mountain

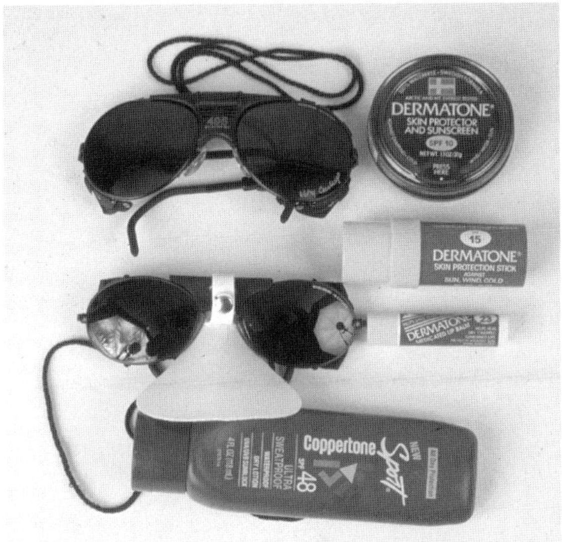

Cebe glacier glasses, Coppertone Sport sunblock and various Dermatone lip and skin protections—all are essential for mountain travel

FOOTWEAR & FOOT CARE

As with clothing and gloves, footwear suited to alpine operations is a "system" and not a single item. Socks, boots, foot-beds, liners, gaiters and crampons must form a mutually compatible system where each component contributes to both function and comfort. It is not sufficient to simply go out and buy the "best" of everything, if in fact they are not designed to work with the other items or could create any discomfort at all.

The boots form the heart of the system and boot fitting is so important that it would not be difficult to write an entire book just on this subject, but that would be impractical for this text and more than a little tedious for the reader.

BOOTS

The first thing to know about boots is that there is no one boot that is suited to all operations in all seasons, so it should come as no surprise to learn that most mountain professionals have several pairs of boots that they rotate throughout the year.

Second, there is no way around the fact that mountaineering means walking—lots of walking; hundreds of miles of walking; often days and weeks of walking. Whether on a mountain rescue team, in a military unit, or just out to bag a few peaks, the alpinist must be prepared to put more than a few miles on his or her boots, and it is often those boots that make or break the aspiring mountaineer, since nothing is more demoralizing than miles of suffering with bad footwear and the resulting painful blisters or frostbitten toes. Without a doubt, more operations have been aborted and individuals med-evaced for no other reason than blisters and tortured feet.

The importance of a good pair of boots cannot be over emphasized since they support the very foundation of our structure and are the interface between our feet and the rock or ice. They are our means of transportation and as important as good tires on a car.

In selecting boots, it is not the color, brand, model, or fashion that rules, but FUNCTION, FIT and COMFORT.

Since there is so much variance in not only shoe size but also individual foot shape, there is no one brand or model of boot that will fit all people. Feet can range from wide to narrow; from low insteps to high insteps; from high arches to no arches; some people pronate while others supinate; and some feet are just plain deformed as a result of old injuries or congenital defects.

For these reasons, boots are best obtained from a reputable store with a staff that has been professionally trained in mountaineering boot selection and fitting. Many outdoors shops also have a policy that they will replace not only defective boots but also boots that simply do not fit the wearer, even after they have been out on the trail.

Not all boots are suited to hiking or climbing. Work boots for instance are designed to protect the feet from falling objects and a lot of standing around, so are generally not suited to long distance working. Some military boots are well suited to walking, but the downside is that they are not of the highest quality since they are usually procured from the low bidder—the company that can build them for the government at the cheapest price.

Just as there is no one boot that will fit all people, there is no one boot for all jobs. For the individual operating in all four seasons on varied terrain, it will be necessary to have at least two or three pairs of boots. Light hikers for summer; sturdy leather boots with good ankle support for rough terrain; insulated boots for winter snow and ice; and double plastic mountaineering boots for prolonged operations at altitude and in wet snow. Additionally, one may want a pair of light hikers or cross-trainers for around camp, and most serious rock climbers will have special friction shoes designed just for rock climbing, or stiffer soled shoes for edging and big wall work.

So boot selection becomes a balance between application, temperature and weight. There is no point in wearing more boot than is needed for the conditions since the added weight will only add to fatigue, and insulated boots in the middle of summer may be uncomfortably hot. However, when selecting boots for a specific application or expedition, they must be suited to the worst conditions expected. For instance, climbing in the Northern Cascades in June, one may leave the parking lot at 2,000′ in hot, sunny conditions facing a five-hour climb up through the forest to the snowline. By 5,700′ you are in wet, knee-deep snow and a thousand feet later stepping onto a glacier. So although lightweight cross-trainers may be suitable in the trees, and heavy plastic boots are less than ideal for approach work, insulated, waterproof mountaineering boots are essential if you plan to spend more than a day on the glacier.

Similarly, when working in an alpine environment, it is not unusual to be traversing talus and boulder fields one hour and then be into snow and ice the next, with temperatures ranging from uncomfortably hot during the day to well below freezing at night.

So, at a minimum, most individuals who work in the mountains will have three pairs of boots. A pair of light hiking boots for approach work and wear around base camp; a sturdier pair of mountaineering boots for four season work up to the snow line; and finally a pair of double plastic mountaineering boots for extended travel or work on snow and ice.

Light Hiking Boots:

All boots must be well fitted and comfortable for prolonged wear over rough terrain but the lightweight hiking boots will undoubtedly be the easiest

Nike Exum lightweight hiking and climbing boots

to find and fit. Several companies like Nike, La Sportiva, Salomon, Merrell, Garmont, Raichle and now North Face all offer a wide range of high tops and light hikers made from a combination of leather and synthetics. But even lightweight boots should be purchased a half size to a size larger than normal shoe size to accommodate a suitable outdoor sock system and the natural foot swelling that may come with heat and altitude.

For military personnel, instead of a light hiker they will use combat boots and if the issue boot is not suitable they should consider purchasing an approved uniform boot that is more suited to mountain warfare. The boot that is most popular with many mountain warfare instructors is the Danner Acadia that comes in both an insulated version for cold weather and an un-insulated version for warmer climates. The Acadia Insulated is an 8" tall boot with full grain, water-proof leather and a Cambrelle nylon lining. There are 200 grams of Thinsulate in this US Marine Corps approved boot that also includes a 100% waterproof Gore-Tex bootie and a Vibram Kletter Sole that resists trapping mud and dirt.

The US military approved Danner Arcadias are available with or without insulation

Matterhorn also manufactures a similar insulated combat boot that is a popular alternative to issue boots.

Mountaineering Boots:

The boots that will get the most use, especially for mountain rescue teams, are the heavier leather boot that is suited to all four seasons below the snow line and for short duration operations above the snow line.

This boot should be of sturdy construction, offer good ankle support, be crampon compatible, and stiff enough for climbing while flexible enough for walking. The amount of insulation will be dictated by the location and anticipated operating conditions.

One of the most popular boots in this category is the La Sportiva Makalu which is a medium volume boot that blends a one piece leather upper with a crampon compatible Vibram sole. Also in this category is the Salomon Super Mountain 9.

For colder conditions, we recommend insulated boots like the Karakoram, K3 or Nepal Extreme also from Sportiva, or the Altitude Plus or Everest XT from Tecnica. Tecnica is using primarily synthetics in the construction of their boots such as Keprotec (Kevlar composit), with a quick drying Cambrelle

Two excellent examples of approach mountaineering boots. Tecnica left and La Sportiva Makalu on the right.

lining and Quattro sole which helps control pronation and supination plus utilizing a thermo-plastic rubber insert in the heel area for further heel strike absorption.

Double Plastic Boots:

Once a team commits to prolonged operations above the snowline then plastic mountaineering boots offer all the necessary warmth and waterproof characteristics. These should be good quality plastic shells with Vibram type soles and removable insulted inner boots. They should also be large enough to allow for a comfortable fit with heavy mountaineering socks.

The two key advantages of double plastic boots are that the plastic shells are totally waterproof and the insulated liners can be taken out and dried in the sun during the day or in the sleeping bag at night. The moisture in the liners coming not from the snow but from natural foot perspiration.

Plastic boots are rigid so it is critical that they fit correctly from the start. Unlike leather boots, there is no break-in period, however, if you have an odd shaped foot or you cannot find a perfect fit, there are boot doctors who can heat the plastic and stretch it away from contact points. Another potential hot spot with plastic boots is under the large hinge rivet that allows the ankle to flex. This area can also be heated and pressed out to reduce pressure and improve comfort. For most people, it is best to simply buy over sized boots and use two or three pairs of socks to get the optimum fit.

The advantage of double plastic boots is that the liners can be removed and dried in the sun or taken into the sleeping bag at night

La Sportiva K3 insulated leather boot and Kolflach Para plastic boot with removable insulated liner on right

Koflach offers a wide variety of plastic boots to fit all needs beginning with the Viva Soft and Degre all the way up to their top of the line expedition boot, the Arctis Expedition. One of their most popular boots with the military is the Kolflach Para which was designed to meet the demands of a high alpine work boot. The Para has a synthetic leather liner with perforated fleece; Y technology for walking comfort; and excellent grip from a special Vibram sole. Kolflach can boast yearly sales of over 70,000 pairs to alpine military units in France, Germany, India, Pakistan, Afghanistan and Chile, plus the USA.

Two other brands of plastic boot that also share an excellent reputation with both recreational users and professionals are the BD Scarpa Inverno with a Pebax upper and EVA insulated inner boot, and the Lowa Civetta Extrem which utilizes a Duratherm liner that resists compression and will not expand at altitude.

BOOT FITTING & INSERTS

The human foot is a complex combination of tendons, muscles, ligaments and 26 odd shaped bones, and even though boots come in a variety of sizes and widths, the probability of finding a PERFECT fit is very low. A boot that feels good when first tried in the store may well become an instrument of torture after humping a pack for several hours in mountainous terrain. Plus the flimsy factory supplied foot beds break-down leaving a flat base that results in a poor fit, with loss of support, comfort and performance.

The first step toward happy feet is to take your time in selecting a new pair of boots and to do it well in advance of a planned expedition or operation. The next step is to be measured and fitted in a store that has a good selection in stock and offers professional boot fitting by a trained staff.

When buying boots, be prepared to take a few hours to try several sizes and styles since it can take an hour or so before a slight pressure spot turns into a "hot spot". Keeping in mind that feet swell with heat and at altitude, plus the fact that you will be carrying a heavy pack, and that a minor hot spot in the store will become a major aggravation in the mountains. By wearing new boots around the store for an hour or so, it will give the trained sales person the opportunity to correct any problems or suggest another size or style. If they do not have the exact size you want, do not compromise. Have them special order the size or sizes that you would like to try.

Plastic boots being heated and jacked out for a better fit on wide American feet

Boots should be tried with the same sock combination that you intend wearing in the mountains. It is not unusual to go a half size larger on lightweight boots worn with a medium to thick sock, or a full size for boots worn with two pairs of socks. There is also variance in European and US sizing and often differences in sizing even within the range of boots from the same manufacturer. European boots are generally low to medium volume, while the average American foot is wider and more suited to medium to higher volume boot fitting.

Heavier leather mountaineering boots and especially plastic boots do not stretch and break in like a soft street shoe or sneaker so they must fit correctly from the start. Boots intended for use in the snow should be on the big side to accommodate additional socks, and it is not acceptable to just jam your feet in with more socks on, since it is the airspace within the socks that contributes to the warmth more than the socks themselves. By jamming more socks into the boot you are only reducing insulation and restricting circulation.

When in doubt about the exact size, it is recommended to go with the slightly larger size and then compensate with thicker socks and a liner.

Another important aspect in boot selection is to ensure that the foot does not slide forward and hit the inside of the boot's toe box when walking downhill. On long downhill treks, this will become extremely painful and can turn the toes black causing loss of toenails. To prevent this, the boot must be correctly laced to hold the foot securely back into the heel pocket but at the same time not restrict circulation.

FOOTBEDS

Footbeds are an essential part of the boot fitting process and after-market ones will add more support than the thin ones supplied with most boots. Good after-market footbeds are designed to add comfort and support, so are not soft and mushy like regular insoles. Most have a well-designed heel cup that stabilizes and benefits balance while absorbing and evenly distributing heel shock. They should also have a biomechanical arch support that will relieve fatigue and realign the foot to a natural position for more efficient energy transmission.

At first, new footbeds will make the boot feel snug, but within a few days they will actually conform to the shape of your foot. Because previous flat or soft insoles may not have supported your foot properly, the tendons and muscles of your feet will need time to adjust to being held in the proper position. By wearing them for a few hours each day they will begin to conform to the foot adding both comfort and performance.

SUPERfeet make their footbeds from a combination of durable, closed-cell foam with a molded plastic stabilizer capsule, and a friction reducing top cover. The Trim-to-Fit feature makes them the easiest and most economical way to improve boot fit and performance and are recommended by most outfitters and retailers.

The people at Surefoot go a step further with their custom orthotics. They have a foot scanner that, through pin registration, takes up to 538 separate measurements of the contour of the foot, computerizes this into a color foot density map and then sends it to an in-store mill center to produce a truly custom machined foot bed in less than 30 minutes. The resulting ethyl vinyl acetate (EVA) orthotic

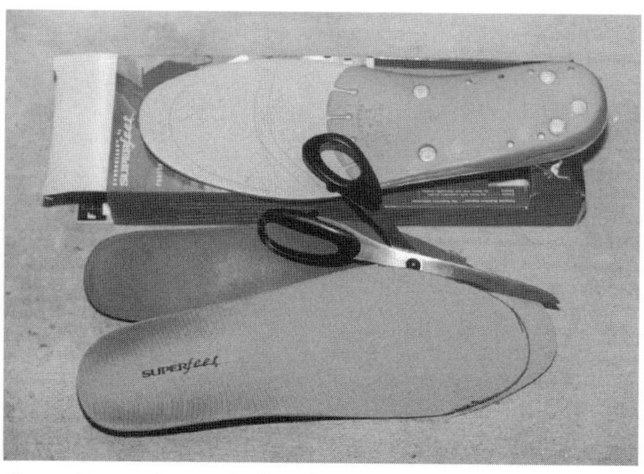

SuperFeet trim-to-fit foot beds are widely used for their support, convenience and affordability

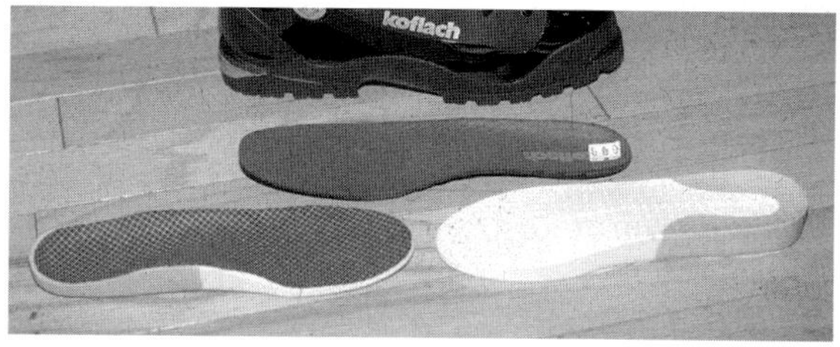

Flimsy stock inserts at back; Surefoot blank before milling on right; and completed custom orthotic footbed on left

Surefoot technician, having just taken pin registration readings of a client's foot, sends this information to the footbed milling machine

exactly matches the sole of the foot, giving essential ache support and positioning the foot for optimum performance. Pressure points are eliminated and previously unsupported areas of the foot are stabilized. These are obviously not cheap but can be used in several pairs of boots and the professionals swear by them.

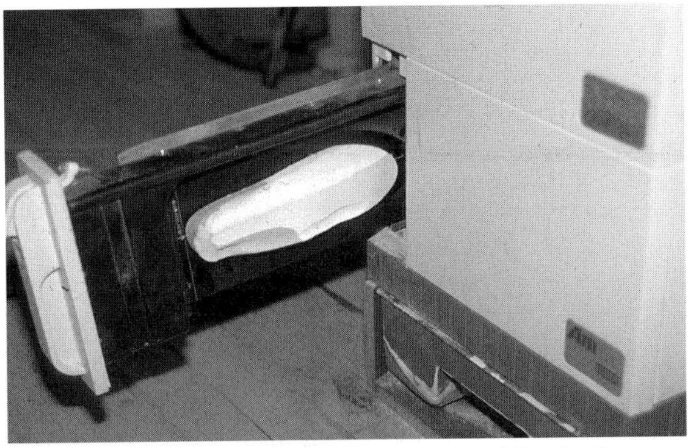

Surefoot mill machines out a contoured footbed designed for maximum support

SOCKS

When confronted with the range of socks on display at many stores, sock selection can be almost as confusing as boot selection. It is not sufficient for manufacturers to just have light and heavy socks, they now have specialized socks for each activity—Light Hiking, Hiking, Light Trekking, Trekking, Skiing, Climbing, Mountaineering and Expedition Weight. In addition, they offer everything from lightweight silk liners to bulky battery heated socks. There is also the big question of wool or synthetic or a blend of both.

Like the clothing system, socks form the Wick and Warmth part of the foot system with the boot being ideally the outer waterproof layer. But the manufacturers also produce Gore-Tex and similar barrier socks (SealSkinz) to form a waterproof layer over the socks and inside the boot.

To simplify all this, the mountaineer essentially needs smooth liner socks to wick moisture away from the feet and reduce friction; under heavier wool or blend socks for warmth and additional foot protection.

Most climbers will carry three pairs of heavyweight wool or synthetic socks to be worn over the liner socks. At any given time you would be wearing one pair, have one pair clean, dry and ready and the third drying. The three pairs of liners should be smooth, thin wool, nylon or Capilene and are also rotated so that a dry pair is always kept next to the skin to reduce the chance of blisters.

The socks that we favor at STTU are the Thorlo Mountain Climbing sock and the Smartwool Mountaineer. Both hold up very well and have cushioned heel and toe areas. For summer and lighter applications, both companies make excellent trekking weight socks and thinner liners.

Socks are an important part of keeping the feet healthy in the mountains and every opportunity should be taken to replace wet socks with a dry pair and to keep the feet equally clean and dry. Socks can be dried on the outside of the pack when moving during the day or taken into the sleeping bag at night. Where it is impractical to wash socks, they can be simply dried in the sun and shaken vigorously to remove any dirt and dead skin. On longer rest periods and around camp it is also beneficial to air out your feet and dry them in the sun.

For extreme conditions, many climbers will opt to add another layer to the sock system and that is a

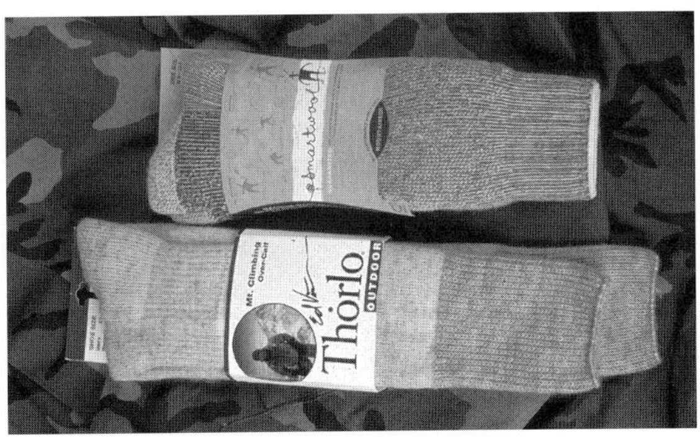

Thorlo and Smartwool Mountaineering socks

VBS—vapor barrier system. These can be worn between the liner and the thick sock and although the feet may become moist from sweat, the heat is contained on much the same principle as a wetsuit. But as soon as one stops for the day they must be immediately removed, the feet dried and dry socks put on.

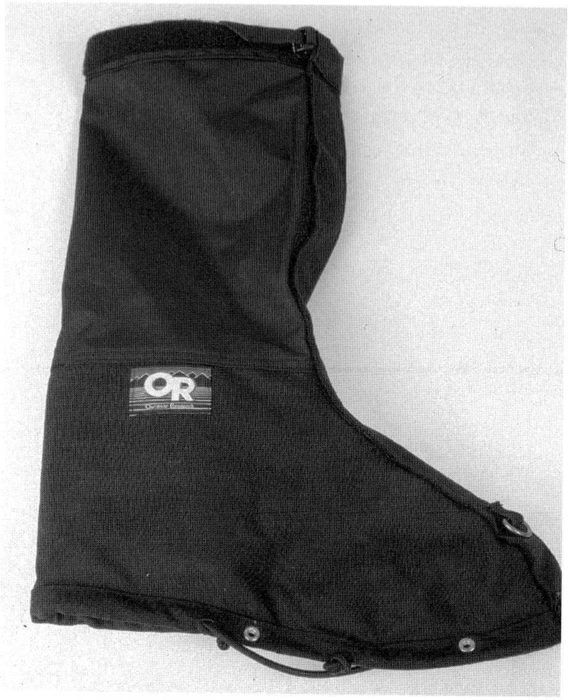

OR's insulated X-Gaiter

GAITERS

The outermost layer of the footwear system is the gaiters which are designed to further waterproof and insulate the boots and feet. The primary value of gaiters is found when walking in deep snow. By extending up the leg and fitting the boot snugly, they prevent snow and water from getting in the top of the boots and soaking the socks and feet.

Outdoor Research "Crocodiles" are the recommended gaiter for most applications and the standard by which most other gaiters are judged. For professional use, we recommend the OR HDs—a heavier duty version of the Crocs and an extremely durable fitted gaiter. They have Cordura Gore-Tex in the leg section and a 1000d Cordura lower section lined with Packcloth. They also have field-replaceable instep straps; cam buckle top closure; 2″ wide Velcro front closure; and heavy-duty lace hooks.

For extreme conditions where more insulation is called for, such as found on Mount McKinley (Denali), Alaska, most guides will recommend supergaiters that are randed, glued or shock corded to the boots. The OR X-Gaiters are one of the warmest Supergaiters available with 3 layer Taslan Gore-Tex upper; 1000d coated Cordura lower section; and a shock cord X under the boot. These come complete with two different forms of insulation—3/8″ EVA foam being the best choice when crampon or snowshoe straps are used since it has a high resistance to compression. When compression isn't an issue, the softer 1/2″ urethane foam provides better insulation.

CRAMPONS

Although technically classified as hardware or climbing gear, crampons are also the final part of the boot system. They are the indispensable sharp spikes that allow safe movement on steep and slippery ice.

One of the most important features of crampons is that they stay firmly attached to the boots when climbing on or traversing steep ice, making boot compatibility a critical consideration.

The preferred crampons for most applications are flexible, as opposed to rigid, and come with "step in" type bindings and flat rather than "cookie cutter" frame rails. The twelve-point configuration is most popular for general mountaineering with several more specialized models available for technical ice climbing (see Chapter 11).

For boots with a crampon compatible toe-bail feature, the Charlet Moser Super 12, the Grivel G-12, and the Black Diamond Sabretooths, which are also compatible with the Atlas Summit Series 33BD snowshoe, all work well. For boots with a rounded toe and no bail slot, the Grivel New-matic will fit the bill. For a crampon that works well for both vertical ice climbing and mountaineering, the more rigid Grivel Rambo with forged front points is a winner.

All crampons need to be periodically checked for cracks, the points sharpened and the screws kept tight. It is also an item that is good to standardize for a team so that one spare parts kit can service the whole team. When not in use, the climber must be careful about carrying and storing crampons as their sharp spikes can cause considerable damage to other equipment and people.

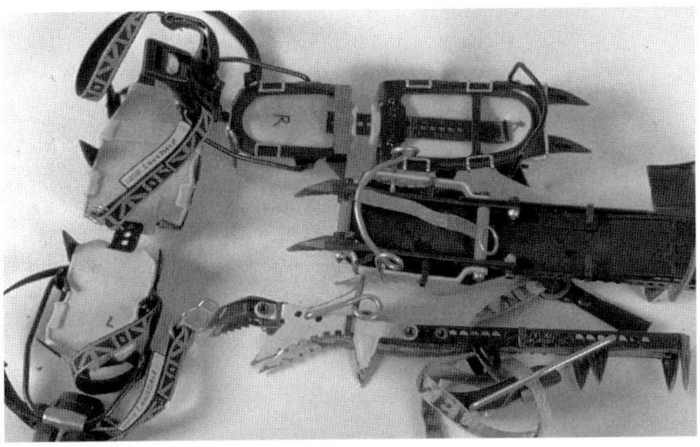

Charlet Moser S12 crampons with anti-balling rubbers on left; Grivel Rambos on right

Grivel Rambos with aggressive toe points suited to more technical ice climbing

BD Sabertooth crampons are compatible with the Atlas BD33 snowshoe

Grivel New-matics, mounted here on Sportiva Trangos, are designed for boots that do not have a slot for a toe bail

FOOT CARE KIT

Blisters can be caused by ill-fitting boots; boots that are too loose or too tight; an irregularity in the boot construction; wrinkled socks, or dirt or pebbles getting into the boot. Since prevention is always preferable and less painful than cure, boots must be well fitted, broken in, comfortable and worn with the correct socks. Similarly, keeping the feet clean and dry and toenails trimmed will also reduce the need for a blister repair kit.

But blisters are a fact of life in the mountains and even if you are fortunate to not get them, others in your party will, so it is essential that certain items be carried specifically for foot care.

The key to foot maintenance is early identification of the problem and to tape the "hot spot" before it worsens into a nasty raw blister, or even worse, becomes infected and debilitating.

Most serious trekkers and climbers will carry some form of Molefoam, Moleskin, Second Skin or Spenco Adhesive Knit and athletic tape for blisters, but in a pinch, duct tape also works pretty well. How much of this one carries will depend on one's predisposition to blisters and the distances to be traveled. Most mountain guides know to come well prepared with ample supplies for their less experienced climbing and trekking clients.

In treating blisters and hot spots, avoid using common Band-Aids since they usually do not stick well to sweaty skin and will roll and bunch until they aggravate more than they help. Where blisters have already formed, a padded "dough-nut" placed around the blister and taped in place may ease discomfort and prevent it from rupturing.

Other useful items for this kit would include antibacterial wipes to clean the blisters and possibly something like iodine or beta-dine to ward off infection.

Equipped for snow travel with TNF Gore-Tex shell jacket and pants, OR gaiters, Leki poles, snow shoes and a good supply of wands

The One Sport Everest boot available through Climb High is the ultimate in warmth for high altitude climbing. Even with Aveolite insulation, a removable inner boot and an integrated breathable Gore-Tex upper gaiter, the Everests at 6.25 pounds a pair, are still lighter than most other boot/supergaiter combinations

4

GENERAL EQUIPMENT

Mountaineering requires not only a high degree of fitness and training, it is also an equipment dependent activity. This means that without the essential equipment, any possibility of success or even survival is limited.

For the purposes of this book, the equipment required for alpine operations can be divided into four basic categories:

- General wilderness camping equipment
- Mountaineering equipment
- Rescue equipment
- Tactical equipment

The last two will be covered in parts III & IV of this book and mountaineering equipment is covered in the following chapter. Therefore, this chapter is an overview of the equipment essential to 4-season hiking and camping in the wilderness.

In developing an equipment list, we come back to the inescapable fact that everything must be carried in a pack, so each item must be essential and should be considered carefully using the following criteria. Equipment should be:

- Essential to the Mission
- Lightweight & Compact
- Quality Construction & Materials
- Reliable & Field Maintainable
- Readily Available & Reasonable Cost

The easiest way to begin developing a list of specifics is to look at what the "big boys" are using—the local professionals with years of experience in the mountains. Failing that, jump onto the World Wide Web and download the recommended equipment lists of reputable mountain guides and guiding operations who are running expeditions in regions and conditions similar to your area of responsibility. More seasoned members of a team should also be encouraged to share their knowledge and experience.

The first ten items listed below are considered essential for survival in the wilderness and no one should venture into the mountains without them. If all hikers and climbers would carry just these ten items, the number of missing persons and stupid, pointless deaths would drop dramatically.

THE BIG TEN

1. Water bottle(s) & Water Purification
2. Knife
3. Compass
4. Matches and/or Lighter
5. Emergency Space Blanket
6. Small First Aid Kit
7. Whistle & Mirror for signaling
8. Flashlight w/ batteries
9. Emergency Clothing
10. Extra Snacks or Energy Food

All of these items will fit into a very small back- or butt-pack, and the minimal weight far outweighs the inconvenience. But these few items are not sufficient to sustain an operator on a multi-day operation, so from here we move into the essentials like packs, sleeping gear and tents.

BACKPACKS

The backpack, like a good pair of boots, must be chosen with care, especially considering the amount of time that it will be on your back and the loads that must be carried. It must be selected primarily for FIT and COMFORT.

Also like boots, there is no one pack that is suited to all purposes. There are expedition packs, day-packs, technical climbing packs, mid-range packs, trauma packs, butt packs, military packs and tactical packs. However, this selection can usually be narrowed down to two general design applications. The first is the large primary multi-day expedition or approach pack, and the second, a smaller one- or two-day assault or summit pack. For a rescuer this would be his light-duty rescue pack and for a military operator a patrol pack.

Gregory Robson Pro is a full sized mountaineering and expedition pack. Note the Ridge-Rest and Therma-Rest sleeping pads strapped to the outside

The large primary pack should have an internal frame with approximately 5,000-6,000 cubic inches capacity and not weighing more than about 7 pounds (lighter is better). It should be comfortable with a 65-80 pound load, even though "comfortable" may be a relative term when suffering up a hill under 80 pounds of gear. It is best to select a pack that is ruggedly simple with no unnecessary zippers, pockets and features which all add weight, however, external cinch straps and attachment points for crampons and ice tools are a necessity.

It is recommended to get as much inside the pack as possible and avoid items dangling on the outside, so the pack needs to be large enough to hold the sleeping bag, tent, food, clothing and cooking gear, along with any mission required equipment. The only items on the outside should be sleeping pads, ice axe(s), snow shovel, and possibly a helmet or wet rope. Even for short duration rescue operations, the operator will still need enough room to carry rope, hardware and medical supplies on the inside, in addition to sleeping bag and bivi sack, should he be caught in inclement weather and have to overnight it.

Once the cubic capacity and weight of the pack have been considered, the next step is fitting. Most modern packs have many more adjustment options than the older models that could be considered torture racks by modern standards, so fitting becomes much easier. The two key measurements needed for pack fitting are back length and waist measurement, both of which should be done in accordance with the manufacturer's recommendations. The goal of correct fitting is to move the bulk of the load off of the shoulders and to transfer it down to the lumbar region. To do this while wearing the pack, the climber should be able to shrug his or her shoulders up-ward, cinch down the waistband, and then lower the shoulders leaving the weight on the top of the hipbones.

A good example of a ruggedly built pack that

meets most needs for both expeditions and rescue work is the Gregory Robson Pro (5,000-5,900 cu in. depending on size), the slightly smaller version of the Denali Pro (6,500-7,000 cu in.) seen on the IMAX Everest documentary. The Robson has also been selected as the foundation of the new special operations pack being considered by the US military.

Like many modern pack systems, the Pro series packs come with an ergonomically molded and shaped shoulder harness and waist pads, along with several adjustment points to better conform to the shape of the human body. However, the Pro series are top of the line packs so if cost is an important consideration for a team, then the Shasta is an excellent alternative in the same size range and equally as rugged.

The second type pack to be used as a lightweight summit or rescue pack, should have a capacity of at least 3,000 cubic inches and be comfortable with about a 35-pound load. Anything smaller will not hold all the necessary rope, hardware and supplies essential for this application. Popular packs in this category are the Dana Bomb Pack and Gregory Gravity which have several features suited to both recreational and rescue work. In addition to two tool loops, reinforced ski loops and hardware loops, the Gravity has a large front compression pocket that is ideally suited to hold a snow shovel, Gore-tex or to stow a wet rope.

The 3,000 cubic inch Gregory Gravity is ideally suited to rescue work and one day operations

Even though weight is an important consideration in almost every piece of equipment, professionals should be somewhat cautious about the ultra-lightweight expedition packs built for hard-core alpinists. By using very light parachute-cloth fabrics such as rip-stop nylon and eliminating the frame, it is possible to make a full-sized pack under 4 pounds, but these are not that comfortable to carry and would probably not hold up to the year round daily abuse of alpine operations.

SLEEPING GEAR

Sleeping warmly and comfortably in the mountains also requires a "system" approach, and like boots, the more time one spends involved in alpine activities, the more sleeping bags one will accumulate. The three criteria in selecting a sleeping bag are:

- Temperature Rating
- Filling / Synthetic or Natural
- Size / Design

First and foremost, a bag should be selected that is temperature rated for the anticipated area of operation and expected climatic conditions. This could range from –40 degrees Fahrenheit (–40 degrees C) in the harshest arctic conditions, to +20 degrees F (–6 degrees C) for most three season mountain activities. Keep in mind that even +20 F is still twelve degrees below freezing (32 degrees F / 0 degrees C).

We have found that bags in the 0 degree to –10 degree F rating fit most of our four season, winter and high altitude needs, especially if used in conjunction with a tent or bivi sack.

The insulating quality of a bag is derived from both the design and materials used, but primarily from the loft or thickness of insulating material packed within the bag itself. A bag with a 775 fill of down like the Marmot Cwm may be rated at –40 degrees F, and a bag with a 600 fill of down will only be rated to +15 or 20 degrees. It is important to note at this point that a warmer bag is considerably thicker, bulkier, heavier and more expensive, so one generally does not want to invest in more bag than one needs or is willing to carry.

It is also important to match the bag to your own personal sleeping patterns. Some people sleep hot while others chill very easily while sleeping. This is directly related to a combination of body fat, muscle mass, diet and metabolic rate. So one man may be very comfortable in a +10 degree F bag while his thinner female climbing partner may require a zero or –10 degree bag for the same conditions.

To further improve insulation, most alpine bags are of a mummy design with a full hood and usually with extra insulation around the foot box where circulation is the poorest. Another desirable feature is an insulating flap under the zipper to prevent heat loss.

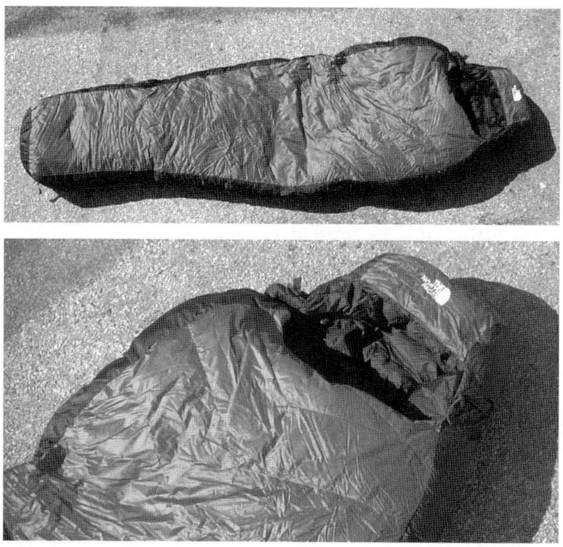

TNF Solar Flare down-filled bag with Dryloft outer shell

The second important decision in bag selection will be Synthetic or Down fill. Goose down is undoubtedly warmer, lighter and packs up smaller, but if wet loses it's insulating qualities and is very slow to dry. Synthetic filled bags like North Face's Darkstar 3D filled with Polarguard, are better suited to conditions where the bag may get wet since the synthetic fill will retain insulating qualities and is much faster to dry. Polarguard is also very durable and compressible.

To make down bags more resistant to moisture from snow and condensation, several companies offer an optional outer layer of Dryloft (i.e., the –20 degree Solar Flare), which not only repels water but also raises the insulating value of the bag by about five degrees. However, if used in conjunction with a Gore-tex bivi sack, this may not be necessary.

The third selection criteria is size. Most bags come in sizes ranging from small and regular to long and extra long depending on the height of the customer. They can also be cut wider in the shoulders for men or wider in the hips for women.

Sleeping bags are designed in either a tapered mummy configuration or cut square to the bottom. The square cut bags are more comfortable to sleep in since one can move around more freely, but that comfort means more pack size and weight. So the mummy bags are the first choice for most applications but a roomy bag like Sierra Design's Paul Bunyon is great for extended stays in base camp.

The roomier Sierra Designs Paul Bunyon synthetic 3D Polarguard bag (35″x88″)

This particular bag has a polyester lining to wick away moisture and extra insulation in the foot box and torso areas where warmth is most needed. The advantage of two bags on expeditions or operations supported by porters or mules is that the warmer mummy bag can be ferried up to high camp while keeping the bulky more comfortable square cut bag at base camp.

SLEEPING PADS

The next layer in the sleeping system is the sleeping pads that both protect the bag and insulate the occupant from the snow and rocks below. Rather than one large thick pad, most climbers carry two lighter pads. One three-quarter length closed-cell foam Ridge-Rest and one three-quarter length inflatable Therma-Rest for both warmth and comfort while sleeping on snow.

The inflatable Therma-Rests offer more insulation and comfort than the Ridge-Rest, but they can also be punctured. The rugged little Ridge-Rests can be laid under the Therma-Rest to better protect it and also make a very handy seat when stopping to rest or preparing food. The Ridge-Rest can also be carried on summit day, in conjunction with a bivi sack to be used in a hasty bivi or if trapped by weather.

BIVI SACKS

A good bivi sack is far more than just a cover for the sleeping bag—it is an essential part of the mountain survival system.

In situations where you would not want to carry a six or eight pound tent, you can at least throw a lightweight bivi in the pack for emergencies. Whether making an early morning summit bid or moving fast and light in rescue mode, the mountaineer must always consider his survival options should he be caught in a storm or delayed by injury or other acts of nature.

Several reputable manufacturers offer a variety of bivi sacks but OR's Standard Bivi really is the standard—at 1.25 pounds (567g) this lightweight, no pole bivi is perfect for that unexpected night in the snow. These are made with a Ripstop Gore-Tex fabric upper; Hyrdoseal-200 coated taffeta floor; fully taped seams to keep water out; No-see-um mosquito netting; a high volume foot section and large zipper opening.

More substantial bivis are available, complete with flexible poles to keep the bivi away from the face while sleeping, and essentially have all features of a one-man tent.

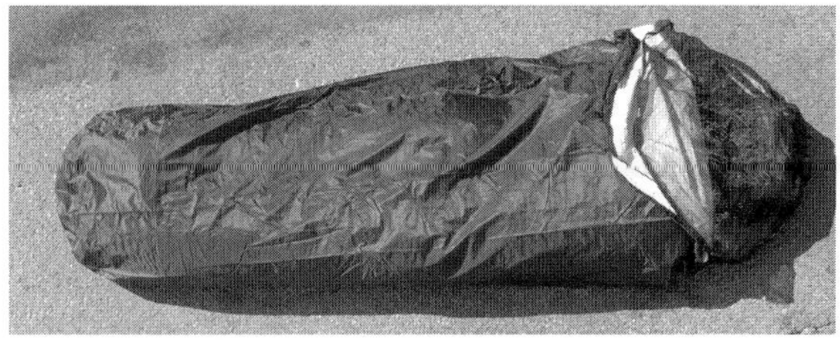

The OR Standard Bivi bag is lightweight, compact and most of all dry

Comparison in pack size of the compact OR Bivi and the Integral Designs one-man tent style bivi

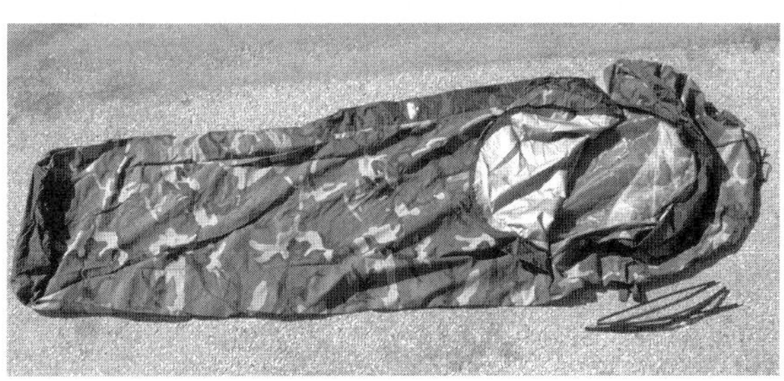

Integral Designs military style bivi complete with aluminum pole to support the hood area

TENTS

As with everything else, weight is a major consideration in selecting a tent since this is probably the single heaviest item, apart from rope (9-10 pounds for a 50 meter coil), that one must carry into the mountains. Fortunately, the weight of the tent can be shared by a two-man team by separating the tent body from the tent fly and poles.

A tent for alpine operations must be a good quality, 2-3 person, 4-season mountain tent that has a proven reputation for handling storms and high winds. One-man tents, although light and compact, do not offer enough flexibility in emergency situations where more people must be housed, and 3-season tents are just not designed to handle rough winter conditions above the snow line.

A look at any expedition base camp will give a good cross-section of the top tent builders that climbers trust with their lives at altitude. The tents most often seen on the mountain are North Face, Mountain Hardwear, Sierra Design and Bibler, however, even the best tent in the world is of little value if incorrectly positioned or poorly secured. More than one climbing team has lost a tent to high winds, and like all nylon products, lightweight tents are sensitive to damage from open flames and sharp points.

The lightest tents are the single walled high altitude summit tents like the North Face Assault at just under 5 pounds, but the most practical and more economical are the double walled, 4-season, 2-man tents like SD's Tiros Assault and TNF Mountain Tent at 7-8 pounds.

For a little more weight and a lot more room and comfort, the 3-man TNF VE-25 at 9-10 pounds is a favorite with many guides and a great investment for any team. In an emergency this tent will handle three with ease and four or five if necessary.

All these tents are of an aerodynamic domed design and have multiple tie-down points on both the tent and the fly to better handle the high winds found at altitude and on the exposed sides of mountains. They derive their shape from lightweight, collapsible aluminum poles and have vestibules for cooking and to store equipment.

Tents, like nylon ropes, should be handled with care, kept clean, inspected regularly and thoroughly dried and aired out before storage. Seams should be sealed with Seam-Grip sealer according to manufacturer recommendations and any small cuts or tears should be repaired immediately with adhesive rip-stop tape before they worsen.

Since conventional tent pegs are of no value in the snow, and even snow pegs will not hold a tent in high winds, an adequate supply of 550 parachute cord or light line should be carried to secure the tent to large rocks, ice axes, pickets and Deadman

The TNF VE-25 3-man, 4-season dome is the standard for many expeditions

The SD Tiros Assault 2-man, 4-season tent has proven to be an excellent lightweight mountaineering tent

anchors. Another alternative is soft "snow anchors", which are small parachute like pieces of nylon that can be placed around rocks or filled with snow and buried to make Deadman anchors.

COOKING GEAR

The rookie, when mountain living, will try to sustain himself on lightweight snacks and energy bars but the seasoned professional understands the necessity of nutritious food and hot, appetizing meals. The individual energy expenditure in the mountains is considerable so must be fueled by an ample intake of food and liquids, and although an energy bar can sustain life, it will not repair fatigued muscles or keep the climber warm or moving at peak performance.

Fortunately we have evolved from cast iron pots and pans and heavy enameled cups and plates to where modern technology has blessed us with a plethora of extremely lightweight alternatives.

The necessities for each person are:

- 1 Cup—preferably 16oz plastic with a snap-on lid that retains heat well and will prevent spills in the tent or around sleeping bags. (Tip: A string should be attached from the cup to the lid to prevent it getting lost or blown away)
- 1 Spoon of good quality and tough plastic or Lexan
- 1 Bowl of deep plastic with a 2-3 cup capacity
- 1 Cooking Pot. Usually 1-quart minimum with a tight fitting lid and rounded bottom corners which make for easier heating and cleaning. Stainless is more durable but aluminum or titanium are much lighter and easier to use. MSR Titanium, Blacklite or Alpine pots are all recommended.
- 1 Pot Grabber to save burning fingers (comes with some cook sets)
- 1 Stove complete with liquid fuel pump and repair kit. The most popular is the MSR Whisperlite Internationale because of its ease of use and the fact that it packs up into the pot. Many high altitude climbers prefer the XGK Expedition for its rugged simplicity, performance at altitude and ability to use almost any fuel, even though it makes more noise and burns more fuel.
- 1 One liter Fuel Bottle (2 for longer expeditions). These should be O ring sealed to prevent spillage and fuel degradation. Preferably MSR Aluminum or Titanium.

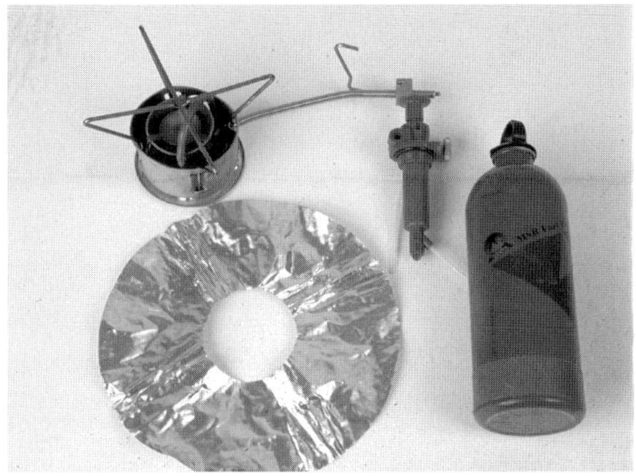

The rugged but noisy MSR XGK expedition stove, pump, bottle and reflector

The MSR Whisperlite is one of the most popular stoves on the market

ALPINE OPERATIONS

Even though MSR stoves are supplied with an optional kerosene jet, white gas is the most popular fuel and the amount needed will vary with the season, access to water and number of hot meals one plans to prepare. The general rule of thumb is to allow 6-8 fluid ounces of fuel per person per day if melting snow for water. Less if water is readily available from a filtered source. If not filtered or iodized, the water will still have to be boiled to kill any bacteria.

Another essential is two good water bottles, which should be of the 1 or 1.5- liter variety with a wide mouth—preferably Nalgene Lexan. The wide mouth allows the climber to pack snow into a bottle during the day to melt for drinking without having to fire up the stove. This works best if snow is continuously added to a half filled bottle of water and then positioned so that it is in the sun or drawing on body heat to help melt the snow.

An additional wide mouth bottle makes an excellent pee bottle that saves having to get out of your sleeping bag at night. This is a luxury under normal camping conditions but an essential when camping or bivouacing in freezing conditions. For the benefit of the entire team, this bottle should be marked clearly as a pee bottle, usually with a large skull and crossed bones.

Use of pickets and a snow fluke to stabilize the stove when cooking on snow

CamelBack type water bladders have become very popular in all areas of outdoor activity and many pack manufacturers now include them in their packs, but they are not a substitute for wide mouth bottles. The bladders work great in warmer regions and are very convenient when packing in but the drinking tubes will freeze solid in colder conditions. They also do not offer the convenience of being able to stuff them with snow to melt for water.

Wide mouth water bottles are an essential

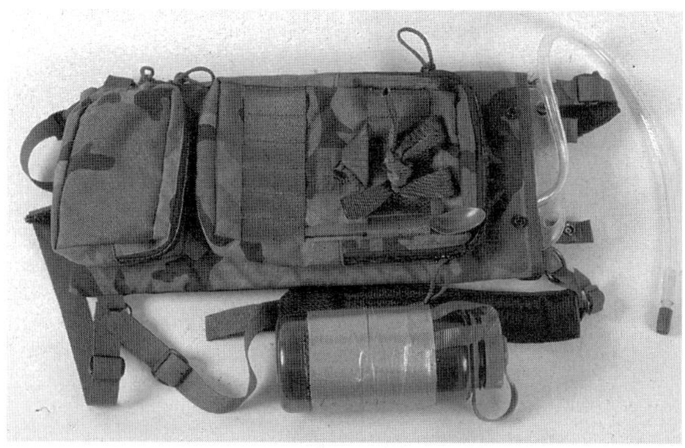

Eagle's Liquidator hydration system and wide-necked lexan bottle

MISCELLANEOUS

The above notes cover all the big items, but there are also a number of smaller items that should be considered essentials.

Many people get lost each year, even experienced climbers who find themselves in whiteout conditions, and all for lack of a good compass. Anyone entering the wilderness should have a compass, and ideally a map and altimeter in the mountains, and at least a rudimentary understanding of land navigation. The compass should have a sighting mirror and declination adjuster (Suunto Professional), but these will be covered in more detail in the navigation section of Chapter 9.

Another indispensable item is a knife or multi-tool. Whether a Swiss Army knife, a Leatherman, or a Buck or Gerber multi-tool, the small tools that they contain and options that they offer for making repairs to equipment are endless.

Every equipment list recommends a cigarette lighter to light the stove but we go a step further and recommend several stashed in pockets, pots and packs. Without a lighter or matches one cannot light the stove, and without the stove there is no melting snow for water or hot meals—and this could be serious if bivouacking or weathering a storm. So rather than one high tech, windproof $50 lighter it is better to buy a half dozen 50 cent disposable lighters and spread them throughout your gear.

To be able to travel at night, descend ropes or work around camp after dark, a reliable headlamp is more efficient than a conventional flashlight. The first choice of most climbers is the Petzl Zoom, which runs on 3 AA batteries, and for extremely cold conditions there is the Petzl Arctic, which has the battery pack in a small pouch that can be carried next to the body to extend battery life.

Another popular light is the lightweight, waterproof Princeton Tec Vortex, which runs on 4 AA batteries, and one of the newer additions to the Petzl line is the Duo, which has the added advantage of two lamps and two bulbs that can be activated independently—one low power and one more intense.

Most headlamps come fitted with a standard bulb for maximum battery life—commonly about 9 hours with a 30-meter range for the Petzl Zoom. For a more intense light for climbing or skiing, the standard bulb can be replaced with a halogen bulb but this will shorten battery life. Extra bulbs and batteries should be carried with any headlamp.

Lastly, everyone should carry at least 4 sturdy plastic bags to be used to waterproof clothing, line packs, or for trash and human waste that needs to be packed out.

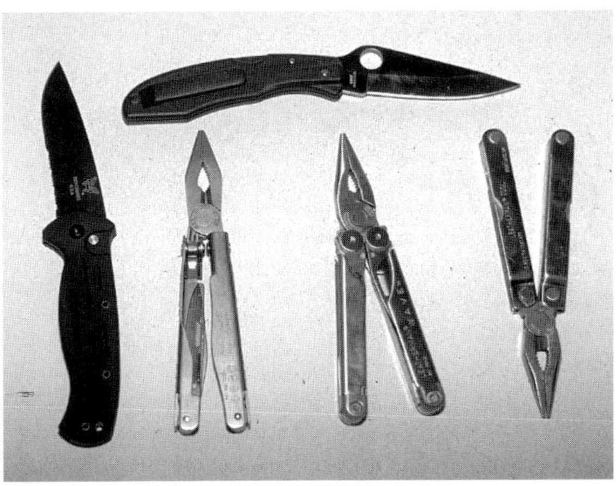

The multi-tool has established itself as the "Swiss army knife" of the 90's (L-R) Benchmade AFO, Bucklite, Leatherman Wave and standard tool, with the lightweight Spyderco folder on top

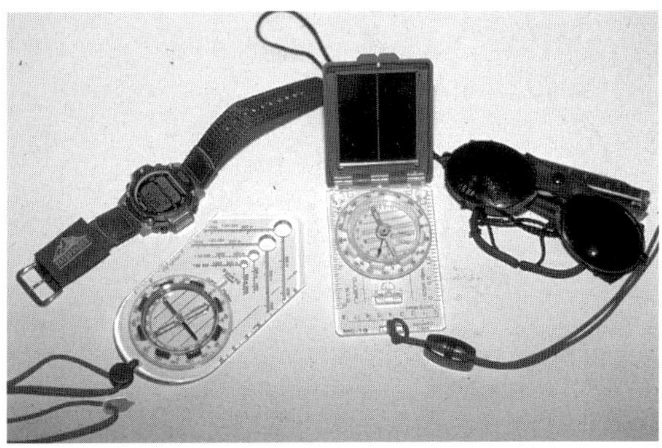

Suunto compasses, Casio wristwatch altimeter and Cebe glacier glasses

Princeton Vortec on Edelrid climbing helmet & Petzl Zoom on Petzl Ecrin Roc

Petzl Duo on Ecrin Roc helmet

PERSONAL HEALTH

First Aid Kits—If a personal first aid kit is not simple, compact and lightweight it will be considered more of a hassle than a help and simply won't be carried. At a minimum, a personal kit should consist of a small waterproof pack containing such items as aspirin, ibuprofen (Advil), moleskin or molefoam, a triangular bandage, a roll of gauze, gauze pads, waterproof first-aid tape, several Band-Aid strips, butterfly closures, a small pair of scissors, a few safety-pins, some form of wound cleanser, Neosporin, and any required prescription medications. If there is any doubt about it being waterproof then it should be stowed in a plastic bag.

A group first aid kit should contain more of all of the above plus athletic tape, Ace elastic bandages, Tincture of Benzoin, Providine iodine swabs, alcohol pads, a compact splint, tweezers, paper and pencil, thermometer, latex gloves, strong pain killers, antibiotics, and even a suture kit. (See Part III—Rescue for additional information on First Aid & Emergency Medicine)

At higher altitudes and on arduous treks, we have found that both Aspirin and Advil are in big demand for headaches and aching muscles, so additional supplies should be packed. Extra Iodine tablets such as "Potable Aqua" are also very important for purifying water taken from mountain streams.

Pepto-Bismol, Imodium and re-hydration salts may also be a necessity if members of the party get diarrhea from the local food or water (especially in third world countries).

Sunscreen—The first rule on sunny days is to cover up and wear glacier glasses (see Chapter 3) but this still leaves skin exposed to the glare and reflection from the snow. Carry a small tube of sunscreen, SPF 30 or higher (i.e., Coppertone Sport), and at least two sticks of lip screen, also SPF 30 or better.

One line of products that has come to our attention for sun protection and skin care, that was tested by ARIEM (US Army Research Institute of Environmental Medicine), is Dermatone. They concluded that Dermatone protects twice as long against frostbite than all other products tested. Their Skin Protector comes in various forms and applicators and combines new non-chemical micro-fine zinc oxide and titanium dioxide to provide broad-spectrum protection.

Toiletries—Keep it light and simple because no one cares that you haven't washed for days, because no one else will have either. Few bother to shave but you can at least bring a small tube of toothpaste and brush your teeth to keep your breath socially acceptable. Toilet paper should be carried in a waterproof bag and anti-bacterial handy-wipes should be used after defecating. A double plastic bag should also be reserved for defecating in areas where one wants to have a minimal environmental impact.

It is however a good idea to leave a towel, biodegradable soap and shampoo in your base camp duffel bag for when you come down off the mountain and want to take advantage of campground showers or a local stream.

DUFFELS

Large durable duffel bags are needed when traveling to and from operational areas to protect all the gear and keep it organized, particularly when flying. The Gregory Long Haul Duffel in XL or XXL is large enough to hold a fully loaded expedition sized pack with room to spare and is ideal for protecting equipment during air travel.

On expeditions, or when moving up to a forward base camp with the support of porters, mules or helicopters, the North Face Base Camp Duffel is the hot set-up. The Large size (27.5″ x 17″) is made from extremely durable waterproof PVC Tarpaulin fabric with a double bottom, triple reinforced haul handles, and has plenty of cinch straps and tie-down points to secure them to mules. One is usually sufficient for most operations but two may be needed per man to carry all the necessary food and fuel to support a team for over a month in the wilderness—especially if the team is ferrying or staging additional equipment at higher camps.

North Face duffles used for soft goods and food and Pelican cases used to transport more sensitive communications and medical equipment

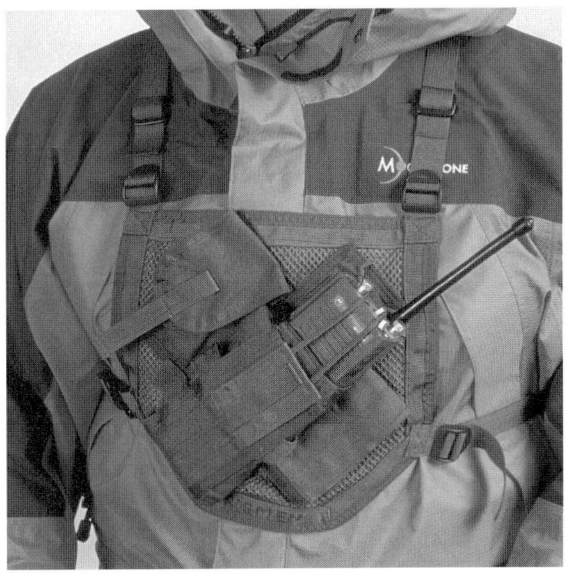

Eagle radio chest harness

BASIC INDIVIDUAL CHECKLIST

MOUNTAIN GEAR

- ❑ Large Backpack (minimum 5000 cubic inches)
- ❑ Sleeping Bag (0 degree F or suitable for coldest local temperatures)
- ❑ Ridge Rest pad—3/4 length
- ❑ Therma-Rest pad—3/4 length
- ❑ Tent & Pegs—4 Season
- ❑ Bivi Sack (optional)

CLOTHING & FOOTWEAR

- ❑ 2 pair Light/Medium weight Long Underwear
- ❑ Shorts (optional)
- ❑ Pile Jacket & Pile Pants
- ❑ Gore-Tex Jacket & Pants
- ❑ Light Hiking Boots (optional—for long approaches)
- ❑ Running Shoes (optional—for camp or river crossings)
- ❑ Double Plastic Mountaineering Boots
- ❑ Crampons
- ❑ Gaiters
- ❑ 3 Pair Mountaineering Socks
- ❑ 3 Pair Liner Socks
- ❑ 2 Pair Synthetic Glove Liners
- ❑ 1 Pair Gore-Tex Mitts w/ Pile Liners (or Modular Gloves)
- ❑ Hand Warmers (optional)
- ❑ Glacier Glasses
- ❑ Spare Prescription Glasses
- ❑ Pile or Wool Hat & Balaclava
- ❑ Baseball Cap & Bandanna

COOKING GEAR

- ❑ Cup w/ Lid: 16oz
- ❑ Spoon
- ❑ Bowl
- ❑ Cooking Pot (1qt.) w/ lid & Pot Grabber
- ❑ Stove w/pump & repair kit
- ❑ Fuel Bottle(s) 1 liter & Fuel
- ❑ 2 Water Bottles—Wide Mouth, 1—1.5 liters

MISCELLANEOUS

- ❑ Headlamp & Batteries
- ❑ Compass, Map & Altimeter
- ❑ Multi-Tool or Swiss Army Knife
- ❑ Lighters (3+)

- ❑ Small First-Aid Kit
- ❑ Sunscreen & Lip screen
- ❑ Toiletry Bag & Toilet Paper
- ❑ Insect repellent
- ❑ 4 Plastic Bags
- ❑ Repair & Sewing Kit
- ❑ Nylon 550 Cord
- ❑ Food
- ❑ Trekking Poles (optional)
- ❑ Snow Shovel (essential for camping in deep snow)
- ❑ Camera & Film (optional)
- ❑ Thermos Flask (optional)
- ❑ Large Duffel Bag with Lock (for base camp)
- ❑ Summit Pack (optional)
- ❑ Leather Mountaineering Boots (where appropriate)
- ❑ Towel, soap, shampoo (for base camp)
- ❑ Travel clothes (left at start point)

ALPINE GEAR (minimum)

- ❑ Climbing Helmet
- ❑ Alpine Climbing Harness
- ❑ Locking Carabiners (2)
- ❑ Non Locking Carabiners (2)
- ❑ Ice Axe w/leash
- ❑ 30 feet of 6mm Prusik Cord
- ❑ 50m Dynamic Rope (10 - 11mm)

See Chapter 5 for more extensive Mountaineering Gear List.

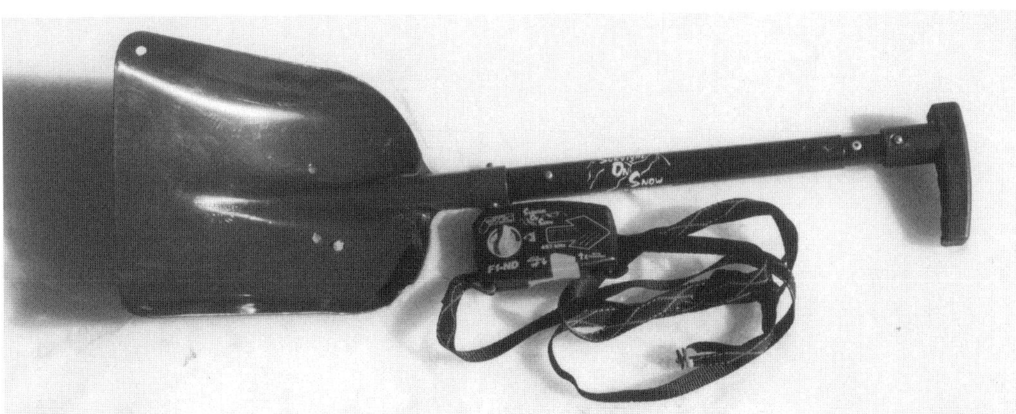

SOS folding shovel and avalanche beacon

Climber moves up the Maestri route, southwest ridge of Cerro Torre in Patagonia (SP Parker)

5

MOUNTAINEERING EQUIPMENT

The whole subject of mountaineering equipment and hardware can be very confusing to the uninitiated, and it doesn't become any clearer when confronted with the ever-widening variety of brands, materials and configurations for the same piece of gear. Take something as simple as carabiners for example—there are steel, aluminum, locking, non locking, screw gate, pear shaped, HMS, self locking, double locking, auto locking, ovals, Ds, offset Ds, straight gate, bent gate, wire gate, hot forged and cold forged—all with their own special features and applications. And the same is true for everything else from ice axes and ropes, to descenders and ascenders, to belay devices and cams.

But if we go back to the principles for purchasing equipment that were detailed in Chapter 4, then it becomes less of a burden. Only get what is essential, buy quality, keep it simple, look at what the professionals are using and get advice from an expert.

Not all mountain operations require technical climbing ability or the related specialized equipment—but there are a few essentials like ropes, harnesses, ice axes and crampons that should be considered standard for everyone on an alpine team. Even if one does not aspire to scale vertical walls of ice or cling to a rock overhang like a human fly, it is still necessary to handle ropes and be able to negotiate some exposure. To simply traverse a glacier or negotiate a narrow snow covered ridgeline, the team will still have to rope up and know how to best utilize ice axes and crampons to remain in balance, or arrest themselves in a fall.

HARNESSES

The harness forms the critical connection between man and rope so should be selected with some care. Obviously one that is of inferior quality or ill fitting is not going to add confidence to the user or be safe in the event of a fall.

There are various types of climbing harnesses, not all suited to alpine operations. At one end of the spectrum is the lightweight, padded sport climbing harnesses, and on the other, heavy-duty rescue harnesses—neither of which are suited to alpine mountaineering.

There are two basic designs of harnesses used for alpine climbing—the sit harness and the full body harness. Sit harnesses are the most common and most popular but they are not well suited to individuals who do not have good hip definition or are overweight. A sit harness needs to fit snugly around the narrow part of the waste and above the hips so that if the climber inverts, he or she will not fall out of the harness. If the individual does not have sufficient hip definition then a full body harness should be seriously considered even though they are less comfortable for everyday use. Small children are also more secure in full body harnesses.

A good alpine sit harness should fit over all

clothing, including the Gore-Tex outer layer; have gear loops; adjustable leg loops; and should be reasonably comfortable to hang suspended in for some time. One should not have to step into an alpine harness like a sport climbing harness since this is difficult with crampons and would require removing the harness to go to the bathroom—not something that is recommended on an exposed ridgeline or icy face.

The alpine harness should be strong but lightweight and made from materials that will not absorb moisture that will ice up. It should be adjustable so as to fit over multiple clothing layers and have drop leg loops to facilitate the call of nature. The waist strap should fit snugly around the waist and well above the hips with a secure double-back buckle system.

The recommended harness for many years, and still an excellent product, is the Black Diamond Alpine Bod Harness, but Misty Mountain's Summit harness has proven to be an even better investment. It is the harness we use at STTU and has also been adopted by the instructors at the Army mountain warfare training center.

Misty Mountain Summit Harness

Correct Use:

The climbing harness should be considered a critical link in the life support system that runs from the climber, through the rope to the anchor or belay point. It is therefore important to not only put the harness on correctly and tie into the rope correctly, but also to stay tied in securely at all times when climbing, belaying and crossing exposed terrain.

The first step in correctly securing a harness is to know how to double-back the buckle that secures the waistband (See Photo). Failing to do this has resulted in the loss of life on several occasions. One Everest climber fell to her death after going to the bathroom and then not re-securing the harness correctly—a fatal mistake attributed to altitude and fatigue.

In addition to paying particular attention to the buckle, the climber should not be distracted while

Incorrectly buckled harness

Correctly buckled harness

Correctly double-backed BD Alpine Bod

donning the harness. Once one begins putting on a harness, continue until the task is finished without interruption.

As stated earlier, the harness must also be worn snugly around the narrow part of the waist above the hips so that one cannot come out if inverted in a fall.

Like climbing ropes and most other nylon products, try to keep harnesses out of direct sunlight and away from nylon damaging substances such as acids, alkalis oxidizing agents and bleach. Retire a harness when it becomes worn and faded or the stitching and tie-in points show signs of deterioration.

HELMET

Protection from falls, falling rocks and ice dislodged by other climbers, along with dropped pieces of hardware, makes the helmet another essential piece of safety equipment.

Even though a climbing helmet is not worn at all times in the mountains, everyone should carry one if they are to be doing any steep rock or ice work. The helmet must be specifically designed for climbing and should be lightweight, comfortable and have sufficient adjustment to fit with or without a balaclava or head warmer. Petzl's Ecrin Roc is the favorite of many teams but the helmets made by Edelrid, Blue Water and HB are also suitable.

ICE AXE

The ice axe is not only the symbol of the mountaineer it is an important tool that has a multitude of uses in the mountains. The mountaineer's ice axe consists of a shaft with a spike on one end and an adze and pick on the other. When it is held like a walking stick, it can be used for balance on sloping ground, going up, down or traversing. When held like a hammer the adze can be used to cut steps in ice, and the pick used as an artificial hold to assist in climbing steep terrain.

The ice axe is also used for self belay or self arrest in case of a fall on steep snow or ice fields, and can be buried in the snow to make a "deadman" anchor. When the handle is driven into the snow pack, it becomes a field expedient anchor to belay others down steep snow, and is also used horizontally to stop the rope cutting into the lip of a crevasse while doing a rescue (See Chapter 12).

Around camp, the axe has several other more mundane uses: like digging latrines, cutting drainage channels, digging holes for deadman anchors and as a solid tent stake in high winds.

When selecting an ice axe for general mountaineering, the shaft length of the axe can be matched to the height of the user. If under 5´ 7″, for example, the axe should be a 60cm tool; 5´ 7″–6´2″ use a 65cm tool; over 6´2″ use a 70cm tool. The SMC Himalayan, SMC Shuksan (70cm, 1.5 pounds/680g) and BD Alpamayo are all simple, ruggedly designed axes that will give years of hard service.

SMC manufacture their axes with an aircraft-quality alloy steel head that is stress relieved and heat-treated for toughness and durability. The aluminum alloy handle is finished with a durable thermo-plastic coating, and is light enough to be carried all day yet strong enough to be used as an anchor. For alpine situations requiring two tools, such as very steep snow and ice, SMC makes an excellent Himalayan Ice Hammer (50cm, 1.6 pounds) as a companion for their ice axe.

If one is looking for a super lightweight alpine climbing axe for high altitude climbing, then the Grivel Air Tech (1.2 pounds in 66cm length) and the sleek Grivel Pamir are worth considering, but keep in mind that these axes are not intended for the rough conditions encountered in guiding and mountain rescue work.

The ice axe should be equipped with a leash, usually made from half-inch tubular webbing, running from the hole in the head of the axe to the climber's harness. This is preferable to a wrist loop and allows the climber to use the axe for self-belay and more easily change hands with the axe when traversing and changing direction. (See photo)

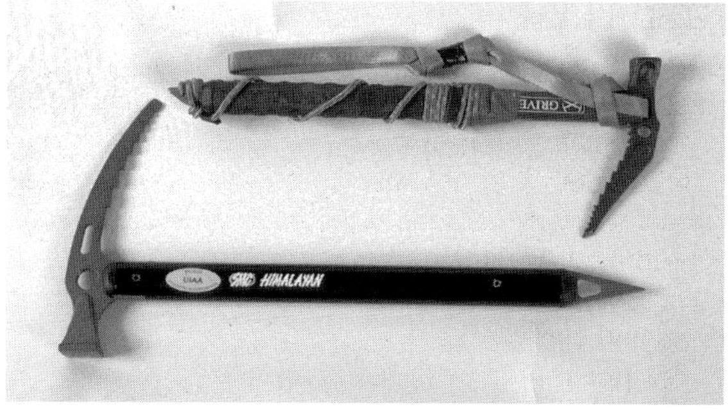

SMC North Wall hammer and Grivel compact Third Tool

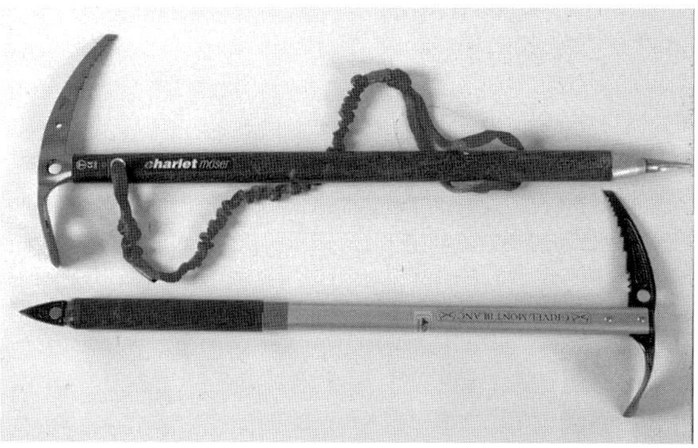

Heavier CM Goulet axe (2.5 lbs) and lighter Grivel Air Tech (1.2 lbs)

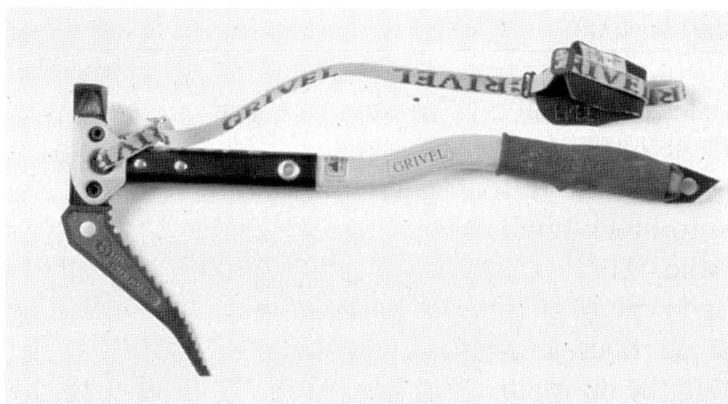

Grivel Rambo technical ice tool usually used in pairs

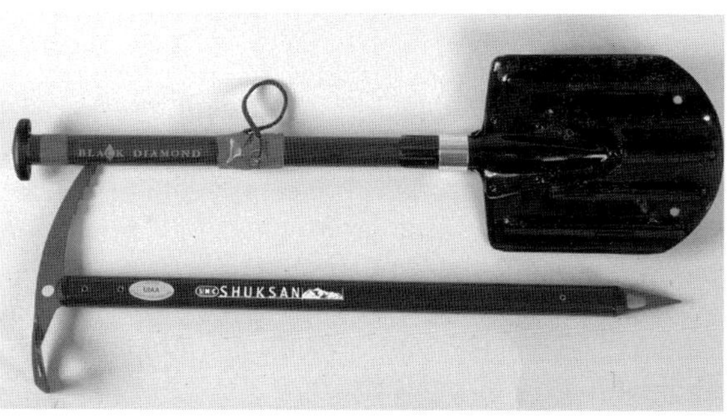

SMC Shuksan (1.65 lbs) and compact BD shovel

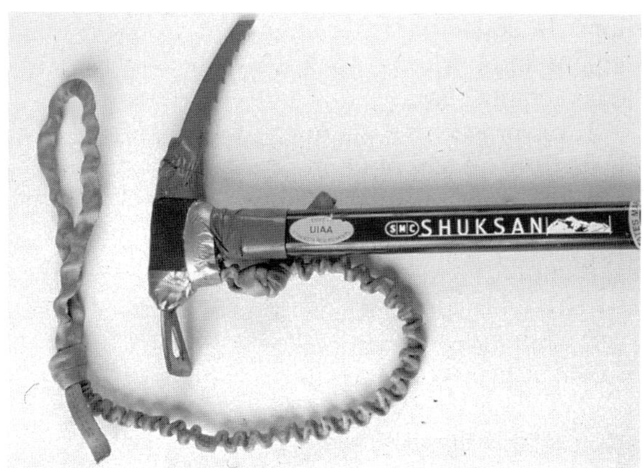

Leash and padded insulation taped on axe head

UTILITY CORD

Light cordage, even 550 parachute cord, has many applications for both climbing and wilderness living, such as guy lines for tents and hauling lines for equipment. It also has numerous repair applications. However, there are also three very useful climbing aids that can be made from 6mm to 8mm accessory cord: prusik loops, Texas prusik ascenders and cordelettes.

A prusik or prusik hitch is used to create an adjustable attachment between a smaller rope, the prusik cord, and a large diameter rope such as the climbing rope. It can be used for a safety / self-belay when rappelling or as a lightweight, non-slip friction device in place of ascenders.

A basic prusik loop has many uses in climbing and rescue work and is made with 5-7 feet of 6-8mm prusik cord, jointed at the ends with triple fisherman's knots. Then by passing the loop around the climbing rope and back through itself two or three times it forms an effective one directional auto-block hitch. (See Chapter 10)

Note: There is some discussion about which size and type of prusik cord should be used with which sized ropes, so some experimentation is called for to find the optimum combination that will best grip the rope that is being used. We have found that 6mm cord works well with 10-11mm ropes for most individual climbing applications and that stronger 7-8mm cord is better suited to 1/2″ rescue rope for most rescue and military rope installation applications. In addition, icy ropes with heavy loads may require more wraps while dry ropes and lighter loads require fewer.

Texas prusiks, shown in detail in Chapter 12, are made from 30 feet of 6mm prusik cord and are used as a lightweight, field expedient, economical alternative to mechanical ascenders.

A cordelette is made like a prusik loop by joining the two ends of a 16 to 20 foot 6-8mm cord with triple fisherman's knots. This large loop makes an excellent lightweight anchor strap that can be used around suitable trees or rocks, or folded to make a two or three leg self-equalizing anchor system. (See photo).

A 6mm cordelette wrapped for convenient carry

Cordelette used for 3-point load sharing anchor

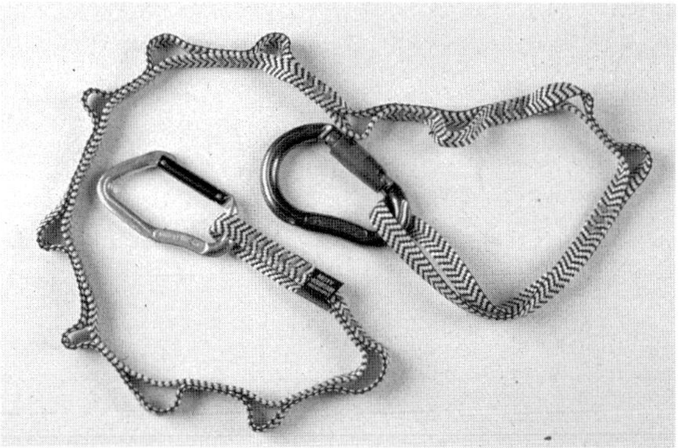

The Daisy Chain has many uses in climbing

MOUNTAINEERING HARDWARE

Hardware is the generic term for the variety of steel and aluminum items such as carabiners, nuts, chocks, pickets and pitons that are used to secure the climber to the rope and the rope to the rock or ice. Descriptions of these items are included in this chapter but many of their uses will be covered in more detail in Chapter 11 on Climbing.

CARABINERS

Carabiners fall into four basic categories, steel or aluminum, locking or non locking. Steel carabiners are primarily for industrial, team and heavy rescue applications, while the lighter weight aluminum ones are designed for most individual climbing and rappelling applications. Locking carabiners have devices, usually screw gates, that stop the gate opening accidentally and are used where accidental gate opening and coming un-snapped could be fatal — are used where speed and convenience is more important than bombproof security

At a minimum, the rookie mountaineer will need two locking carabiners that are the larger pear or D-shaped lockers such as the Omega Jake HMS Screw-Lok or Quick-Lok. Screw lock gates are manually screwed down while quick lockers have spring-loaded gates that lock automatically when released. Some schools and individuals prefer the auto lockers because they are somewhat idiot proof, but they can be awkward to manipulate when frozen or with thickly gloved fingers. If in doubt, you can always carry one of each and see which you prefer.

The next most commonly used type of carabiner is the aluminum oval. This type of snap-link has been around for years and is the workhorse of the climbing community. A non-locking oval is usually used to connect a short sling to the available protection while another is used to connect the other end of the sling to the rope. This configuration is used in not only rock and ice climbing but also when using snow pickets or snow flukes on exposed snow slopes for running belays.

Ovals are a very versatile piece of kit with most climbers carrying between two and six for various uses, even just for clipping off a water bottle or camera. Two ovals can also be used to replace a locking carabiner when placed with the gates opening in opposite and opposing directions to form an X.

Following in the concept of the oval, the D shaped non-locker was developed for its stronger basic design, but even these are now being replaced by more high-tech innovations like the Omegalite 4.0. The Omegalite is a lightweight, low profile cold forged carabiner that is stronger than an oval and with an ergonomic design that makes them faster and easier to manipulate—both important considerations when hanging by one hand and trying to clip in with the other.

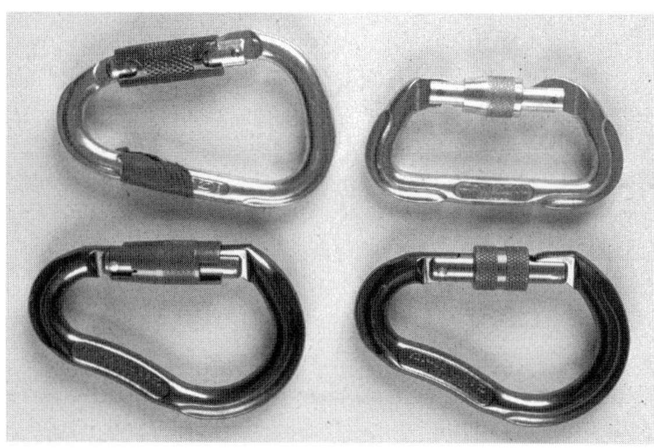

Auto-locking carabiners on left. Screw gate locking carabiners on right

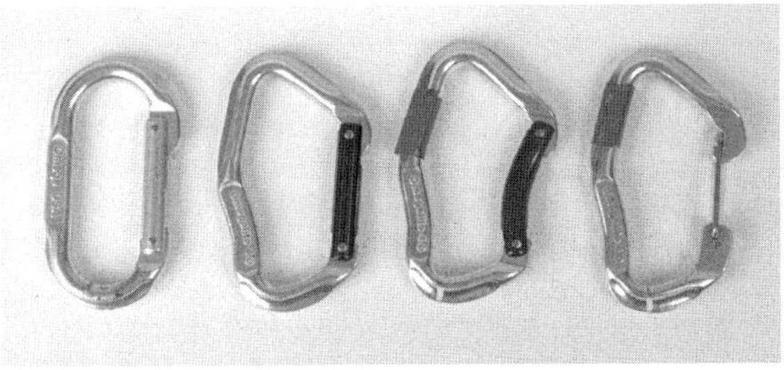

Non-locking carabiners—oval, Omegalite 4.0 straight gate, bent gate and wire gate

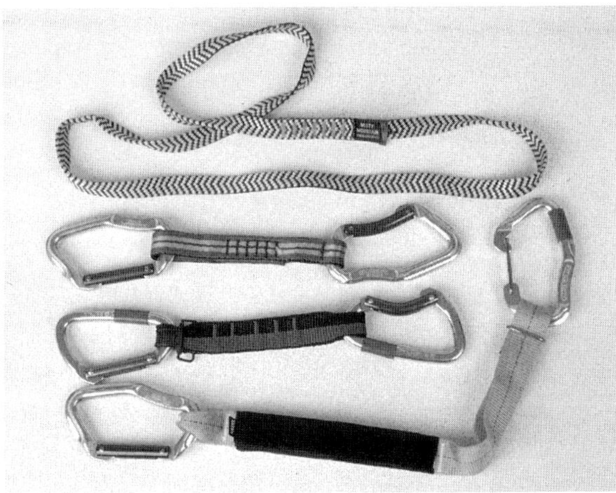

Top to bottom—Misty 24″ spectra sling, Omega quickdraw, BD quickdraw and Yates screamer

BELAY DEVICES

The first rope skill that a novice will learn after mastering the essential knots is rappelling—a technique that allows an individual to descend a rope in a safe and controlled manner. The next important skill to master is belaying which is a process of controlling the rope as another person climbs and then being able to arrest a fall if the climber should lose their grip or footing.

Both these skills take advantage of small aluminum devices that connect to the belayer's harness and increase friction on the rope when used correctly. They are marketed under a number of names and range in price from ten to seventy dollars. For example: the ATC, Jaws, Pyramid, Variable Controller, Sheriff, Sticht plate, belay plate, Rope Controller and Grigri.

With time and practice, the mountaineer will become familiar with most of these, but at the start, a simple ATC (Air Traffic Controller) type belay device will meet most all requirements.

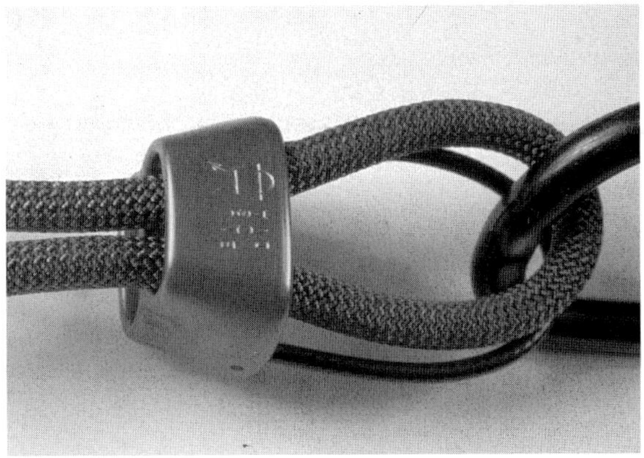

Black Diamond ATC belay device

DESCENDERS / RAPPEL DEVICES

Although belay devices like the ATC can be used to make a rappel, there are other devices more suited to just rappelling—the most popular being some form of figure 8 rappel plate.

The standard sport version of the figure 8 is a small smooth device that can be used with most static and dynamic ropes. However, it is possible to accidentally jam one of these up with a girth hitch when sliding over ledges. To prevent this, units like the CMC Rappel 8 and larger Rescue 8 have ears to stop the rope sliding around and locking up unexpectedly.

For lowering or rappelling with heavier loads, a brake-bar rack can also be used (see Technical Rope Rescue Chapter 20) and to have more control over longer rappels, some operators and spelunkers will use a mechanical brake like the Petzl Stop.

BD figure 8 rappel device

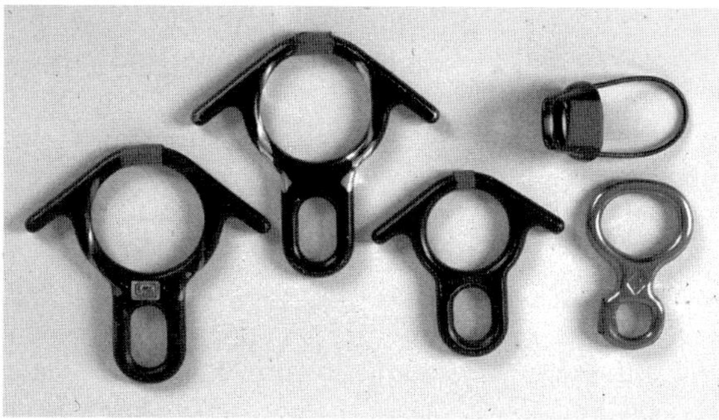

(L to R) CMC steel Rescue 8, aluminum Rescue 8, aluminum Rappel 8, BD Sport 8 and ATC

Trango Jaws belay and rappel device gives very smooth, positive braking

The Trango Pyramid has large cooling fins to help dissipate heat build-up

ASCENDERS

Mechanical ascenders consist of a matching pair of handles with thumb operated camming devices that grip the rope and allow the climber to jug or jumar upward without sliding back down. The term "Jumaring" comes from the brand name of the original high quality ascenders.

Since then many other companies have joined the market with Petzl's Ascensions setting the standard. These ergonomic designed ascenders are both CE and UIAA certified and color coded to help learning impaired big wall rats tell the difference between left (yellow) and right (blue).

Ascenders make climbing a fixed rope much easier and more efficient than using prusiks, and allows a weaker climber to follow a lead climber or guide who has moved ahead to fix ropes over difficult pitches. But for all these advantages, the extra weight of the ascenders, as with everything else, must be considered before they are automatically added to the mountaineer's gear rack.

A more compact form of ascender is the ruggedly built Gibbs Ascender which has several uses for climbing, caving and rescue work. The big wall climber or spelunker will use a pair of light-weight #1 or #2 Gibbs ascenders (6 oz) to rig a ropewalker type climbing system; or to rig a 3:1 mechanical advantage non-slip pack hauling system. In rescue work the heavier duty #3 and #4 Gibbs (8-11 oz) are used as mechanical prusiks for quickly adjusting tensioning systems and for one directional safety on load raising systems (See Chapter 20—Rope Rescue). These are available in two sizes to fit 7/16″–1/2″ ropes or 5/8″–3/4″ ropes, and with either aluminum or stainless steel shells depending on the anticipated load and use.

To best utilize a set of ascenders for rope climbing, the climber will also need a set of etriers or aiders. These are a series of nylon loops sewn into a ladder that allows the climber to "walk" up a rope. The Easy Step aiders made by Misty Mountain Threadworks are some of the better ones that we have evaluated.

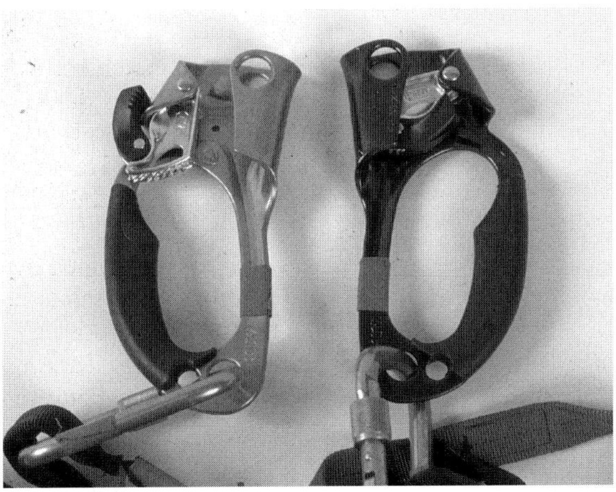

A matched pair of Petzl Ascension ascenders

Compact Gibbs ascenders are widely used in rescue work and for rope climbing systems

PROTECTION

Protection is a broad term to cover all the methods used to secure ropes to the rock or ice that will prevent climbers from falling or catch them before they fall too far. Where natural anchor points and protection are not available, artificial protection devices can be used. These include a variety of protection used on rock, such as cams, nuts and pitons; ice screws for hard water-ice; and pickets and flukes for travel on snow.

Rock protection can be divided into three basic groups:

- Temporary protection that can be removed without damaging the rock—Cams, chocks, nuts and hexes
- Temporary protection that may damage the rock when removed—Pitons
- Permanent or fixed protection—Bolts

Cams, also known as SLCDs (spring-loaded camming devices), marketed under a number of names such as Friends, BD Camelots, HB Quadcams, Flex Cams, and Metolius 3 Cam & 4 Cam, are designed to be used in sets made up of various sizes to fit most cracks. The cams have an expansion range from 8mm-14mm, for the smallest cams, all the way up to 115mm-190mm for a Size 6 Technical Friend. In inches that could be 0.35″-0.53″ for a Metolius 00 up to 2.80″-4.20″ for their 4 Cam #10.

In principle cams all do the same thing, in that they have diametrically opposed spring-loaded camming surfaces that can be manually retracted when inserted into a crack and then open out gripping the inside surfaces of the crack when released.

A cam works best when correctly positioned in a crack that is close in size to the mid-range of the particular cam—for example, a cam with a 1″-2″ expansion range will work best in a 1.5″ crack.

When pulled on or loaded, cams should lodge themselves into the crack even tighter and yet can be easily removed by compressing the trigger release mechanism, providing the cam hasn't walked itself

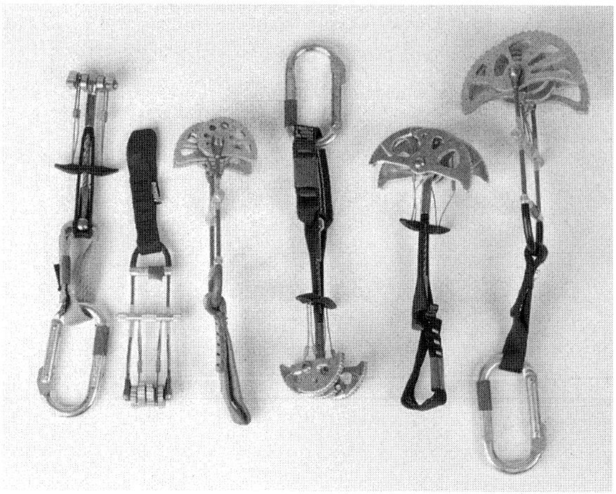

Assorted Cams from Metolius and Wild Country

too deeply into the crack.

Nuts and chocks come in an ever-widening variety of shapes, sizes and materials and are designed as a more passive jamming device. Mostly made from aluminum and pre-threaded with a swaged wire loop to clip into, nuts and chocks should be placed so as to wedge into the rock and grip more tightly when loaded in the potential direction of pull.

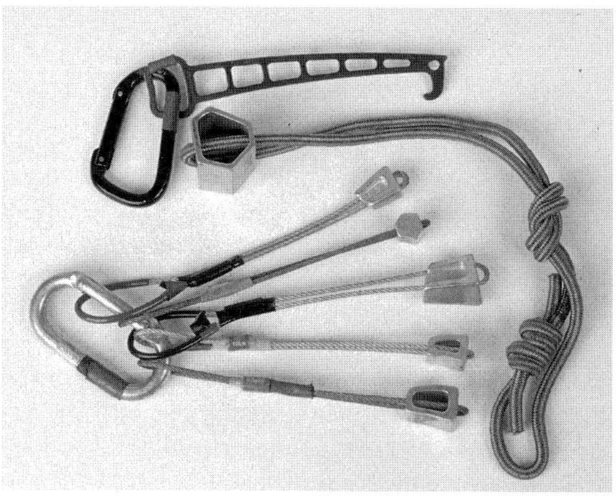

Nut pick, large hex with accessory cord, and assorted wired nuts and wedges

Hexentric chocks (Hexes), because of their six surfaces can be used passively as a chock or with a camming action to secure them in a crack. The smaller sizes come pre-rigged with wire loops, but the larger sizes come with two holes, designed to be threaded with 5mm to 8mm accessory cord.

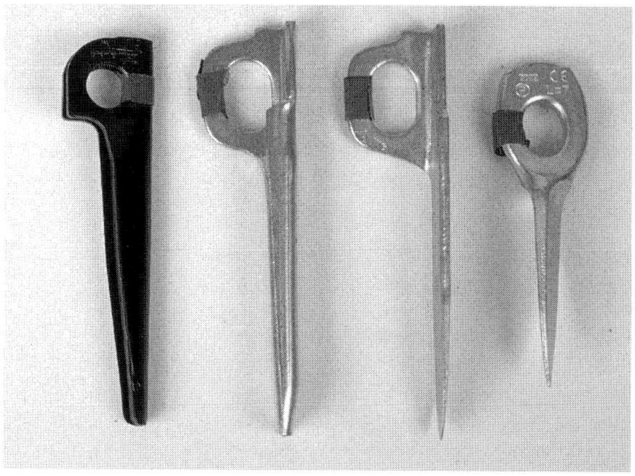

Assorted pitons

Pitons are steel or aluminum alloy spikes that are driven into small cracks with a hammer and come complete with an attachment eye for a carabiner and sling. There are various types of pitons such as blades, angles, Tops, Nailers, Zs, Wedge Hooks, Birdbeaks, lost arrows, and Crack Smacks, each designed for slightly different applications.

Most pitons are hardened steel and designed to be reused many times. Two-thirds of their length should fit into the crack before hammering and when hammered the piton should ring, rising in pitch until the eye comes in contact with the rock. Soft iron pitons are more suited to irregular cracks and are generally left in place since they are difficult to remove without damaging them.

Pitons should always be placed with the eye horizontal or sideways in vertical cracks, and oriented downward in horizontal cracks.

Pitons are relatively cheap and easy to use, but they can be a problem to remove and can cause permanent damage to the crack and rock face, something that is frowned upon by climbers trying to preserve the quality of a climbing route. However, carrying a small selection of pitons is justified since they are useful and expedient anchors when faced with no other options, particularly narrow cracks in horizontally bedded rock.

Bolts also come in several forms and sizes, but are usually 3/8″ Rawl or Fixe expansion type bolts that are drilled into the rock, inserted and expanded for a permanent anchor point.

There are both hand drills like the Pika, and power drills made by Boche, that are suited to rock drilling. However, even though hand drills are relatively light and easy to use, drilling rock is a slow, tiring and tedious task.

Bolts are used in conjunction with hangers which are angled steel plates with two holes—a small one to attach the hanger to the bolt with a nut, and a larger one that will accommodate carabiner and sling or quick-draw.

Bolting can be used for an emergency bail-out where no other natural protection is available, and on rescue operations where raising and lowering systems need to be deployed directly above the rescue site. Since the bolt is only as good as the rock that it is drilled into, multiple bolts should be used at anchor points and belay stations for greater security.

Pika 3/8″ rock drill with expansion bolts and Metolius hangers

When placing bolts, a climber must keep in mind that these are permanent anchors that will most likely be used by other climbers in the future. Because of this responsibility, and to be confident in their strength and safety, bolts should be placed only by experienced individuals.

GEAR RACK

Although a considerable amount of equipment can be carried on the climbing harness gear loops, long climbs such as those found on the big walls require far more protection than can be carried on the harness alone. To augment the belt harness, some climbers will rig one or two diagonal chest slings made from tubular webbing, but a more substantial chest harness is preferable for big wall work.

Misty Mountain's Bigwall Gearsling is built to handle a comprehensive wall rack and also doubles as a full strength chest harness to prevent accidental inversion. The foam padded shoulder and chest straps adjust to fit a wide range of sizes and the reinforced clip-in point centers with the use of double buckles. Four racking slings carry and organize even the biggest wall rack. (see photo)

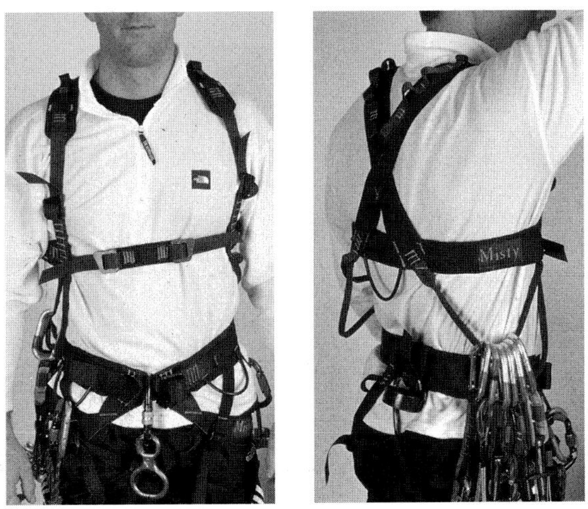

Misty Mountain's Big Wall Gear Sling also doubles as a full strength chest harness

SNOW & ICE PROTECTION

Snow protection comes in three forms—Deadman Anchors, Flukes and Snow Pickets, all of which work best in densely packed snow but do not work well in dry, unconsolidated snow.

A Deadman is a generic term for anything that can be buried in the snow, with a rope fastened to or around it, to form an acceptable anchor. This could be anything from an ice axe to a pair of skis and even a backpack or a rolled up sleeping bag. In most cases, a girth hitch is used to secure a sling or prusik loop around the middle of the deadman anchor and the climbing rope is then carabinered into the sling.

Flukes, such as the BD Deadman (8.9ozs) and SMC Snow Anchors, are shovelhead shaped aluminum alloy plates that come in various sizes and are driven into the snow for a hasty deadman anchor. They are most effective when angled back from the direction of pull by 40 degrees, and wire cables attached in such a way as to make the fluke dive deeper into the snow when loaded.

Snow pickets are large aluminum stakes with a T-shaped cross-section, such as the SMC 24″ and 36″ units, that have multiple holes for carabiner attachment points. Snow pickets can be either driven into firm snow on an angle (40 degrees), used in a series connected together, or buried horizontally to form a deadman. They are also used by rope teams for a running belay over exposed low-angle, snow-covered terrain.

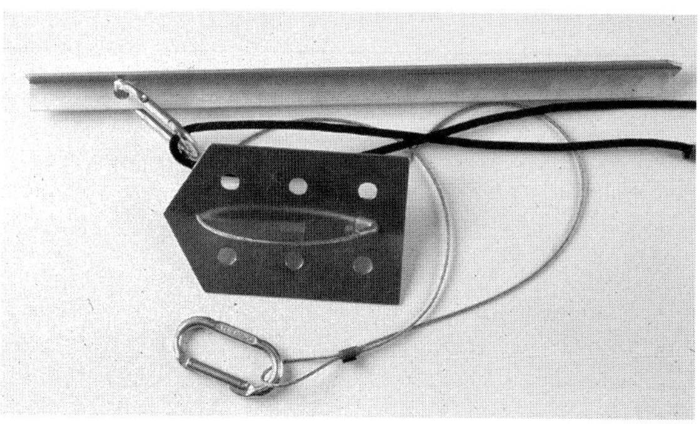

SMC Snow picket and Deadman fluke

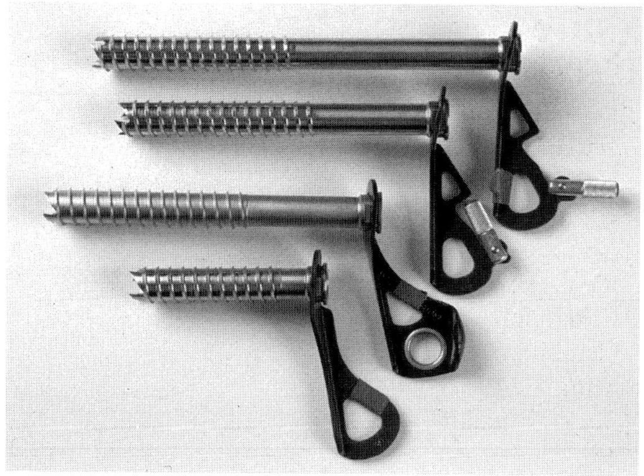

Note the small crank handles on the BD Express Ice screws

Any and all of the above can be combined to create a multi-point anchor and a more secure system.

The Ice Screw is designed to serve as a more solid anchor in hard water- or glacial ice, in place of a conventional piton that would only shatter the ice. Most modern ice screws are aluminum or titanium tubular devices that are hand drilled into the ice to a depth of 10 to 22cm. When the screw is drilled in completely, the head should be flush with the ice leaving the hanger exposed to accept a carabiner and sling.

Several companies manufacture quality ice screws, such as Black Diamond, Charlet Moser, and Grivel, but by far the most convenient are the BD Express Screws since they have a small integrated handle that allows them to be more easily cranked into or out of the ice. They are available in 17cm and 22cm lengths with a hanger strength of 22kN (4945 pounds).

BASIC MOUNTAINEERING EQUIPMENT CHECK LIST

INDIVIDUAL GEAR
- ❏ Helmet & Headlamp
- ❏ Alpine Harness
- ❏ Crampons
- ❏ Ice Axe w/Leash
- ❏ Carabiners—Locking (2)
- ❏ Carabiners—Non Locking (2)
- ❏ Sling—24″
- ❏ 30′ of Prusik Cord—6mm (to make prusiks)

- ❏ Belay Device—ATC/Jaws
- ❏ Rappel Device—Figure 8 Plate (optional)
- ❏ Rescue Pulley—Small (optional)
- ❏ Cordelette—16′-20′ (optional)

GROUP GEAR (PER ROPE TEAM)
- ❏ 50m Dynamic Rope 10-11mm
- ❏ 150′ Static Rope—3/8″ or 7/16″ (for fixed lines)
- ❏ Carabiners—Non-locking (10)
- ❏ Webbing—Tubular 1″—15-20 feet
- ❏ Slings & Runners—12″ & 24″

SNOW & ICE
- ❏ Wands (12)
- ❏ Snow Pickets (3)
- ❏ Snow Flukes (1)
- ❏ Ice Screws (4+)
- ❏ Slings—Spectra (5)
- ❏ Ice Tools (2)
- ❏ Tool Loops

AVALANCHE COUNTRY
- ❏ Snow Shovels
- ❏ Avalanche Beacons
- ❏ Avalanche Probes

ROCK CLIMBING
- ❏ Chocks, Nuts & Hexes—Assorted Sizes
- ❏ Nut Pick
- ❏ Cams—Assorted Sizes
- ❏ Quick Draws (for Bolted Routes)
- ❏ Slings & Runners—Assorted Lengths
- ❏ Chalk Bag & Chalk
- ❏ Rock Climbing Shoes

- ❏ Pitons—Assorted Sizes
- ❏ Big Wall Hammer
- ❏ Rock Drill—3/8″
- ❏ Bolts & Hangers—Assorted
- ❏ Chest Harness (Big Wall)
- ❏ Ascenders (Fixed Line or Big Wall)
- ❏ Etriers (Big Wall)

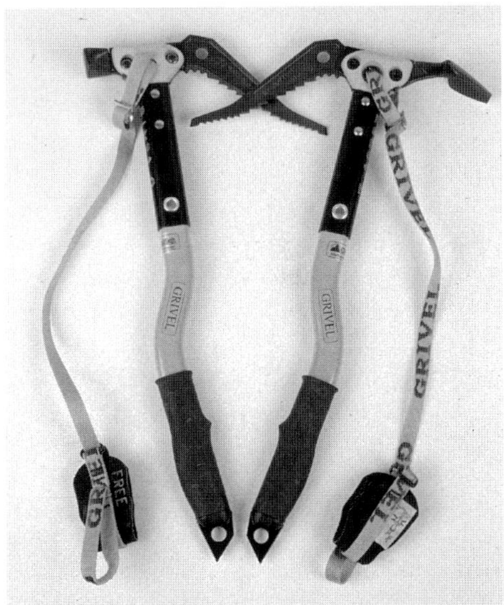

ALPINE STYLE MOUNTAINEERING—"LIGHT IS RIGHT"

There are two approaches to mountaineering—expedition style and alpine style. In expedition style climbing, the team plans on spending numerous days or weeks laying siege to the mountain and equips accordingly. But this means a considerable amount of food and supplies that must be moved up to base camp by porters, mules, yaks or by ferrying loads between camps.

The alternative is alpine style climbing that is popular with purists and is founded on the creed of, "light is right." Now although this is a good policy in all mountaineering, the alpinist is a true minimalist who will take it a step further.

The climbing philosophy is to travel light, move fast and get on and off the summit before weather closes in, avalanches break loose or snow conditions change for the worse. Traveling lighter means less fatigue and faster movement but it is also taking a gamble on one's own physical ability and the unpredictable conditions that nature can throw at a climber.

The alpinist will study every single piece of clothing and equipment with a view to lightening his load.

- A lightweight Nylon Rip-Stop pack will be used in place of a standard expedition weight pack.
- Packs will be trimmed of any excess straps, buckles and gadgets.
- A heavy sleeping bag will be traded for a lighter bag with the intent of sleeping fully clothed and in a bivi sack.
- Instead of a tent, climbers may opt for a bivi sack and a snow cave.
- A heavy 10.5mm rope will become a lighter 8.5mm or 9mm and in place of two ropes for longer rappels, 6-7mm accessory cord will be used as the second line.
- Heavy Gore-Tex outer clothing will be replaced with a lighter Activent type shell.
- Plastic climbing boots will be customized and the liners lightened.
- The ice axe will be a lightweight, racing model.
- The snow shovel handle will be replaced with the shaft of an ice axe.
- Food will consist of a few energy snacks.

By making just the modifications and substitutions listed above, an alpinist can easily cut 10-15 pounds from his load, which is a significant amount of weight when climbing and at higher altitudes. However, keep in mind that lighter gear is not as strong or durable so is less suited to professional applications, but for situations where a team must move fast and light, it is possible to do so without leaving essential items behind.

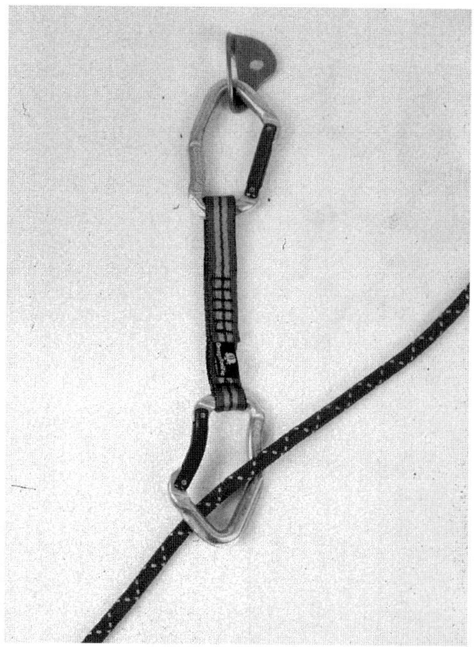

The correct method of clipping the rope into a sling or quick draw

6

ROPE & ROPE CARE

It is not surprising that climbing rope is depicted in the logo of most alpine clubs and mountain rescue teams since rope is the single most important piece of equipment to the mountaineer and truly is the thread that connects all forms of alpinism. It is rope that allows us to travel and work in a vertical environment, literally on the edge of certain death, with both safety and confidence.

Rope, its construction, selection and use, is a subject that has filled volumes and can be quite confusing to the novice mountaineer since there are different sizes and types of rope for all different applications. Ropes that are designed for climbing are not suited to rescue and tactical rappels, just as static rappel line could cause serious injury to a climber taking a thirty foot whipper (fall). Yet both look very similar.

For the sport climber or SWAT team member who may use only one type of rope, the whole subject of rope can be greatly simplified. But for the military mountaineer or rescue team member, the correct application of these various types of ropes is critically important making a more in-depth study warranted.

TYPES & SIZES

There are primarily two types of construction found in ropes—Laid and Braided. Laid ropes are recognized by their distinctive "twist" form and are made by twisting fibers into small bundles and then further twisting those bundles around one another. One draw back of this design is that when put under load, the lay wants to unwind causing the rope to twist and the load to spin. This is not a pleasant experience for a victim or rescuer hanging on the end of that rope so this type of construction has been replaced by braided rope.

Braided ropes are constructed in several different ways but are generally recognized by their tightly woven threads braided into a much smoother format than the laid or twisted ropes. There are single braid ropes, hollow braid and double braid, but the form used in virtually all mountaineering ropes is kernmantle construction.

Kernmantle ropes are designed with a strong central core (the kern) covered by a woven outer sheathing (the mantle) that supports a lesser portion of the load. However, this tightly woven covering does protect the core from abrasion, dirt and environmental effects. The result is a rope that is strong, flexible, easy to handle and resistant to damage.

In addition to strength and abrasion resistance, kernmantle ropes are constructed with two uniquely different characteristics to suit different applications. Dynamic rope is designed for climbing and fall protection, while Static rope is best suited to rescue and tactical operations.

Dynamic rope, as its name indicates, has a lot of stretch and elasticity built into the rope to cushion a climber's fall and to reduce shock loading on the protection or anchors. Static rope has minimal elongation (less than 2%) and is used more for fixed lines, rescue, caving and tactical rappelling. It should not be used for climbing since the un-cushioned shock loading from a fall could both injure the climber and tear out any anchors or protection.

Both ropes come in a variety of lengths and diameters for various applications but there are two sizes used more than any other. In dynamic there is the 11mm climbing rope in a 50-60 meter length; and in static the 7/16″ (11.1mm) rappel line or 1/2″ (12.7mm) rescue rope in 100′-150′ lengths.

Note: Since 11mm dynamic and 7/16″ static are almost identical in diameter and appearance, it is important that teams using both types of rope purchase them in different colors to avoid confusion. Although it is perfectly safe to rappel on dynamic rope, using static rope for climbing and fall protection could result in serious injury or death.

DYNAMIC ROPES

Most guiding, mountaineering training and rock climbing is done with single 11mm dynamic ropes since these have good handling characteristics and will withstand the abuse of regular use. More experienced climbers will opt for a lighter rope in the 10mm or 10.5mm range to save a few pounds in their pack and for the reduced drag while climbing. Lighter single ropes, like 9.5mm can be used for glacier travel but are not recommended for vertical climbing where a high impact fall is a possibility. To use any lighter rope means going to a Half Rope or Twin Rope system.

Half ropes are usually 8.5mm-9.5mm in diameter and are identified by the symbol "**1/2**" enclosed in a circle on the CE/UIAA label. They are used where the climber wants lighter rope and the added security of two separate ropes, especially when climbing in areas with numerous sharp edges. With half ropes, each rope can be clipped alternately into staggered protection points on either side of the climbing route.

Where weight and drag are critical, some sport climbers will go to twin ropes (climbing floss). The diameter of these is down in the 7.6mm range and can be identified by the infinity symbol **∞** on the CE/UIAA label. Being even smaller in diameter than half ropes, they must both be clipped into each piece of protection.

Half and twin ropes are particularly popular in ice climbing and when doing crampon work where there is a very real chance of punching one rope with a crampon spike or ice tool pick.

The use of half and twin ropes also has the advantage of facilitating longer rappels on the retreat since it is not necessary to halve the rope, like must be done with a single rope, so that it can be retrieved at the bottom of each pitch.

50 meters of two color 11mm dry PMI dynamic climbing rope in a mountaineer's butterfly coil

ROPES—COMMON SIZES & USES

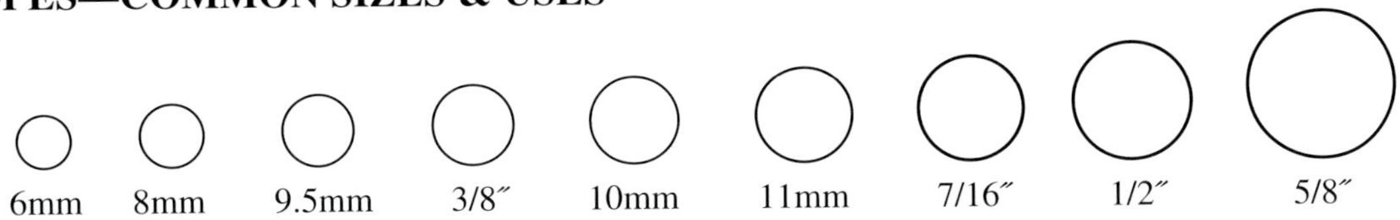

DYNAMIC ROPES:

11 mm:	General climbing, training and mountain guiding
10 mm	Light Single rope climbing and mountaineering
9.6-9.9 mm:	Lightest Single ropes used for mountaineering
9.5 mm:	Primarily used for glacier travel
9 mm:	Half-Rope used for double-rope climbing
9 mm static:	Static line used for hauling systems
8 mm-8.8 mm:	Half-Rope used for double-rope climbing, used in pairs
8 mm static:	Static line used for fixed ropes in expedition-style climbing
7.6 mm-8 mm:	Very light Twin-Rope climbing—must be used in pairs

Smaller line ranging in diameter from 5mm-8mm is generally static accessory cord used for prusik loops, slings, cordelettes and hauling line.

Dynamic climbing ropes are usually packaged in metric lengths of 50 to 70 meters (1 meter = 39.4″); 50 meter = 165′; 60 meter = 197′; 70 meter = 230′

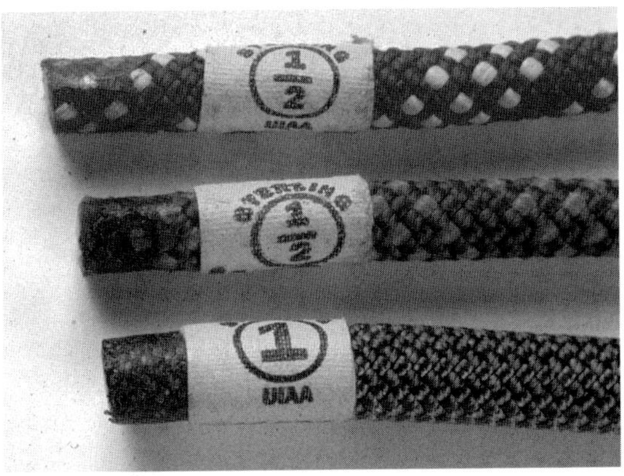

The UIAA 1/2 marking on a pair of Sterling Dry 8.8mm Half ropes and the 1 symbol indicating a Single rope.

Static 7/16″ rappel rope clearly marked "Not for Lead Climbing"

STATIC ROPES:

Static line does not have the stretch of dynamic rope so is ideally suited to rescue work, rappelling, hauling lines and fixed rope installations. Even though climbers rappel on their dynamic ropes rather than carry a static rope just for that application, the advantages for tactical rappelling and rescue work are obvious. The elongation in a long dynamic rope could cause the SWAT operator to pass the window that he intends to enter, or the rescuer to pass the victim. Similarly, when a climbing party has to ascend up a fixed line with the aid of ascenders on an exposed piece of mountain, it is considerably more efficient if the rope does not stretch and elongate every time it is loaded.

Static Rope comes in several diameters each suited to different applications:

5mm-8mm static line is routinely used for utility cord and prusik loops.

8mm static line is commonly used for fixed lines on expedition-style climbs.

9mm is often used as a hauling line for big wall climbs.

3/8″ (9.5mm) line is rated for approximately 5,000 pounds and can be used as a lightweight one-man tactical rappel line, fixed line or hauling line. 3/8″ Rescue Rope is also approved by some teams for light-duty mountain rescue applications with each team member carrying a 100′ length (minimum).

7/16″ (11.1mm) rope, rated for 6,500 pounds, and known as tactical assault line (black) or rescue rope (colored), is the most popular for rappelling, and is NFPA rated for One-Person rescue applications. 7/16″ Rescue Rope is used by many teams working in remote locations as their primary rescue rope since it is lighter to carry than half inch.

1/2″ (12.7mm) CMC Rescue Lifeline is rated for 9,500 pounds and is the workhorse of the rescue community since it is NFPA rated for Two-Person rescue applications such as pick-offs. New England KM III Static Kernmantle and Wellington Rhino are other excellent rescue ropes in this category.

150-foot coil of CMC 1/2″ Rescue Rope

5/8″ (15.9mm) CMC Rescue Lifeline is rated for 12,500 pounds and is best suited to heavy-duty rescue applications and litter recoveries. 5/8" rescue ropes are favored by fire departments, USAR teams and other truck based rescue teams.

Static ropes can be purchased by the foot but are usually purchased in standard lengths of 150′ and 200′, or 300′ and 600′ spools.

Another static cord that is getting considerable attention is New England Ropes Maxim Tech Cord used in their Micro Rappel System (See Chapter 30—Tactical Rappelling). 5mm Tech Cord is a lightweight, super strong cord (rated at 5,000 lbs) that can be used for emergency access, escape or belays. The cord is a uni-directional Tehnora, one of the next generation of Kevlar-like fibers, covered with a polyester sheath. Tech Cord has very little degradation from UV and good resistance to most acids, alkalis, bleach, and solvents.

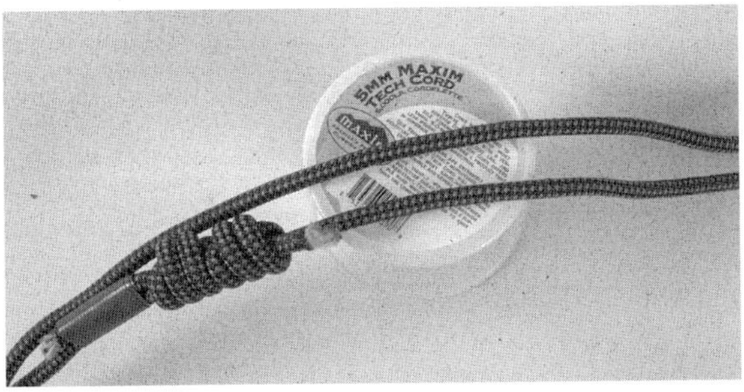

Maxim 5.0mm tech cord

ROPE CARE

Climbing ropes can be damaged in many ways, with the main culprits being UV light, chemicals, sharp edges and general abuse. It is important to visually inspect ropes regularly and get into the habit of running the rope through your hands as you coil and uncoil ropes to feel for irregularities such as bulges in the core or damage to the sheath.

Quality ropes are not cheap, and every time one is taken on the mountain someone's life may depend on it being in good condition, so all climbing, rappelling and rescue ropes should be considered "life support equipment" and treated accordingly.

Malcolm Daly of Trango preparing to rappel into the Uncompahgre Gorge in Ouray, Colorado. He is using Dry ropes that will not absorb water or freeze up.

The following are a few tips to get the best performance out of your ropes:

- Avoid sharp objects and edges (use edge protection)
- Avoid abrasion on rocks and rough surfaces
- Try to keep gravel and sand out of the fibers
- Do not walk or stand on ropes (especially wearing crampons)
- Take special care when working with crampons, ice axes and ice tools
- Avoid chemicals which contain alkalis, oxidizing agents, bleach, acids, gas, oil, battery acid and acid fumes
- Avoid impact, hard falls and shock loading
- Avoid fast rappels which generate excessive heat and damage ropes
- Store rope in a rope bag or pack for better protection
- Store in a clean, well ventilated, dry area
- Avoid cement floors since concrete contains destructive compounds that could be harmful to ropes
- Avoid UV light which will deteriorate nylon over time. Do not hang ropes in direct sunlight to dry.

8mm accessory cord knotted for ease of use as a prusik loop or cordelette

So when should a rope be retired?

The general rule is to retire a rope when you no longer have confidence in it, but this depends primarily on usage so it is important to keep a log on each rope. Obviously a rope that shows signs of being cut or abraded to the core, or one that has sustained a severe fall should be retired immediately. However, there are several other indications that a rope should be retired that can only be detected with careful inspection.

Rope inspection should be a routine procedure performed before and after all operations and training. Each rope should be run through the hand and visually inspected for irregularities in diameter, deformities, soft or mushy spots, hard spots, glazing, dirt, abrasion, cuts, or any discoloration or smell that may indicate exposure to chemicals. Any of these could be grounds for retirement.

The entire length of the rope should be inspected each time the rope is put away whether the entire length was used or not. Each rope should have its own log page and usage should be logged by an experienced member of the team so that there is consistency in the procedure.

Regular usage alone will shorten the life of a rope, especially when working in the sun, running the rope over rough surfaces, constantly loading and unloading or being used for training purposes. Some teams will designate specific ropes for training and save their newer ropes for actual rescue deployment. However, any rope to be used operationally should be first inspected and used in training to stretch it out and remove any of the kinks found in brand new ropes.

Any rope over 10 years old should be retired regardless of history and usage, even if it has just been sitting in the rope locker. For rescue and tactical teams, rope retirement will be based on training and operational tempo, but a general guide to rope retirement would be:

Professional daily use/Guiding	3 months-1 year
Regular weekend use/Training	1-2 years
Occasional use/Rescues	2-4 years
Storage/minimal use	5-7 years

Some teams will routinely rotate their ropes by using them for a year as an operational rope, then making them training ropes for the next year, and then finally cutting them into shorter lengths for utility lines in the third year. By cutting them up, the more worn or frayed sections of the rope can be cut out. These are usually the sections that are ten to twenty feet from the ends of the rope where they have laid over the edges of cliffs and buildings.

DRY COATINGS

One feature in dynamic climbing ropes that is an asset when moving above the snow line is Dry Coating. This treatment gives much better abrasion resistance and prevents the rope from taking on water weight or freezing in icy conditions. This is important since wet rope is cumbersome to carry, has a lower tensile strength (up to 30-50% less) and compromised knotability at temperatures near freezing. Some ropes with loosely woven sheaths and poor dry treatment can absorb enough water to virtually double their weight and make them unusable when they freeze.

After manufacturing, using Maxim rope as an example, the rope is run through a bath of fluoropolymer which allows the coating to fully penetrate to the core yarns. The rope then proceeds through a low temperature oven that sets the dry coating. Other manufacturers will go a step further and treat the individual strands before the rope is woven together.

PMI literature recommends never using soap on dry treated ropes, since the dry coating begins to degrade after 3 washings and is gone after 7. However there is a product called Proarid that is an aqueous fluro-chemical product that when applied to a rope provides excellent water repellency and one bottle should dry treat up to four ropes.

MARKING ROPES

It is often useful to mark the ends and mid-point on a rope. On the end one can note the length of the rope and the date it went into service, and the middle can be marked for doubling the rope for rappelling purposes.

There are several brands of marker that will not harm the rope: Carter Marks-A-Lot, Sanford Sharpie, and Binney-Smith Majic Marker.

ROPE COILING

Most rescue ropes are stored in throw bags or deployment bags that both protect the ropes and make rapid deployment more efficient. However, climbing ropes are usually coiled into a Butterfly coil that can be carried like a backpack in the center of the back (see Photos). But since the mountaineer will invariably be wearing a pack, it is best to stow the rope in the pack to better protect it from the elements.

Some climbers will use a combination bag and ground cloth (Rope Ranger) to facilitate rope handling and keep the rope cleaner. With this system the rope can be loosely flaked onto the ground cloth, which is then rolled around the rope and stuffed into the bag that is secured with a draw-string.

Every effort should be made to keep ropes clean and dry so a ground cloth is very useful when working in the snow, but a wet rope can also be carried on the outside of a pack to help drying, as long as it is tied securely since ropes have a tendency to slip out and become a big mess.

When climbing as a rope team on multi-pitch climbs, it is important that everyone on the team can handle the rope or ropes efficiently. To do this requires a system of taking in and letting out rope when belaying that will avoid unnecessary tangles. The rope coil or pile should also be positioned so that it is out of the way yet still convenient to work with. If there is any possibility that the rope has become tangled in the pile, then the entire rope should be quickly run through the hands and re-flaked into a loose pile to ready it for the next pitch.

20 meter coil of 9.7mm rope

ROPE CLEANING

Ropes can be machine washed by placing the rope in a mesh bag or pillow-case then washing on gentle cycle with cool water and a mild chemical-free soap such as Ivory Snow or Woolite. Commercial rope washing compounds are also available such as Lifeline Cleaner which is biodegradable, nontoxic, non-acid, non-alkaline, non-flammable and mild on the skin.

The empty washing machine should be run through a rinse cycle with just water to first remove any residue from previous uses. Then, to avoid tangles, first chain the rope and agitate for a few minutes adding diluted Lifeline cleaner (1/2 cup for a full-sized washer). Stop and let soak for about 30 minutes then resume normal wash cycle.

The rope can then be dried by hanging out of direct sunlight, and leaving it chained makes it easier to handle. Once dry and before coiling or packing, the rope should be re-inspected and a record of the wash made in the rope log.

UIAA TESTING OF DYNAMIC ROPES

Since most climbing ropes carry a UIAA certification and the test data is frequently published in company literature, the reader should have some understanding of what all these numbers mean.

Any climbing rope carrying UIAA certification has met all the minimum safety standards set by the UIAA (Union Internationale Des Association D'Apinisme). This series of tests is used to evaluate a climbing rope's performance characteristics under laboratory controlled conditions. Energy absorption, elongation, and flexibility are only a few of the properties tested.

UIAA laboratories test the ability of a single rope to survive, without breaking, five successive impacts of an 80kg weight dropped 4.8 meters. The stress of that drop must be absorbed by 2.8 meters of rope, a test designed to simulate the fall of a lead climber. The result is seen in the number of UIAA falls and the Impact Force number. A high Number of Falls and a low Impact Force number is preferable.

Impact Force: Probably the most important single statistic on the hang-tag, this figure is an accurate indication of how much force a rope is able to absorb to protect the climber in a fall. Since the Impact Force is the maximum load transmitted to the climber and his protection in a fall, the lower the number is the better. The UIAA standard for maximum impact force is 12 kN (2640 lbs.) for single ropes and 8 kN (1760 lbs.) for half ropes.

Falls Held: The UIAA test drops an 80 kg weight 5 meters on 2.8 meters of rope repeatedly at approximately five-minute intervals until the rope breaks. To meet the UIAA standard a rope must survive a minimum of five falls. This test is a factor 1.78 fall, simulating a short, severe climbing fall. It represents a theoretical "worst case" fall combining impact forces and the effect of a standard carabiner edge. Since all ropes carrying the UIAA mark are tested in the same manner, this number (UIAA test falls held) indicates a performance comparison that can be made between two different climbing ropes.

Static Elongation is the percentage of a rope's length it will stretch in a static loading situation. The UIAA test hangs an 80 kg (176 lbs.) weight from one meter of rope and measures the percentage the rope stretches. This gives an indication of how much the rope will stretch in a static loading situation (rappelling, lowering, or ascending). The maximum static load stretch allowed by the UIAA is 8% for single ropes and 10% for half ropes.

Tensile Strength is the same as Static Breaking Strength. The rope is stretched until it breaks and the force required to break the rope is measured. Although this is not a UIAA specification, static breaking strength can be used as a comparison of the ultimate strength of the rope.

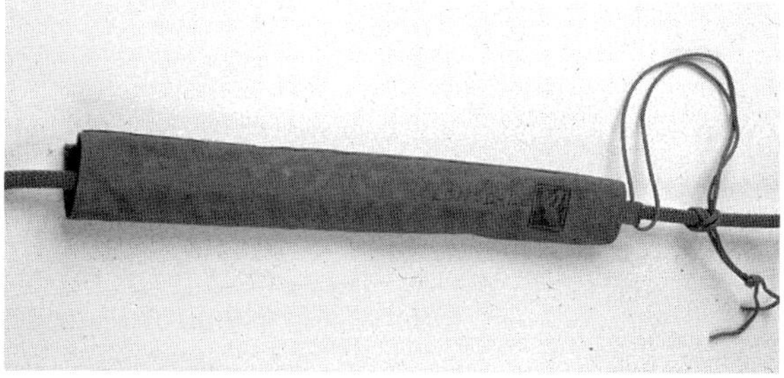

Eagle Industry's edge protection sleeve can be Velcroed around any rope

MANUFACTURERS & ROPE SPECIFICATIONS

Over the years we have used and evaluated a number of ropes at STTU for rescue, tactical and climbing applications. Since selection of a rope usually begins with a look at the manufacturer's reputation and published specifications, we have included some additional information that may be of interest to the reader.

NEW ENGLAND ROPES

New England has been a leading manufacturer of high-quality, specialized ropes for over thirty years and has the distinction of being the first American rope manufacturer to earn UIAA and CE approval. New England manufactures a range of both dynamic (Maxim) and static (KM III) kernmantle construction climbing and rescue ropes.

Maxim rope uses continuous, high tenacity strands of premium nylon fiber in the core and sheath with tension and twist that optimize flexibility, durability and long life. For maximum abrasion and moisture resistance, the core is covered in a 48 carrier high-fiber density sheath and bonded with a proprietary dry coating.

DYNAMIC ROPES

Maxim	11mm	10.8mm DC	10.5mm	9.8mm	9.0mm
UIAA Falls	10-12	9-10	8-10	7-9	9-11
Impact Force	9.6kN	10.2kN	9.8kN	8.7kN	7.2kN
Elongation	6.3%	6.3%	7.3%	7.3%	7.4%
Grams/Meter	74.6	75.8	67.1	62.4	52.0
Type	Single	Single	Single	Single	Half

KM III - STATIC

KM III is exceptional static rope for rappelling, caving and rescue applications. It has a unique continuous filament polyester sheath that absorbs less moisture, is a lower stretch fiber and more abrasion resistant than nylon. For balanced construction, the polyester sheath, which makes up 50% of the weight of the rope, is braided over a continuous filament Nylon core to limit rotation and elongation and improve rope handling.

These ropes provide high-tensile strength with low stretch and are torque-balanced for no spin when rappelling. They are also UL certified to meet NFPA 1983 (1995) standards for rescue work.

STATIC ROPES

KM III	3/8″	7/16″	1/2″	5/8″
Weight / 100′	4.5 pounds	6.0	7.2	10.4
Tensile Strength*	6,600 lbs	8,000 lbs	9,700 lbs	13,600 lbs
Elongation @ 450 lbs	2%	2%	2%	2%
@ 900 lbs	3%	3%	3%	3%

*Tensile strengths are the average of three tests conducted according to Federal Test Specification 191A, Method 6016. For 1/2″ diameter rope, all test results exceeded 9,000 pounds as required by NFPA 1983, 1995 edition.

MAMMUT ROPES

Mammut ropes, marketed by Metolius, are renowned for their high quality and ease of handling and have long been a favorite of many climbers. The two that we use and have been impressed with are the Galaxy and the Flash.

The **GALAXY** is a 10 mm Dynamic Super-Dry. In a 70-meter length it has a coil weight of about 10 pounds (64 grams/meter); is UIAA rated for 7 falls; with an elongation of 7%.

The **FLASH** is a 10.5 mm Dynamic Duodess Super-Dry. In a 60-meter length it has a coil weight of 9.5 pounds (70 grams/meter); is UIAA rated for 9 falls; with an elongation of 6.5%.

Both have a soft, subtle feel that makes them a pleasure to handle and climb with.

PMI ROPES

STATIC—PMI's Low-Stretch Kernmantle ropes have an average elongation of less than 2% at 200 lb, 17% at 75% of maximum breaking strength and less than 20% at failure. Their core uses a continuous, block and creel construction with parallel nylon fiber bundles for low stretch and no spin.

DYNAMIC—PMI's UIAA tested and labeled 11mm Dynamic UltraDry, 50-meter climbing rope has a coil weight of 9.5 pounds. The 10.5mm is the best balance of performance and weight, is considerably lighter than the 11mm and yet extremely durable.

	11mm	10.5mm	10mm	8.6mm
Weight	78g	69	62	48
UIAA Falls	16	12	7	17 (Single/55kg load)
Impact Force	7.2kN	6.8kN	6.8kN	4.8kN
Static				
Elongation (80kg)	6.8%	7.2%	7.8%	8.8%
Sheath Slippage	0mm	0mm	0mm	0mm
Type	Single	Single	Single	Half

STERLING ROPE COMPANY

Sterling is a US company that made its first ropes in 1993 and now manufactures a full range of both static and dynamic ropes. In addition to the black static line that has been included in Marine Corps MACK boxes, Sterling also makes rope in OD green and camouflage, and ropes that are Infra-Red (IR) neutral for night applications.

STATIC—Sterling's 7/16" Static, 50-meter, has a coiled weight of 10.5 pounds and comes clearly labeled "Static Rope—Not for Climbing" to prevent military assault climbers grabbing the wrong rope for climbing purposes.

(Note: It is good policy to color code different types of rope—for example, all static ropes Black and all dynamic ropes OD green or colored).

This static rope has a 48 carrier sheath and dry core construction, a tensile strength of 8000 pounds; an elongation of 1.5% with a 200 pound load; and a weight of 6 pounds per 100 feet.

Sterling's black 8mm Accessory Cord is used by the military for prusik and tensioning systems and has a tensile strength of 3300 pounds / 1500kg. A 40′ coil weighs 1.1 lbs.

DYNAMIC—For a pack-weight comparison, Sterling's 10.2 mm Dynamic Dry, 60-meter, has a coil weight of 10 pounds.

The 8.8mm Dynamic Dry Half-Ropes, 60-meters (Blue & Red), have a coil weight of 6.1 pounds each. The 7.6mm Dynamic Twin-Rope, 50 meters (Camo), has a coil weight of 4.1 pounds each.

	11mm	10.6mm	10.2mm	9.8mm	8.8mm
UIAA Falls	11	9-10	10-11	8	11-12
Impact Force	9.4kN	7.4kN	8.8kN	8.9kN	6.5kN
Elongation	6.6%	8.0%	6.5%	7.0%	8.9%
Grams/Meter	78.1	75.2	74.4	69.3	54.2
Type	Single	Single	Single	Single	Half

Army National Guard Mountain Warfare School

PART II
BASIC ALPINE TRAINING

(Atlas © 1999 John Claussen)

7

BASIC TRAINING & FITNESS

After two decades of developing training programs for a wide variety of law enforcement and military special operations teams, STTU has not only had time to perfect an approach to new program development, but has also learned a few things about why more traditional programs are ineffective and how institutionalized thinking can be the progressive instructor's nightmare.

Knowing all this, we have not only developed and launched numerous highly effective training programs, we have also created instructor and command level courses on how to develop more effective in-house training programs. This has been included as a guide for instructors and team leaders seeking direction in alpine program development.

STTU approach to Program Development:

- Study the Mission Requirements of the team or agency
- Review existing budget, personnel and equipment resources
- Study prior related operations and mission debriefs
- Study the equipment and training of several similar US and foreign teams
- Break with institutionalized thinking and bureaucratic inertia
- Seek out the best available civilian and military subject matter experts
- Test and Evaluate a broad range of mission related equipment
- Integrate existing in-house knowledge and experience
- Develop operational protocols and essential equipment lists
- Develop command, control and communication protocols
- Re-define the team selection and individual fitness criteria
- Develop training aids and training evolutions
- Establish training conditions and certification standards
- Validate training and equipment selection through realistic FTXs
- Launch the program with a view to periodic revision and up-date

TRAINING PROGRESSION

Once the individual skills have been identified, the next step in launching a specific program is to establish a training progression based on the principle that each subject forms the foundation for the next. For example, there's no point in jumping into mountaineering training if the students are not competent in rappelling and belaying, and the foundation for both of these skills is knots and rope management. Similarly, before one can aspire to be an alpine mountaineer, one must have experience in basic winter hiking and camping skills. So the progression for an Alpine Operations program could be divided into Four Phases.

PHASE ONE—SELECTION
- Physical Conditioning
- Psychological Evaluation
- Team Selection

PHASE TWO—WILDERNESS FUNDAMENTALS
- Outdoor Living / Camping Skills
- Back-Country Travel
- Cold Weather Operations
- Knots and Rope Management
- Basic Rappelling and Belaying

PHASE THREE—BASIC ALPINE OPERATIONS
- Basic Mountaineering
- Rock & Ice Climbing
- Glacier Travel & Crevasse Rescue
- Avalanche Training (where appropriate)
- High Altitude Operations
- Expedition Planning & Execution

PHASE FOUR—MISSION SPECIFIC TRAINING
- Mountain Rescue
- Technical Rope Rescue
- Mountain Warfare
- Tactical Operations

PHYSICAL FITNESS

Mountain fitness is different to basic physical fitness in that the mountains are a unique environment in terms of climate, terrain and altitude that bring a whole new meaning to the word endurance. Many individuals with exceptional levels of fitness, even champion tri-athletes, have found the rugged terrain, altitude and steep exposure more than they were willing to endure. Flat ground is a luxury in the mountains; the weather is always unpredictable; days can be hot and nights freezing cold; preparing food can be difficult and sleep does not come easily at altitude. All contributing to a miserable work existence for some, yet there are those who thrive in this environment, and those are the ones one must seek out for alpine operations.

In addition to being physically fit, the mountaineer requires a positive mindset, the right attitude, a firm commitment to the mission, and the technical skills to not only survive but to be able to work with relative comfort.

The training for all this is a balance between aerobic exercise, strength training and mission specific conditioning. Increasing levels of aerobic exercise for the long uphill hikes and climbs, leg and back strength to handle the heavy loads and arm strength necessary for some climbing moves.

SPECIFICITY IN TRAINING

There is a term used in human physiology studies that is very applicable to alpine training—"S.A.I.D"—Specific Adaptation to Imposed Demands. In brief, it means that the human body will adapt based on the demands placed upon it.

At STTU we are staunch proponents of specificity in training, in that if you want to be a combat swimmer then you have to get fins on and swim long distances. If you aspire to greatness as a combat shooter, then you have to put in the trigger time on the range. So it is only logical that to be strong in the mountains, there can be no better training than time in the mountains—not at some plush ski resort but humping a pack up and down every possible type of terrain available.

However since this is not always practical or convenient for many teams, there are several related recreational activities that will condition the climber and help maintain existing levels of fitness. Beach running, stair climbing, inclined treadmills, hiking, mountain biking, alpine and cross-country skiing, free weights and sport climbing are all beneficial to the aspiring mountaineer.

Whatever the pursuit, a regular aerobic workout, with a good cardio-vascular output for over an hour each day is essential. The first 20-30 minutes are often hard but after that there will be the rush of endorphin release giving the ability to continue. It is important to stay within your comfort zone and it is recommended to train at a moderate level, not pushing the body to exhaustion. As soon as you over-do it, the endorphin release tapers off and your body moves into a distress response. It will also stop burning fat and revert back to burning the sugars and glycogen that the brain needs to remain alert and coordinated.

Personal fitness for the mountain rescue or mountain warfare specialist must be more than just an occasional trip to the gym or a jog around the block—it must be a life style.

There is no question that to feel good, perform better and live longer, one needs—

- A positive mental attitude with realistic expectations of improvement
- An exercise regime including both aerobic exercise and strength training
- A diet high in complex carbohydrates and plant proteins
- Drinking adequate amounts of water every day—more when exerting heavily, in heat and at altitude
- A lifestyle that incorporates self responsibility, self-reliance and an avoidance of excesses—no smoking or drugs; with sweets, desserts and alcohol in moderation

PSYCHOLOGICAL FITNESS

On the subject of fitness it is also interesting to look at the psychological fitness and attitude of many climbers and mountaineers. In studying mountaineering, the regularity with which climbers are killed, even big names in the sport, becomes immediately evident. In reading the accounts and memoirs of many great climbs and climbers we find

Military mountaineers going through the Austrian Army's International Rock & Ice course feel the exposure of the European Alps (Sgt Clearman)

an interesting thread running through many of these stories—and that is the need to summit regardless—regardless of extreme discomfort, fatigue, hypothermia, frostbite, lost equipment, dead team members, inclement weather, lack of daylight, or lack of supplemental oxygen. It is as though they are oblivious to self-preservation. On the side of that mountain, and at that point in time, their world is reduced to one single-minded task—making the summit.

Taking a step back from that position, it is interesting to sit in the comfort of base camp and listen to climbing teams discuss and agree on contingency plans and turn-around times. They'll agree that it is important to stay together; that they will take care of each other; that the mountain will always be there; and how much they want to return to their families after they summit. But then on the final push for the summit, after several days on the mountain working their way up to high camp, and several hours of hard climbing, and within striking distance of the top, they will begin making irrational decisions.

Hours after the agreed upon turn-around time, they will try to rationalize that "there's still time"; that the "storm front is still hours away"; that "my toes no longer feel cold and painful"; not realizing that the altitude, fatigue and their own need to "beat the mountain" have skewed their judgement. They will literally climb to their deaths unless calmer heads prevail.

World-renowned mountaineer Ed Viesturs, put it best when he said, "Getting to the summit is optional—getting down is mandatory". And yet all too many climbers get to within sight of the summit too late in the day, under inclement conditions, or without their climbing partner who may have wisely turned back, and push on regardless. For some the gamble pays off but for many their destiny is to become a frozen lump on a wind swept ridge, often not found until the following season if at all. The only eulogy being, "He died doing what he loved—but what a waste. His poor family."

But I ask you now, try to find me a climber laying in the snow exhausted and frostbitten, who says, "Leave me to die, I am doing what I love."

It is understood that mountaineering is a high-risk activity, and if you are not willing to assume that risk and take full responsibility for your actions, you shouldn't venture into the mountains. But it is also important to realize that no summit bid is worth the loss of your life or permanent disability.

Sudden deceleration (falling) is the number one cause of death for climbers, but in analyzing a number of mountaineering accidents, the following are some of the common mistakes that climbers make that ultimately contribute to their deaths.

- Poor judgment driven by an over active ego
- Lack of training and experience
- Climbing un-roped
- Moving through known avalanche areas without due care
- Ignoring the signs of altitude sickness and not going down immediately
- Continuing a climb when missing critical pieces of equipment (rope, crampons, ice axe, gloves)
- Trying to summit late in the day and after the agreed upon turn around time
- Ignoring the signs of approaching bad weather or low visibility
- Failing to turn the climb when oxygen supplies are insufficient to continue
- Guides competing to summit the most clients
- Continuing to climb with known frostbite or other injuries
- Making a challenging dynamic climbing move knowing it could be the last
- Continuing a climb with sketchy or no protection

Granted, it takes determination to overcome many of the obstacles that nature presents us with on the mountain, but too much determination will get you killed. Determination must be tempered with common sense combined with training and experience.

8

WILDERNESS LIVING

The mark of an experienced outdoorsman is the level of comfort that he is able to maintain utilizing only what he carries in his pack or can find in the wilderness. It is interesting to note that many people venture into the mountains only to have a miserable time because they were cold, wet, hungry or unable to sleep, whereas others can go into the same conditions and be totally content. The same is true for veteran soldiers or Marines in a war zone. In the early days of an overseas deployment, the young bucks are wholly consumed with the tactical aspects of the conflict and load themselves down with ammunition and grenades. They are, however, surprised to see that the veterans of previous conflicts have brought in an array of items needed more for comfort and good health than for killing the enemy.

In time, as the deployment inevitably drags on and the first timers become seasoned veterans, they too will become more concerned with creature comforts and begin working to improve their base camps and personal living quarters. They will also come to realize that they must pay particular attention to personal hygiene and nutrition if they are to maintain any level of fitness and combat readiness. In a tropical combat zone for example, even the smallest wound can become quickly infected, and in a cold weather warfare environment minor frost-nip can result in amputation if left unattended.

The lesson to be learned here is that being relatively comfortable, like being healthy, is important and a process that begins long before entering the mountains. It begins with education and research on the area of operation, careful gear and food selection, equipment preparation, and additional training if necessary. Shots will also be required if operating in countries that have a history of disease such as Cholera, Malaria, or Hepatitis—any of which can debilitate a man and cause an operation to be aborted.

OUTDOOR FUNDAMENTALS

The fundamentals of camping and outdoor living are a combination of the following:

- Attention to personal Health & Fitness
- Careful selection of Boots, Socks & Clothing
- Systems appropriate to the region
- An understanding of the importance of Nutrition & Hydration
- Pack selection and sensible Packing limits
- Selection & use of suitable Tents & Camping Equipment
- Careful preparation of Water & Food in the wilderness
- Good Navigation & Back-country Travel skills
- A knowledge of First Aid & Basic Survival

ALPINE OPERATIONS

For the professional, wilderness living is more than just the ability to simply survive, but the need to be able to function with a high degree of efficiency for a prolonged period of time. This requires a level of comfort and confidence and the ability to maintain a high degree of health and fitness for the duration of the operation.

While doing research for this book, I have seen teams that after as little as five or six days in the mountains looked like they were part of the Bataan death march and incapable of any significant work other than putting one foot in front of the other. Blisters had gone unattended; they became too tired to prepare hot food or melt snow for water; their personal hygiene deteriorated and they lacked the energy or motivation to improve their situation. Without attention blisters become infected; without water climbers become dehydrated; and without hot food they become cold and listless—and if disaster hits under those conditions the team is doomed.

Bad weather, high winds, bone chilling cold, deep snow, thin air, dehydration and lack of nutrition conspire to drain away the morale and energy of the unprepared. Similarly, exertion over long distances, steep climbs, heavy loads, compounded by dehydration, headaches, a hacking cough and lack of sleep at altitude can turn strong athletes into walking zombies. This makes it important to establish practices and procedures, like forcing oneself to eat and keeping water bottles filled that will head off these ugly problems before they disable the individual or an entire team.

BASIC CAMPING SKILLS

The following is a list of basic camping skills that must be taught in a comprehensive alpine program:

- Compass navigation and map reading
- Packing & waterproofing gear
- Foot care & blister repair
- Locating & purifying water
- Selecting & preparing a suitable camp site
- Setting & anchoring tents
- Stove operation & cooking
- Tent & sleeping protocol
- Washing & drying clothing
- Bathroom procedure & human waste disposal
- Low impact camping & environmental considerations

US military current issue 4-man tent made by North Face in a well-established campsite carved our of deep snow

COMFORT & HEALTH CONSIDERATIONS

The first line of protection for an outdoor enthusiast is the clothing system along with basic camping gear—tent, sleeping bag and ground pad. These should be selected for the season, anticipated terrain, expedition duration and expected weather conditions—all of which will require some prior knowledge of the area of operation.

Since clothing is such a critical factor in maintaining both comfort and health, the novice mountaineer should be constantly working to improve his basic clothing system and layering practices.

A useful acronym for this is, "Think C.O.L.D. to stay warm"

- Keep clothing CLEAN
- Avoid OVERHEATING
- Wear clothing loose and in LAYERS
- Keep clothing DRY

One of the most miserable experiences in the mountains is to have essential gear like spare clothing and sleeping bags get wet, since this greatly accelerates heat loss and can lead to hypothermia. Spare dry clothes, socks and sleeping bags should be stowed in plastic bags and lightweight, waterproof stuff sacks. In this way, no matter how heavy the rain or how deep the rivers, you will have dry clothes to change into when you make camp. After making camp each afternoon, clothes and sleeping bags should be placed on the tops of tents to air out and dry in the sun, but being sure to tie them down if there is any possibility of wind.

Being dry and relatively clean are important steps toward maintaining health for an indefinite period. It is not sufficient to say, "Hey, its only a week," and then come off the mountain at the end of the 7 days looking and smelling like shit. The professional pays meticulous attention to personal hygiene when in remote locations and especially in third world countries. He should be able to come off the mountain after several weeks almost as healthy as when he went up—except probably several pounds lighter. Any health problems, such as diarrhea or blisters need to be treated immediately and before the problem worsens.

Note socks tied to pack and drying during the day

Without a good degree of personal hygiene, the body becomes a breeding ground for bacteria that will infect the smallest cut or scrape. Bacteria thrives in warm damp areas so if socks are not rotated and washed, blisters will become infected and walking will be slow and painful. Rest days around camp and even long stops on the trail are a good opportunity to get your boots off and air your feet in the sun to both dry-out the sweat soaked skin and give blisters the opportunity to dry and heal.

Although it is normal practice to wear one set of underwear for two to three days before changing, every opportunity should still be taken to clean and dry underclothes. Anytime weather and conditions permit, an undershirt or pair of sock liners can be rinsed out and hung on your pack to dry while walking, as long as one dry set is always maintained in the pack for evenings. If washing is not practical, then just airing them in the sun and shaking out the dirt and dead skin will help.

THIRD WORLD COUNTRIES

In the past two decades I have worked and trained in over 40 countries, including considerable time in the Himalayas and Andes doing research for this book, and have been fortunate to have almost no health problems. But in contrast, almost everyone on my teams has had serious bouts of diarrhea or some form of infection that has totally incapacitated and even hospitalized some members because of their lack of attention to personal health and hygiene.

Everyone knows, or should know, not to drink the water in third world countries, but it goes much further than that. Some semi-third world countries will claim to filter, irradiate or chemically purify their water—but who wants to take that chance. If it is not personally boiled or iodized, don't risk it!!

In some towns the local street venders have a devious habit of re-filling supposedly "purified bottled water" bottles with local water and then selling it to the tourists and trekkers under the pretense that it is bottled, pure and sealed. If in doubt about the seal, iodize it!!

A full range of shots and inoculations (especially hepatitis and tetanus) are essential when traveling in third world countries, and if the health warnings indicate that rabies is a problem, then spend the money and get the rabies shots as well. This is preferable to the treatment that would be necessary after a bite from a rabid dog, which must be done whether you know the mangy mutt is rabid or not. The treatment is both painful and unpleasant—if even available at the local clinics.

Medical assistance in remote regions is often slim to non-existent; clinics that are available may have marginal hygiene standards themselves; essential medicines may also be in short supply, so don't count on any local aid. (In one Eastern European country I had to have surgery on my foot with no anesthetic and no antibiotics because the hospital was out of both.). So it is recommended to get advanced levels of training in emergency medicine and to take all your own supplies. (At the end of some expeditions or operations we will give a lot of our medical supplies to local village clinics that are strapped for funds).

The following are a few basic health and hygiene rules for under-developed countries:

- DON'T DRINK THE WATER—Beer and Coca Cola can be bought almost everywhere and tea and coffee made with boiling water should be safe
- When in the towns, even in hotels, keep a good supply of bottled water on hand
- Don't buy bottled water from street vendors or if the seals have been broken
- Don't eat salads or anything that may have been rinsed in water
- Don't brush your teeth with anything except bottled, boiled or iodized water
- Run the showers hot and keep your mouth shut
- Only eat fruits that can be washed and then peeled
- All food should be adequately cooked
- Know what is up-stream of drinking and bathing areas—it may be a village toilet facility

Most rivers in third world countries are heavily polluted. This one running through Kathmandu is particularly filthy and yet the locals consider it sacred and bath in it. The concrete platforms to the left are for burning the dead—and you can guess where the ashes go

Civil unrest and political instability is not uncommon in many of the countries that are home to the world's highest mountains

- Carry a good supply of anti-bacterial handi-wipes
- Keep your hands away from your mouth
- Clean your hands with handi-wipes after defecating
- Don't allow anyone else to touch your food
- Don't allow others to dip their dirty hands into your candy or trail mix bag
- Don't allow others to drink from your water bottle or drink from theirs

- Get all necessary inoculations well before traveling
- Carry remedies for diarrhea, head aches and muscle pain
- Treat small cuts and scratches immediately
- Repair blister damage at the first inclination of a hot spot
- Prevention is always preferable to Cure

HYDRATION

Fact 1—While the human body can go for weeks without food, it can only survive a few days without water.

Fact 2—A man's body is 60-65% water; a woman's 50-60%; and the human brain is about 75% water.

Fact 3—Water regulates body temperature, carries nutrients and oxygen to cells, removes wastes and protects organs and tissue.

Conclusion—Water and Hydration is important.

Medical experts recommend 8 ounces approximately 8 times a day (4 pints/2 liters) under normal living conditions, but we use and lose that just through normal elimination, sweating and breathing. With exercise, humidity and altitude we may lose another quart or more. So in the mountains that intake needs to be increased to 3-4 quarts/liters per day depending on altitude, temperature and exertion.

Water is also a simple remedy for a multitude of health problems and can aid digestion and ease muscle soreness. Water helps people engaged in physical activity perform more efficiently since it lubricates the joints, helps maintain proper muscle tone, cools the body and helps replace fluids lost through perspiration.

To satisfy this demand, most mountaineers will carry two 1 liter water bottles at all times and try to top them up at every opportunity. Water should be consumed constantly throughout the day and only conserved when absolutely necessary.

The simple test to see if you are adequately hydrated is to observe your own urine stream, which should be clear, copious and odorless. If not then you may be dehydrated, but also be aware that some foods like multi-vitamin pills will turn the urine yellow even when fully hydrated.

Melting snow for water becomes a full-time chore above the snow line. During the day climbers should be stuffing snow into half-filled, wide-neck water bottles to make more water. This will melt in the sun, especially when placed in the top of the pack or near the body. Around camp, snow can be kept in large black plastic trash bags (preferably new clean ones) right outside the tent. This will also melt in the sun during the day and make snow more accessible at night without having to pull on boots and go dig for it.

The author uses and XGK to melt snow for water at 17,600′ in the Andes. Note the team's Nalgene water bottles lower right waiting to be filled

At night and in cold over-cast conditions snow will have to be melted in a pot over the stove. All water bottles should be routinely filled in the evening and kept in sleeping bags at night to prevent them freezing. At lower elevations water can be more easily pulled from streams and then run through a filter, boiled or sterilized with iodine tablets.

Hot liquids are an excellent way to start the body's furnace in the morning and pick-up morale on cold and arduous multi-day operations. One luxury that many mountain guides and rescue teams will allow themselves is a lightweight stainless steel thermos (OR) which will retain the heat in a liquid, losing only about ten degrees over a five hour period, while a plastic or poly bottle will lose sixty degrees in the same amount of time.

FOOD & NUTRITION

One essential ingredient for the body while in the mountains is carbohydrates and lots of them since they are easily absorbed for energy. Some fat and protein are also necessary but both are more difficult to digest and better eaten in the evening to repair tissue torn down during the day. Digestion also draws blood to the stomach and away from the extremities where it is needed to maintain circulation and warmth and help prevent frostbite. For quick energy and convenience most climbers will carry a bag of their favorite candy and trail mix, along with Snickers Bars, Cliff Bars and jerky.

To simplify food preparation, several companies like Alpine Aire and Mountain House market a variety of pre-packaged freeze-dried menus suited to camping and climbing, but it is also not difficult to make up your own menus from the local supermarket. Bulk buying for the team is cheaper and allows one to repackage everything into more convenient plastic bags eliminating a lot of unnecessary individual packaging that must be carried down off the mountain.

The daily menu for many climbers usually consists of instant oatmeal and hot chocolate or coffee in the morning; snack foods and drink supplements for lunch; and then something more substantial for dinner such as a rice or pasta based meal, along with more hot chocolate. In reality, lunch can begin shortly after breakfast and continue throughout the day with a number of small snacks, candies, cookies, sardines, jerky, energy bars, and drink supplements.

Where appropriate, climbing teams should try and pack in as much fresh food as possible, at least to base camp, especially if taking advantage of mules, porters or helicopters for the initial move onto the mountain. Nothing is more appetizing and more of a morale booster than eating real meat, eggs, fresh vegetables, fresh fruit or bread in the mountains. "Fresh" also supplies essential vitamins and nutrients not present in processed foods.

When calculating pack weight, one should allow about 2 to 2½ pounds of food per man per day, so a 6-day operation will add 12 to 15 pounds to each pack. A three-week expedition could require upwards of 40-50 pounds of food per man so menu selection and packaging become critical factors in minimizing weight and garbage.

From a purely weight and bulk consideration, military MREs (Meals Ready to Eat) are the worst at over 1½ pounds per single meal serving with the bulk of this being packaging. If MREs are to be used, several meal packs should be stripped of the heavy plastic external packaging and cardboard inserts and repackaged into daily meal plans. However, the advantage of being "ready to eat" is that they do not require cooking or even heating so there is a saving in fuel, but then one is denied the hot meal which can be so important for extended operations in colder conditions.

For eating utensils, most mountaineers will carry no more than a large Lexan, plastic or aluminum spoon, a plastic bowl and a 16oz plastic mug. Cooking is done in a single aluminum or titanium 1-2 quart/liter pot complete with a lid and pot grabber. A second pot can be used just for melting snow and boiling water.

After cooking or heating food, pots should be scrubbed out so as to remove any food particles that may decay or attract bacteria. A pot scourer is handy but snow and gravel can also be a field expedient alternative.

The stove is another critical part of the alpine cook set in that it not only cooks food and makes hot drinks, it is essential for the daily routine of melting snow to make water. Most climbing teams will carry one stove between every two or three climbers, but when operating in remote locations, a spare stove may be deemed essential, even if it is just kept at base camp.

One of the most popular stoves is the MSR

Whisperlite International—a lightweight, compact and quiet unit that is run primarily on white gas but comes with a jet for kerosene as well. Many high altitude climbing teams, venturing above 20,000 feet (6,000+ meters) prefer the MSR XGK Expedition stove for its rugged simplicity, high heat output and ability to utilize almost any fuel including kerosene. Fuel should be carried in aluminum or titanium fuel bottles with O-ring sealed screw stoppers such as MSR's 33oz (975ml) units.

At lower altitudes a climber may need to carry 4-5oz (118ml) of fuel per day for heating meals, but at higher elevations where one has to melt snow for water, fuel consumption could jump to 8oz (237ml) per day, or even up to 15 ounces for places like Antarctica.

Be sure stoves are used in well ventilated areas since 4-season alpine tents are double zippered and almost air tight, resulting in deaths from carbon monoxide poisoning.

Alpine Ascents guide uses a flat rock and aluminum heat reflector to increase the efficiency of her MSR stove

SNOW SHELTERS

Even though most nights in the mountains will be spent in a good tent (see Chapter 4) there will be times when a snow shelter may be necessary. In situations where high winds are expected, tents should be located in the lee of a large rock or snowdrift, just below a ridgeline or in a natural depression in the snow for shelter. This can be augmented by building a wall of snow blocks or by digging down three or four feet to create a pit for the tent. Both methods will lessen the force of the wind on the tent and make for a quieter and more peaceful nights sleep.

In really exposed areas however, building a snow cave has several advantages, and even though sleeping in one appears to be somewhat like trying to live in a refrigerator, they offer more protection and can be many times warmer and quieter than a tent.

The temperature in a tent in gale-force winds will only be about 10 degrees F above the outside temperature and when the outside is –20 F, the inside is still damn cold at –10 F. However, no matter what the wind, the inside of a well constructed snow cave will be a sheltered and quiet 32-35 degrees F, even though a little damper than a tent.

The disadvantages of snow caves is that it can take 2-3 hours to construct one, and all that toil will leave the builders wet from both snow and sweat. In addition, where a tent will heat up to forty or fifty degrees above outside temperature in the sun, the snow cave will remain cold and wet making it unsuited to drying clothes and sleeping bags during the day.

Knowing how to building a snow cave is an essential mountaineering skill that can be used for an unexpected bivouac if caught out in a storm or in situations where tents are burned or blown away. Some climbers will go to the added trouble of building a snow cave near their tents, knowing that if things turn nasty in the night they have a prepared bolt-hole to weather out a storm.

A lightweight, wide-mouth shovel is the best tool for digging a snow cave but a cooking pot, climbing helmet or snow fluke could also be used in an emergency.

The easiest way to build a snow cave is to dig into the side of a steep snow bank (30+ degrees), starting with a small hole (1½´x 5´) for the entrance and then burrowing in and opening out the interior. As one climber digs, the other can be clearing the snow away from the entrance, all the while maintaining a moderate work pace so as to avoid sweating profusely.

The interior should be scooped out to make a domed ceiling for strength, carefully smoothed to prevent dripping, and a vent hole punched out with a ski pole. The snow cave should be just big enough for the team and their gear and built with raised areas for sleeping platforms.

Wands should be placed outside on the roof to mark the position of the shelter to prevent anyone from accidentally dropping in, and in the evening, the entrance can be blocked with packs or snow blocks to keep the wind out and the warmth in, but sufficient space must be left for ventilation and to avoid suffocation.

Other forms of hasty snow shelters can be built by digging down around the base of a large tree to take advantages of the overhead protection afforded by the branches, or by simply digging a narrow trench and covering it with branches, a tarp and snow.

Climbers at the 14,000´ camp on Denali build a wall of snow blocks for protection against the elements. (Steve Dundorf—1998)

SURVIVAL TIPS

Poor planning and bad judgement can both lead to a survival situation in the wilderness with two of the most common causes being getting lost or becoming trapped by bad weather. The degree of severity of the situation will be dictated by the amount of supplies that a party is carrying when they find themselves in this predicament.

For the plane crash survivor or individual who finds him- or herself separated from the group or stranded in the wilderness, the situation can be quite serious. However, if one stays calm, thinks it through and has even elemental training and initiative, the problems may not be insurmountable.

The most important response to finding oneself lost or stranded is to STOP and THINK about the situation calmly. Next, YELL or blow a whistle—3 blasts and then a pause, to all four points of the compass. Some comfort can be derived from having been part of a group, or knowing that when the plane fails to arrive at its destination, people will begin looking for you in a matter of hours.

After stopping and thinking, the next order of business is to identify the essential life support priorities:

- Evaluate and stabilize injuries
- Seek warmth and shelter
- Find water and food

In evaluating injuries, life-threatening conditions take priority—the basic ABCs of Airway, Breathing and Circulation; followed by serious bleeding and shock. Once these are stabilized, then fractures, sprains, burns and abrasions can be attended to.

Warmth is a critical need in a cold weather environment since hypothermia can quickly take affect, draining away any will to struggle for survival. Apart from just feeling cold, the first indication of hypothermia is shivering but this is soon followed by loss of coordination and stumbling. Without intervention and re-warming, the individual will continue to deteriorate. Victims have described hypothermia as a gradual slowing down and a dimming of the light of life, until it finally burns out. This can only be prevented by immediate donning of sufficient warm clothing for the conditions, seeking shelter and if necessary climbing into a sleeping bag with another warm body.

Even with the right clothing, exposure to the elements, particularly icy winds, will continue to have a debilitating affect on a survival victim. If no other shelter is available, then the victim can at least dig a snow cave or trench, and even though this may take a few hours to build, it will afford shelter from the wind. With little more than a candle, the inside temperature of a snow cave will be considerably warmer than the exposed slopes outside.

Once the essentials of first aid, warmth, shelter and water have been considered then one can re-evaluate the situation and decide whether it is best to stay put and wait for assistance or try to walk out. This is best answered by the following questions:

- Do people know your planned route and itinerary?
- How long before they will begin to worry?
- Will a search and rescue effort be mounted?
- Are there local search and rescue assets?
- How far are you from the nearest road or habitation?
- How many days could you survive in the present location?
- Do you know exactly where you are?
- Do you have the skills, equipment and level of fitness to walk out?

If the decision is to stay put and wait for search teams to find you, which is usually the best plan, then you must prepare signaling devices and fires; get more food and warm clothing; collect anything from the crash site that may be useful; improve the shelter; and gather more fuel for the fire. Only in situations where nobody knows where you are, or the region is so remote that assistance is not available will walking out even be an option.

SIGNAL DEVICES

A radio is the most valuable tool to have in an emergency in the mountains as long as someone is within range and on the right frequency to hear a call for help. On Denali most climbing teams use CB radios on Channel 19 to contact the rescue ranger camp, and on many overseas expeditions teams use marine band VHF radios or the newer compact Motorolas developed just for the outdoor market. But with the proliferation of cellular phones and cell sites, these have now become almost standard equipment for many teams and individuals. Even in remote areas where one would not expect to get a signal, one can often get a call out by climbing higher onto a hill or ridgeline. In really remote areas expeditions are now using lightweight portable satellite phones often complete with internet connections.

For military and governmental teams, the Motorola Sabers and MX360s are commonly used for inter-team communication and the AN/PRC-112 multi-mission transceiver for C-SAR applications.

However, in a survival situation without a radio or cell phone, there are still numerous ways to signal a search aircraft or ground team with items that are normally carried in the mountains.

- Whistle (3 blasts) or Shouting
- Flashing mirror in daylight
- Flashlights or lamps at night
- Fire and Smoke
- Waving arms to passing aircraft (usually considered a distress signal)
- Colored clothing, tents, sleeping bags or material
- An X in the snow made by walking, laying out rocks or branches
- Colored rope laid out on the snow
- Crossed skis and poles
- Gun shots (3 shots at 5 second intervals each hour)
- Signal Flares
- Explosives or Pyrotechnics (if available)

For a climbing, rescue or military team who has moved into the mountains on an operation, and is hopefully well equipped, getting lost or delayed should be less of a critical situation. Even for an individual who has become separated from his team but is still carrying a full pack, the situation is initially not too serious.

The problem solving process will be:

- Stop—Think—Communicate—Orientate—Navigate
- Communicate with the rest of the team if you have a radio
- Calculate how far you have come since the last known location
- Use a map and compass to identify land marks and get oriented
- Move to higher ground to identify significant features if necessary
- Attempt to navigate back to the last known point
- If it is late, make camp, have a hot meal and sleep on it

The more serious situation is the team or individual that leaves camp traveling light and then becomes trapped or disoriented because of low clouds, fog or poor navigation. Now survival may depend on the ability to navigate back to camp in whiteout conditions, or being able to endure a night on the mountain.

To navigate back to camp will require one of the following:

- A compass and a known bearing to follow back to camp
- A trail of footprints and wands laid on the way out
- The ability to communicate with someone at base camp who can come out and guide them back
- Someone making enough noise in camp to guide them in

To survive a night in the snow without a tent and sleeping bag will depend on just how sensible the individual was about loading his small summit pack. With little more than a down jacket, a bivi sack or space blanket, an iso-mat and a summit pack (all weighing less than 6-7 pounds total), one has a good chance of survival. By finding a sheltered area, putting on the down jacket, climbing into the bivi sack and putting one's feet into the pack, one will probably survive the night.

Survival has a lot to do with attitude and determination, but knowledge and training also play an important role, as illustrated by this climber extricating herself from a deep crevasse with prusik cords

Add a shovel and stove (another 3- 4 pounds) and survival becomes even easier since one can then dig a snow cave and have both warmth and drinking water. So at tops, a 10-11 pound pack could mean the difference between survival or freezing to death. Even if injured and unable to walk, with the above minimal equipment, one could last several days until found by a search team.

Obviously radios add a significant level of safety to mountain operations (if batteries are strong), but simple signaling devices can also make the difference between being found and freezing or starving to death in the snow. The two essentials are a whistle and a signal mirror (also found inside the lid of a compass). The whistle can be heard much further than a shout and is internationally recognized as a distress signal. Similarly, a flashing mirror can be seen by search aircraft from many miles more than an individual waving on a mountain or in the snow. Another small lightweight item that has merit is a pencil flare launcher, but even colored clothing, rope, material or drink mix can be used to mark the snow with a large X to signal search aircraft.

If the decision is made to move or try to walk out, then it is a good policy to leave notes of your intentions and direction of movement, dated and timed, for the searchers. Also leave trail markers in the form of arrows made from rocks or sticks and discard nothing, especially those items that can be used for shelter from the elements, warmth or to make or hold water.

With experience and physical fitness comes the confidence to handle most situations, but in the face of true adversity there must also be a high degree of resilience and mental toughness—the attitude that "I will not give up." One needs to have the personal self-discipline to focus attention, analyze the situation, organize resources, accept the discomfort and then not give up.

Wall built from blocks of snow to shelter tent from the wind

Soldiers setting out onto the ice in Greenland (SFC Gordon)

Approach to Cerro Torre (SP Parker)

9

BACK COUNTRY TRAVEL

Although it is more efficient to be inserted and extracted by helicopter, this is not always possible in the mountains. For one thing, even military helicopters have a peacetime ceiling of 10,000 feet and many cannot operate much above 14,000 feet. A few high altitude rescue helicopters can effect rescues up to 18,000 feet but not if they are carrying any significant load, or if visibility is limited by low cloud or fog. In addition, no aircraft can operate in the mountains in storms, high winds or bad weather.

So when high-tech fails there is always low-tech—and that means walking, with some snow shoeing and skiing where possible, but once we eliminate the more technical skills like skiing and glacier travel that are covered in the next few chapters, backcountry travel can be broken into three basic categories:

- Navigation
- Travel below the snow line
- Travel above the snow line

NAVIGATION

When working in the mountains, it is essential to stay oriented always know one's exact location. Whether it is to bring in a rescue helicopter, coordinating military assets or just getting from A to B, an experienced operator should be able to plot his route and identify his current location on a map.

In both rescue and tactical operations, just knowing you are "not lost" is unacceptable. You must be able to convert your position into a grid coordinate so that this information can be radioed to other responding assets. Even though a good pilot, familiar with the area, may be able to find your location by a description of terrain features, it is significantly more efficient and professional to be able to transmit exact map coordinates.

To do this requires the melding of several skills such as terrain following, map and compass reading, altimeter interpretation, terrain feature recognition, and in recent years, familiarity with GPS (global positioning system).

NATURAL NAVIGATION

Although having natural navigation skill or "mountain sense" is a very desirable ability, this is only acquired with considerable time and experience in mountain and wilderness terrain. However, nature does provide us with several natural common sense clues for estimating our direction of movement.

- Rivers flow down out of mountains and more often than not, moving down river will take us back to habitations or at least roads and bridges.
- The ridges in a particular mountain range will often run in the same direction, for example, the Sierra Nevadas run up through central California in

a north-westerly direction and parallel to the 395 highway to the east. Further up into Oregon they turn straight north. Similarly, the Rocky Mountains run from Colorado clear up to Canada in a north-westerly direction.

• If one can see the ocean, then that will be another navigational constant, usually to the east or west.

• We all know that the sun rises in the east and sets in the west, but in the northern hemisphere, the sun arcs across the sky to the south and in the southern hemisphere it is to the north.

• Some areas have prevailing winds and weather patterns that are quite predicable, with clouds coming from the same direction almost every day. Very often this is dictated by the nearest ocean or the jet-stream.

• Animals will usually make trails in the easiest terrain and on the more sheltered side of a ridgeline. By being near the ridgelines they also stay warmer by getting the last sun in the evening and first sun in the morning.

• When in doubt, it is always recommended to move to the high ground to get a better view of the surrounding terrain. Valleys are also colder and more prone to fog in the mornings.

Even a glimpse of an identifiable peak can be a valuable navigational aid

Animal trails can also be used for navigation but not all will be as identifiable as this well used yak trail

No matter how good you think you are at natural navigation, it is still recommended to carry a map and compass and have some familiarity with the planned area of operation. Before traveling to any new area that one is not familiar with, it is a good practice to do some research on the local terrain and significant features. This can be done with the help of maps, books and aerial photographs and through talking to the local guides, rangers and climbers.

One essential piece of navigational information that should be calculated before venturing into any wilderness is which direction to head in an emergency. For example, if there is a major north/south highway to the west, then west would be a good direction to head; or if civilization is down valley to the south of your planned area of operation then south would be the bail-out plan.

Neither of these emergency plans requires either a map or compass since general direction navigation can be done by use of the sun during the day or by knowing the direction that rivers or ridgelines tend to run in that region. However, more precise navigation, lacking any significant features, in low visibility conditions or at night, will require some degree of expertise with a compass.

COMPASSES

Hundreds of people get lost each year when they venture into the mountains without a compass, many probably thinking that they simply won't need one, and the rest just not thinking at all.

The simplest scenario that justifies a compass for route finding occurs when one cannot see the trail or any significant landmarks that are routinely used for navigation. This can include situations where you are delayed and caught on the trail after dark, or lose sight of camp in whiteout conditions. When the clouds settle around a peak or the wind whips up the snow, and when caught in the open as the fog rolls in. Teams caught in any of these situations would benefit and derive comfort from knowing which direction to head.

It is the compass that for years has allowed boat captains and pilots to navigate at night and during the day in a world devoid of reference points. Even with modern technology like the VOR, DME, GPS, Loran and radar, it is still the compass and the altimeter that give the pilot the confirmation that all the other electrically dependent instruments are functioning correctly.

Note: I learned my lesson about compasses at age thirteen, while participating in a YMCA Junior Adventure Camp. On a 30-mile bush hike we became lost when clouds descended onto the ridgeline that we were following, but we pushed on using our trusty compass. Several hours later, when we caught sight of the coast and realized that we had walked many miles in the wrong direction, we also discovered that the needle in our cheap compass had come loose, rendering the compass inaccurate. Since then, and to this day, I now carry two compasses at all times: one on my person and one in my pack.

Compass Use Simplified

A compass has three primary uses:

- To give a basic direction of travel and for route following on a known bearing.
- To shoot bearings of two or more fixed objects such as mountain peaks and thereby finding your location by resection (triangulation).
- To aid in transferring this information to or from a map.

A compass consists of a circular dial divided into 360 degrees with a north-pointing needle that rides

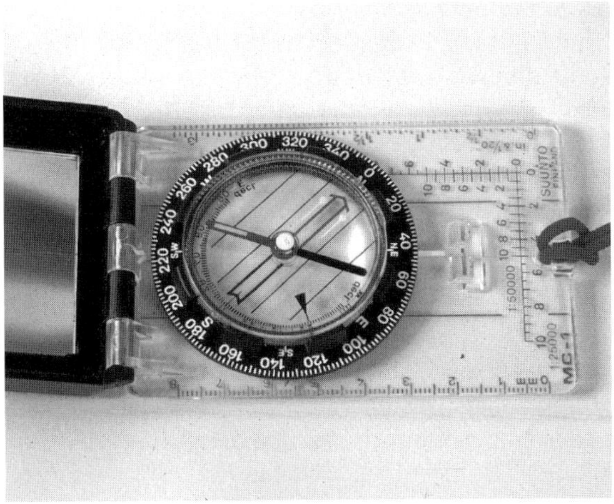

on a near frictionless axis and is dampened by oil. The cardinal directions of North, South, East and West are shown in degrees—0, 180, 90 and 270 correspondingly.

(Note: Another unit of angle that is used in the place of degrees in European countries, and also within US military artillery and tank units, is the mil—with a compass being divided into 6400 mils).

To use a compass, one must first hold it flat. Then when the north-seeking needle is oriented with the magnetic north indicator in the compass, the user can shoot a bearing (azimuth) of a fixed object or move off in a required direction.

There is however, a difference between True North, also called Grid North on a map, and Magnetic North, which is not a constant for the entire planet and will vary with geographic location.

This magnetic declination, or Grid-Magnetic angle (G-M angle) will be shown in the margin of most maps as either an easterly or westerly angle ranging from 25 degrees East in Alaska to 15 degrees West in New Hampshire. The line of zero declination runs from Wisconsin down through Mississippi; the California Sierra-Nevadas have a 15-degree East declination and the Colorado Rockies 11-degree East.

This angle must be added to or subtracted from a

magnetic compass bearing to be able to apply it to a map that is laid out in Grid North. To convert a magnetic azimuth with a West declination to a grid azimuth, add the G-M angle value. To convert an East declination subtract from the magnetic azimuth.

To eliminate the need or confusion of having to make these calculations, it is best to have a compass with a magnetic declination adjustment built into the dial.

Compasses:

There are several companies that manufacture compasses suited to backcountry navigation—Suunto, Nexus, Silva, Brunton are all companies with excellent reputations that offer a wide variety of compasses.

Apart from rugged reliability, the desirable features in a compass include: rotating bezel; a north-indicating needle on a jeweled bearing; liquid filled to steady and dampen the needle; a base-plate with straight edge and map scales; a sight for shooting a bearing; a magnetic declination adjustment; and a nylon lanyard.

Other desirable features would be a mirror that could be used for signaling; an inclinometer to measure up and down angles; and a magnifying lens for reading the fine print on maps that could possibly be used for starting a fire. The very small novelty compasses found on key chains and coat zippers are inaccurate so should not be relied on for precise navigation.

The following compasses have been included as an example of the various models and features available.

• Suunto MC-1G Global Compass—Has all of the above mentioned features plus an inclinometer to measure up or down slopes; an extra large luminous dial; an adjustable declination correction scale; scales in inches and centimeters and for four different map scales.

• Suunto A-30 Woodsman II—Has a rotating bezel; magnifying lens; sapphire-jeweled bearing; liquid filled to steady the needle; 2 degree increments on the dial; temperature range of –40F to +140F; individually balanced; luminous points for night use; declination correction scale; inches, cm, 1:24,000 scale; and easy to read clear numbers.

• Suunto GPS Plotter Compass—This new compass is used in conjunction with a GPS (Global Positioning System) to plot a location on a map. It can be used as a conventional compass and has four map scales: 1:24,000, 1:25,000, 1:50,000 and 1:65,000.

On right, the professional level Suunto MC-1G with mirror inside the lid that can also be used for signaling. In the center is the GPS Plotter compass and upper left a military signal mirror

The GPS plotter requires the use of a UTM map that is always read in meters. UTM coordinates are determined by Northings and Eastings that are marked on the sides of the map. After taking a lat/long off of the GPS, the triangular base of the compass is over-laid on the nearest grid-square on the map. At the intersection of the appropriate plotter scale is a round hole. This represents the 200-meter circle in which you are located. The 200-meter circle is a graphic representation of the 100-meter radius degradation that the military randomly generates in civilian GPS units.

For those looking for a high-end compass, Brunton recently added the Eclipse 8099 to their line of survey quality compasses. Best known for the Pocket Transit, Brunton has been around since 1894 and also has a full line of smaller and more affordable backpacking compasses. The Eclipse has a patented orienting system; field reference cards for knots, navigation and survival; three inclinometer systems; and magnified readings—all housed in a folding rubberized case.

STTU staff evaluating the Brunton Eclipse multi-featured compass

Topographical maps show various features and contour lines needed for mountain route planning

MAPS

Maps are a two dimensional pictorial representation of the earth's surface drawn to a very specific scale. However, not all maps are the same and there are a number of different maps that the professional should be familiar with since foreign countries, the military and the civilian world all operate with different systems.

Some recreational maps show only trails and significant features without giving much indication of elevation. These may be useful for hikers who stick to the well-marked trails but are of little value to teams needing to operate away from these trails.

The type of map most suited to alpine and off trail operations is the topographical map (topo) that depicts the shape of the earth's surface, or topography, by a series of contour lines that represent elevations above sea level. Topos also show numerous other features such as roads, railways, trails, rivers, lakes, peaks, ridgelines, valleys, structures, snowfields and cliffs, all overlaid with lines of latitude and longitude.

To use a topographical map, the user must refer to the legend in the margin to decipher all of the symbols. On USGS maps, all major roads are in red; rivers and lakes in blue; minor roads and trails in black and forests in green. White with blue contour lines will indicate permanent snowfields or glaciers and the standard contour lines are in brown indicating elevation above sea level.

Maps are made to a scale that can range from 1:1,000,000 up to 1:24,000. The 1:24,000 scale means that 1 inch on the map represents 24,000 inches on the ground or roughly 0.38 of a mile, and on the one million map, 1 inch would be almost 16 miles.

Smaller scale maps can be used to get an overview of a region and to identify distant peaks and land marks, but maps in the larger 1:50,000 scale are more useful for approach work and the 1:25,000 scale for getting an even closer look at the local terrain and features. Both are suited to mountaineering and operational planning.

US Geological Survey (USGS) topographical 7.5" maps are the standard now for all 49 states, except Alaska, and the ones most commonly used by mountaineers. They have a 1:24,000 scale or roughly 2.5 inches to the mile. The USGS 15″ series of maps, still standard for Alaska, use a 1:62,500 scale or about 1 inch to the mile.

MGRS—Military Grid Reference System has a 1:25,000 scale ruled at 1000-meter intervals, and has rectangular metric grid systems adopted by the military after WWII that is a modification of the UTM system. The MGRS map is designed to be used in conjunction with a protractor that has coordinate scales of 1:100,000, 1:50,000 and 1:25,000 etched into its surface. With these scales, a soldier is able to give a coordinate accurate to within 10 meters of a position.

UMS—Uniform Mapping System was devised in Washington State and became the official system used for SAR operations in that state. UMS uses letters and numbers to identify points and is tied into Sectional Aeronautical Charts (1:500,000) that are broken down into a grid used by Civil Air Patrol (CAP) and US Air Force (USAF). UMS is also keyed into 15-minute topographical maps that are more widely used. The advantage of this system is that it allows the exchange of map points between ground SAR teams and air support assets. Some find it difficult to apply in the field and attribute this to trying to adapt air search procedures to ground operations.

UTM—Universal Transverse Mercator system is used by the US military and has become very popular with SAR teams since the maps correspond to the data output from many GPS units. UTM maps have a metric grid system that is the foundation of other maps and is easily scaled down to common measurements of kilometers and meters. Military maps are marked along the border with blue ticks that are connected to representing 1000-meter grids. USGS maps have black hash marks in the border that also represent 1 kilometer grids but are not connected with grid lines on the map. Each tick is then represented by two numbers and UTM coordinates are given in six digit numbers—for example 467589.

This is actually a coordinate of 46.7 and 58.9. To find this point on the map locate the upper case 46 tick on the bottom of the topo map and measure 7/10ths to the next tick. Now locate the 58 tick on the left or right of the map and measure 9/10ths to the next tick mark. Where these two marks intersect is a point accurate to within 100 meters.

GPS—Global Positioning System is a navigational system that utilizes the precisely timed signals from several of the twenty-hour DOD satellites in geosynchronous orbit above the earth to identify a position on the earth's surface. By drawing on the signals from up to 12 satellites, a GPS receiver has the ability to find itself, give an accurate reading of its current location and then calculate the distance to other locations that have been stored or entered as way points.

Handheld civilian GPS units are now very affordable, portable and easy to use, but they only receive the Coarse Acquisition (CA) signal from the satellites, not the military precision codes that allow a positional fix accurate to within one meter. The DOD has deliberately built in a variable, random accuracy error of 15 to 100 meters to prevent their signals from being used by the enemy for navigation or weapon's targeting systems. But there are plans to remove this degradation by the year 2000 which will improve accuracy to about 10 meters.

Once the GPS receiver has fixed its position, it then displays the position in a format chosen by the user such as UTM grid coordinates or Lat/Long degrees, etc.

As wonderful as GPSs are, they still have a few drawbacks so should not be considered an alternative to a compass. In a weight conscious activity like mountaineering the GPS is one more item that may or may not be considered an essential; they also require batteries and that means more weight and the possibility of battery failure. The GPS is a somewhat delicate, battery powered electronic device that must be handled accordingly and will stop working at temperatures below about 10 degrees F. There are also times when the GPS may not be able to get a signal such as under the canopy of heavy forests, or in deep valleys and gullies.

The basic compass on the other hand is lightweight, simple, requires no batteries and is unaffected by temperature.

Marines waiting for their GPS to find itself—but beginning to doubt the batteries

ALTIMETERS

Most maps, for instance standard road maps that people are used to seeing, are two-dimensional. But navigation in the mountains has a third and very important dimension—elevation. A topographical map will show elevation with contour lines, but to be sure of a position on the correct mountain or ridgeline, the mountaineer needs to be able to confirm elevation with an altimeter.

The altimeter is valuable to the mountaineer in a number of ways. First, it can be used to calculate rate of climb, for example 1000 feet per hour, and thereby indicate if there is sufficient time left to complete the climb or make it to a planned destination.

The altimeter can also be used to find a specific location. If for example, the climbing team came out of a couloir and onto a summit ridge at 14,000 feet, or chose to stash gear at a certain point, they can record that elevation for the return trip. On the return the team will know to be looking for the same trail or stash site as they descend through 14,000 feet. (This is also a useful application for a small GPS).

However, there is one factor that is both a help and a hinder with altimeters. Since the altimeter works on air pressure, an error in altitude reading can occur when the weather changes and the pressure rises or drops. A low pressure front moving in will be interpreted by the altimeter as an increase in altitude causing the altimeter to read higher than the actual altitude.

But the benefit of this is when at a constant or known altitude, for instance high camp at 18,000 feet, and the altimeter reads higher, then it is a good indication that barometric pressure is dropping and a bad weather front is moving in. The same is true if the altitude suddenly decreases, then a high-pressure front and better weather are on the way.

Another error can occur if the temperature of an altimeter is not kept constant. For this reason, most mountaineers will wear their altimeter watches on the front of their jackets or pack straps rather than on their wrists where they will be affected by body heat.

The not inexpensive but very accurate Thommen altimeter is the standard for mountain guides, but many people who spend time in the mountains will wear some form of altimeter wristwatch, such as the popular Casio Pathfinder, which is smaller and more convenient. The new addition to the list is the easy to read, digital, multi-function Suunto Altimax which records not only altitude & ascent/descent rate but also barometric pressure, time and date and can function as a stop watch.

BELOW THE SNOW LINE

When planning an expedition or operation, it is important to first study the maps and get a clear mental picture of the route to be taken. You should know which general direction you will be heading, for example north or northwest, etc, and how long each leg of the journey should take.

A long approach or trek should be broken into a series of manageable legs with planned rest stops at easily recognizable land marks such as bridges, river crossings and ridgelines. This gives the opportunity to gauge the speed and progress of the team and provides the stragglers with a place to regroup with the rest of the team.

Some thought should also be given to packing and which items need to be kept out. Water bottles, candy, compass, map and altimeter should be carried in such a manner that there is no need to stop or remove the pack to get at them. Sunscreen, lip block, a hat, bandanna and camera should also be kept easily accessible.

If there are marked trails then these should be used both for convenience and to reduce impact on the environment. In many areas the trails will start out following a river giving easy access to water but also requiring numerous river crossings. If bridges have been installed or the water level is low this may not be a problem, but in others, team members will want to carry a pair of tennis shoes or Teva sandals to wear while crossing to keep boots and socks dry. In deeper or swift flowing areas the team may also need to rope up to make the crossings safely.

There are several other disadvantages to river walking in areas that do not have an established trail, such as log jams, debris, huge boulders, loose rocks and steep gorges that all make for hard going.

As the trail leaves the river and snakes upward, the team should fill all bottles with boiled or purified water and plan their clothing for the increased exertion of the climb followed by the cooler tem-

Frequent rests are required for a steep approach through the boulder fields

River crossings are a common part of the approach, however bridges like this are a rare luxury

peratures on the ridgeline.

There are several advantages to ridge walking but the most important is ease of navigation. In a riverbed it is easy to take the wrong fork in a river or miss a crossing, but on a ridgeline the direction of travel is more obvious and it is easier to shoot bearings on other peaks.

One note worthy piece of equipment for hiking is a good set of trekking poles, like those made famous by Leki. Although not an essential piece of equipment, many consider them indispensable for long approaches since they aid balance on rough terrain and reduce strain on the knees on long downhill descents.

Moving above the summer snowline in the Cascades-6,500'

ABOVE THE SNOW LINE

Operating in the snow brings a whole new set of problems to be negotiated. The obvious ones are colder temperatures, high winds and wet snow, but under the relentless beating of a noon-day sun it can also be uncomfortably hot and the glare unbearable without glacier glasses.

Avalanches, glaciers, crevasses and frostbite also await the unprepared and the unaware. Hundreds lose their lives each year in those picture postcard high mountains, a place where even world-class climbers perish at a sad but alarming rate.

Glaciers and crevasses are covered in Chapter 12, and avalanches in Chapter 13, so we will concentrate here on the more routine but essential aspects of traveling on snow.

Before moving above the snow line or crossing onto the first snowfield, the team will need to stop and re-evaluate their clothing systems. The first items to be added will be gaiters to keep the snow out of boots and to keep socks dry, along with additional sun block under the chin and nose to protect against the reflected energy of the sun.

The next consideration will be the clothing system. Moving in the heat of the day with a heavy pack, one may be wearing no more than a synthetic T-shirt or long sleeved poly-pro for sun protection, but as clouds pass over or the team moves into the shadowed side of a ridge, temperatures will drop rapidly and layers will need to be added. This is where clothing with full front and armpit zippers is best suited to allow both convenience and ease of ventilation.

Moving in deep snow and breaking trail is a very slow, methodical and tiring task for the team leader, and a task best done at a comfortable pace trying not to sweat unduly. When the sun goes down that sweat will severely chill the individual who has not had time to dry out in the sun or change shirts.

The pace set by the team leader or point man is also important in that it should be comfortable for the slowest team members, especially if roped-up to traverse a glacier, exposed slope or ridgeline. This is generally not a problem in deep snow since the first man must actually wade through the snow breaking it down into a passable trail. The rest of the team has it much easier following directly in the leader's tracks and working to improve the trail for the next in line. If moving with another climbing team, it is even a good tactic to let them leave camp first in the morning so they will have the task of breaking trail to the summit, and your team simply has to follow their beaten tracks.

When planning movements on the snow, it is recommended to start out early in the morning when the snow is firm and before the sun hits it. Once the long shadows of morning have receded and the sun has blazed on the snow for a few hours, the snow will become wet and slushy, and unstable in the steeper areas prone to avalanche activity.

Another advantage of setting out at dawn is that the next campsite is made earlier in the afternoon and the team can then enjoy the remaining daylight while they make camp and relax. Since most climbers are limited by the amount of altitude they can gain in a single day (1,000′ or risk AMS), the next campsite is usually reached by two in the afternoon leaving ample time to pitch tents, dry clothes and relax.

Making camp above the snow line is actually more pleasant than most would think since the snow, although cold, is far more comfortable and easier to level than rocky, uneven ground. A few minutes with a snow shovel and a flat spot can be prepared for the tent; a snow wall can be built to block the wind; and a cooking pit dug in the vestibule of the doorway. In addition, sleeping pads do a good job of insulating the climbers from the cold, and snow is abundant right outside the tent to melt for water.

(Photo - SP Parker)

SNOW CLIMBING

Following a broken trail on snow is relatively easy but on steep exposed areas, one misstep can still mean a several hundred-foot ride on the "snowball express." When climbing steep inclines, it is important to kick deep pockets into the snow while using the shaft of the ice axe for a self-belay. The lead climber should kick each step and then press down with a flat foot to make a firm shelf for the next climber. The rest of the team should work to improve the steps by keeping their feet horizontal and not letting their heel drop, which could tear out a well-formed step. This could result in a fall into the next climber trying to follow on the same trail.

Although guided groups will rope up when climbing or traversing steep snow that has no convenient run-out at the bottom, many teams will opt to move un-roped but hopefully taking a little more care with foot placement and self-belay techniques.

Whether moving roped or un-roped, every member of the team must be proficient in self-belay and self-arrest techniques on snow. In fact it is good policy to practice these techniques the first day on the snow to make sure everyone is current and confident. If there are areas of significant exposure where a fall would mean serious injury or death, then the team may opt for a running belay using snow pickets and snow flukes. This is another skill that should be practiced for both safety and efficiency.

Moving in rope teams over exposed terrain (SP Parker)

MOVING ON ICE

As one moves through the snowfields and into the high alpine zone, it is possible to move above the snowline and into a rocky desolate area swept clean of snow by high winds. There will also be large patches of ice, especially in deep valleys and on the shaded side of ridgelines that do not see the sun, where crampons will be required to move safely. A fall on steep ice is a very serious situation since it is extremely difficult if not impossible to self-arrest, especially if wearing a heavy pack.

If crampons are required, they must be checked for both fit and function. If not correctly adjusted to match the boots, they can come loose at a critical time and actually cause a fall. Points should be sharp, straps in good shape and buckles secure.

*The first rule on ice is "**Don't Fall**". Note the use of a flat foot "French" technique for descending this slope on Mt Baker (SP Parker)*

Note the lower crampon pressed flat against the snow (Mike Goldsmith)

The key to walking with crampons is to look before you step; use determined foot placement; get as many points as possible in contact with the ice (flat footing); and maintain balance especially when wearing a pack.

If moving on really steep ice, the climber must practice a self-belay technique of always having three points of contact with the ice. This is best done with two tools—a primary ice axe and a second technical ice tool or North Wall type ice hammer. The climber then has four usable points of contact—two tools and two crampons. This then allows one point to be moved at a time leaving the other three in place. For example, the climber reaches out and places one tool in the ice, then moves his feet up or across one at a time, and then lastly the trailing tool.

In the event of a fall, it will be the two tools that are used to try and arrest the movement since digging in with the crampons in a feet-first slide will only cause the climber to flip and cartwheel out of control. But keep in mind that the first rule on ice is DON'T FALL! (See Chapter 11 for more on ice climbing techniques)

Climbing steep ice using crampon toe points and two tools (SP Parker)

SNOW SHOEING

Where snow is too deep and too soft to support much weight and the mountaineer is posting up to his crotch, skis or snowshoes are required to better "float" on the snow. The science behind snowshoes is quite basic—the bigger the surface area, the more the float, but there must also be a balance between size and convenience. There is a limit to just how big a pair of snowshoes can be before they become too cumbersome to walk in.

Earlier snowshoes were made from wood, skins and fiber bent and lashed into a large tennis-racket like shapes. Modern snowshoes make greater use of aluminum, stainless steel plastics and synthetics like Hypalon, to produce a shoe that is more efficient, lighter, stronger and requires less maintenance.

The efficiency of snowshoes is directly related to the snow conditions, the weight of the user, the amount of load being carried and the size of the shoes. Larger individuals, or backcountry travelers carrying heavier loads, will require larger snowshoes, but in firmer snow conditions, a smaller snowshoe may be appropriate. Where loads are significant, it may be necessary to pull a sled with the

Soldier packing older style aluminum "tennis racket" snowshoes

extra gear and then take advantage of the convenience of smaller snowshoes.

Snow shoeing is an aerobic activity so there is again a need to layer clothing to prevent over-heating or excess sweating. Although not essential, ski poles with medium to large baskets will also aid balance and be of assistance when moving over uneven terrain.

Almost any boot can be used with snowshoes but they should be suited to the worst conditions expected on the climb. As one moves above the soft snow and into a higher alpine region, the snowshoes will be replaced with crampons for movement on ice. Only one snowshoe, the Atlas Summit 33BD, developed in conjunction with Black Diamond, is designed to work with the BD Sabertooth crampons for step-in convenience. This is the first ever crampon compatible snowshoe designed specifically for mountaineering. Most other mountaineering snowshoes only have blunt aluminum or stainless steel cleats to aid uphill movement and prevent sliding backwards.

The technique for snow shoeing is not difficult to learn but it is quite tiring. The alternative is skis but they are more expensive and considerably more difficult to master.

When maneuvering with the older design snowshoes that have long tails, extra care must be taken in stepping over logs and obstacles. Newer snowshoes have eliminated these tennis-racket tails altogether making them more user friendly when maneuvering in trees and confined areas.

Atlas 33s with OR Crocodile gaiters

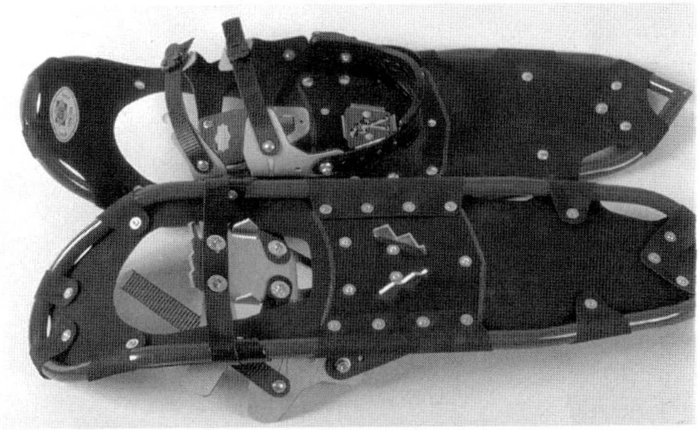

Note the climbing cleats on these mountaineering snowshoes

To climb steeper snow banks, it is necessary to kick the front of the snowshoe into the snow and then press down to make a step, much the same as without snowshoes. The leader should space and form steps so that it is easy for others to follow.

When descending, it is best to keep the knees slightly bent and lean back keeping weight on the heel cleats to maintain control. On really steep descents it will be necessary to remove the snowshoes and use a heel plunging technique to form more secure steps.

When traversing, one can use an edge technique by kicking the side of the snowshoes into the slope making use of the cleats. To do this, swing the heel hard towards the uphill slope, and then stomp down to secure the snowshoe edge into the slope.

On deep virgin snow it is best to walk in a single file behind the leader who is breaking trail and compressing the snow. When leading, one should take consistent even steps that are easy for the rest of the group to follow.

SNOW SHOES

As with most other pieces of mountaineering equipment, there are several companies that manufacture snowshoes, but not all are of the same quality or have the same features. An example of snowshoes that are well suited to professional use and abuse is the Atlas Summit Series. These shoes have strong, lightweight TIG welded frames that are tapered for a more comfortable stride; a spring-loaded binding with 4 point floating cleats; a free-heel harness design with a glove-friendly ratchet buckle system; contoured footbeds; one piece Hypalon decking with double reinforced heel strike; and dual All-Trac stainless steel climbing cleats. They also pack flat for ease of carry, transportation and storage.

Since one size of snowshoe does not suit all conditions, the Summits come in a range of sizes that are matched to the weight of the user.

The Summit 22 is 8″ x 22″, weighs 3.8 pounds, and is suited for up to 190 pound load on firm snow.

The Summit 33 is 9″ x 29″, weighs 5.1 pounds, and is designed to work with bigger and heavier plastic boots and will handle loads up to 275 pounds on firm snow.

The Summit 33BD is also 9″ x 29″, weighs 4 pounds and designed to work exclusively with Sabertooth crampons.

Because the military has to be able to field a 200-pound fighting man with another 100 pounds of gear, they are always interested in the maximum

Boot bindings should be secure and easy to operate with thick gloves

Stainless steel climbing cleats on the Atlas mountaineering models

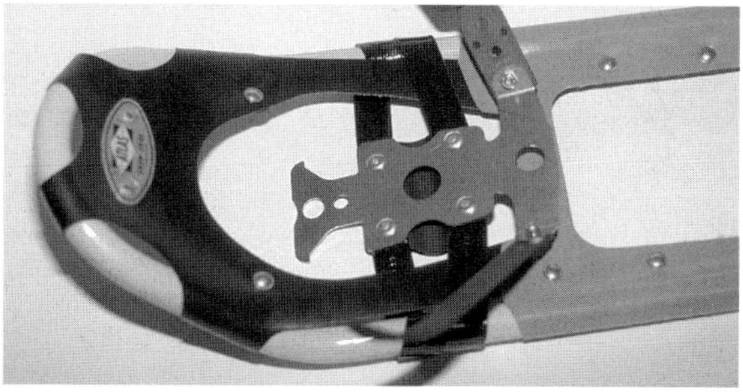

Stainless plate that mates to the BD Sabertooth crampon on the BD33 snowshoe

float they can get from a snowshoe. So it is no surprise that the Seals have been ordering the Atlas Back-Country 1044 for their training and operations in Alaska. The 1044 is 9.5" x 36", weighs 4.8 lbs (2.2kg) and is recommended for loads up to 300 pounds.

For more comfort and hold with more flexible hiking or combat boots, the Back-Country uses a softer padded, asymmetrical switchback harness, and is also available in shorter 22" and 29" lengths.

POLES

As with skiing, snow shoeing with ski type poles has many advantages. Primarily used to aid balance on steep or difficult terrain, poles also reduce stress on joints and lower back, improve respiratory functions much the same way pumping the arms in running works, and work to stabilize the upper body when carrying heavy loads.

Poles should be lightweight, compact and indestructible with large baskets that will give adequate push against the snow pack. And for work in avalanche country specially designed poles like the Leki Extreme and Atlas AVI Probe Poles serve a useful dual function. By removing the handles and fitting the two poles together, the poles convert to a 290 cm (9´6″) avalanche probe.

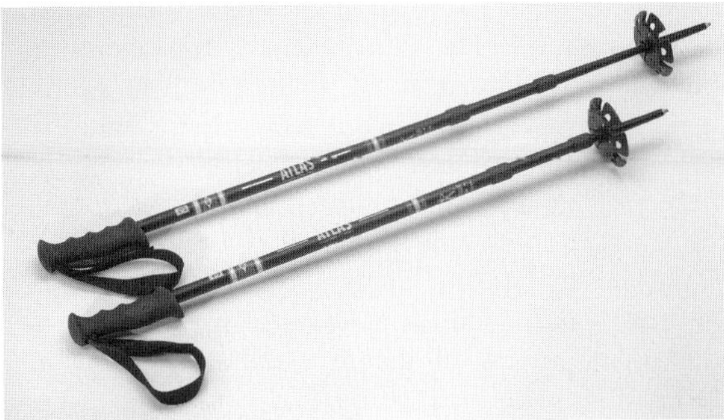

Atlas 3 piece Avi poles that can be joined at the handles to make an avalanche probe

SLEDS

A lightweight plastic sled allows the backcountry traveler to move a greater amount of gear over soft snow by spreading the load over a larger surface area. A sled also gets the load off the shoulders and reduces back and leg fatigue.

One of the best sleds is Wilderness Engineering's Base Camp Sled System used by the USMC in their winter warfare program and now by FBI-HRT.

The Base Camp sled is very durable because of its rotationally molded cross-linked polyethylene hull that has been tested in temperatures below –109 degrees Fahrenheit. It also moves very well across the surface of the snow since ice or snow will not stick to the hull.

This sled comes complete with an 11,000 cu in duffel bag of 1000 denier DuPont cordura nylon that releases from the sled with two buckles. In emergencies it can also be used as a field expedient shelter.

To pull the sled, there is a harness and two powder-coated, two-piece chrome-moly steel poles that are stronger than any aircraft aluminum poles found on other units. The harness is fully adjustable to keep the pulling points right at the sides where they should be and the sled has exceptional side-hill stability.

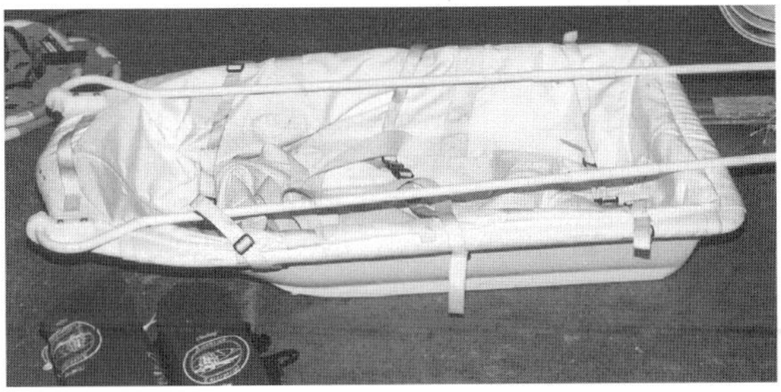

USMC fire team sled made by Wilderness Engineering

"ACCIDENTS IN NORTH AMERICAN MOUNTAINEERING"

Edited by Jed Williams, and published by the American Alpine Club and Alpine Club of Canada.

AVALANCHE—Washington, Mount Rainier—1998

This accident occurred as a RMI guided team was descending from the summit. Two rope teams were clipped into the same fixed line when the avalanche occurred. The avalanche caught the first rope team, which pulled two of the anchors on the fixed line. The slide continued unattested also pulling the second rope team down the hill. Finally one anchor (a picket) held at the other end of the fixed line as a few climbers became entangled at the top of the cliff bands. What was left is detailed. One guide and one client were caught on the fixed line above the cliff. Three clients and one guide clung to the top of the cliff, tangled in the rocks and ropes. Three clients dangled below them on a cliff of ice and snow, while the solo client (Nestier) hung below a second cliff band in a waterfall of snowmelt. Nestier died as a result of this exposure.

The Park Service assembled climbing rangers from Camp Schurman and Muir, Mountain Rescue Volunteers, Rainier Mountaineering Guides and helicopters assisted with the rescue. On scene, Gauthier along with Rainier Mountaineering Guides Randolf and Eieshner worked to assess the situation. The location was extremely hazardous with 40-degree icy slopes, 20-foot vertical rock bands, exposure to avalanche hang-fire and a 300-foot drop to the glacier below. The danger made it necessary for rescuers to secure the exposed climbers with new ropes and reliable anchors. One of the distressed climbing teams was pendulumed over a refrigerator-sized rotten rock, the other clung to the cliff or dangled on a rope which was frayed to the inner strands and pulled tight over a sharp rock held by one picket! Once new anchors and ropes were established, on scene rescuers negotiated the cliff securing the injured and triaging the patients.

Teams of climbing rangers and guides were inserted with US Army and private helicopters at Ingraham Flats. Some of the rescuers climbed to the accident site to assist with the raising evacuation while another team headed to the base of the cleaver to assist with the lowering of one climber. That climber, Patrick Nestler (29), had fallen substantially farther down the Cliff than the others. The fastest evacuation was to lower him off the mountain rather than raise him back to the accident site. New anchors and ropes were set to assist Nestler. However no one had heard from him for over an hour. As the injured were being raised off the cliff above, Nestler was quickly lowered, taken across the bergsehrund and evacuated to the helicopter, where he was pronounced dead.

At 1:45 p.m. on June 11th, an independent climber camping at Ingraham Flats overheard screams of distress coming from the Disappointment Cleaver. The climber, using a cell phone, alerted Mount Rainier communications and reported that a snow avalanche had swept two rope teams off "The Nose" of the Cleaver. The initial report indicated that many climbers may be dead and the accident was extremely serious. Off duty climbing ranger Mike Gauthier overheard the emergency announcement on the Park Service radio and responded to the accident from the summit by riding his snowboard down the climbing route. On scene, Gauthier reported that Rainier Mountaineering guided teams had been hit by an avalanche and up to ten climbers (two rope teams) were unsecured on the cliff or unaccounted for.

Climbing rangers remained to clean up and conduct the accident investigation on the following day while additional guides stayed to escort the remaining clients back to Camp Muir.

Efforts to raise the other nine climbers off the cliff and up the slope were hustled as rescuers raced against nightfall. The injuries included: one guide with a severely injured hand, a client with an injured leg and hand, three hypothermic clients, another client with an injured hand, a climber with an injured leg and two shaken but ambulatory climbers. The Chinook helicopter hovered at Ingraham Flats till darkness when the last of the most hypothermic climbers was loaded on board in a liter. All of the injured and a few of the rescuers were flown to Madigan Hospital in Tacoma.

Analysis:

This avalanche was described as a "wet, loose snow slide." Released on a 40-degree slope at 11,600 feet, it ran on a layer of isothermal melt-freeze grains when it hit the rope teams at 11,200 feet. The width of the slide when it hit the teams was 38 feet, at a depth of 6-10 inches. Warm temperatures and clear sky (solar radiation) are the most significant weather factors in its cause. At the time of the avalanche the snow pack was in the melt stage of the melt-freeze cycle and the snow grains lacked cohesion. Only a small trigger was needed to start the snow mass moving.

No definite trigger was positively identified at the starting point. However, boot prints and climber activity mark the area on the slope above the traverse. This location is notorious for rock and icefall. Guides fix the traverse because the exposure is great should a climber fall. The guides observed no evidence of any avalanche activity that day. Senior guides commented that the area had no avalanche activity for 20+ years.

Avalanches are not just winter phenomena. Big mountains like Rainier create their own climate and conditions. Different slopes, elevations, angles, and aspects mean new conditions and circumstances. Always consider the possibility of an avalanche, particularly on suspect 25 to 50 degree snow slopes on warm days. Hazards can be assessed by digging a snow pit and checking the slide potential. One can also minimize exposure by moving quickly through hazardous areas. Also consider that humans cause many avalanches. In dangerous areas, make sure your teammates or others are not above-or below! (Source: Mike Gauthier, SAR Ranger, Mount Rainier National Park)

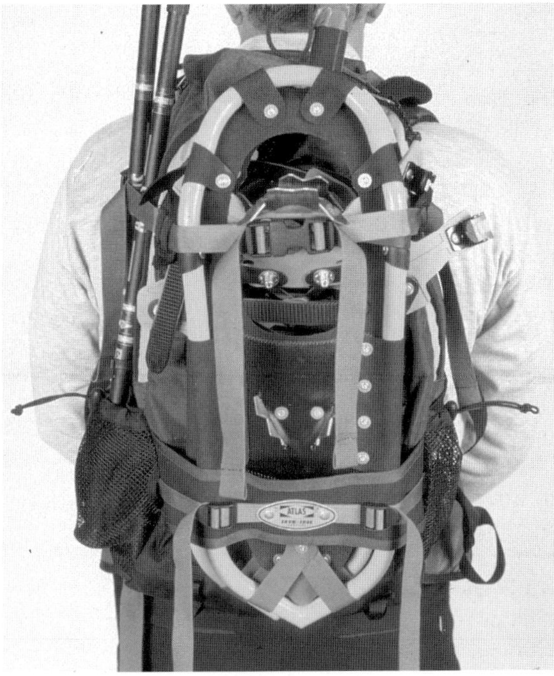

Right: Atlas day pack with hydration system and snowshoe retaining straps

(Photo - Mike Goldsmith)

(SP Parker)

10

ROPE SKILLS & BASIC MOUNTAINEERING

The foundation of all climbing and mountaineering, and a good indicator of a climber's experience, is his or her rope management skills. The ability to handle ropes efficiently, safely and in a manner suited to alpine operations. Where the rookie will invariably be able to turn a perfectly coiled climbing rope into a tangled ball of string in a matter of seconds, the experienced climber will move rope smoothly and efficiently even when perched precariously on an icy ledge.

Rope management training is probably the best place to start any climbing or alpine operations program, beginning with learning about the types of ropes and their applications, then moving into efficient handling, deployment, coiling, care, cleaning, inspection and storage (see Chapter 6—Rope). From there, the student must learn and master the essential knots, basic rappelling and belay techniques before being taken on the mountain. Rappel and belay training is also a good time to introduce students to anchors and anchoring systems.

KNOTS & HITCHES

The best rope in the world is of little value if the operator does not know how to secure it to an anchor or to his harness. Learning knots and hitches is considered so important in military recon and mountain warfare schools that the trainees must carry a length of rope with them at all times and are drilled mercilessly by the training staff until all the knots and hitches become second nature and can be tied blind-folded (literally).

Each type of knot or hitch has a particular application and is designed to minimize the strength loss in the rope that is inherent with all knots, and at the same time, it is important that the knots can be easily undone after being loaded.

Although there are dozens of knots and hitches that are useful to know, there are only a few that are used routinely in mountaineering and MUST be learned. They are:

- Figure 8 loop—used as an end loop on a rope
- Figure 8 Follow-through (Rewoven)—used to tie into a climbing harness; also used to join two similar sized ropes
- Bowline—used as an end loop and to tie into a climbing harness
- Bowline in a Bight—used to tie into a harness in the middle of a rope
- Overhand Knot—used to secure the ends of two similar sized ropes
- Clove Hitch—used as an adjustable attachment at a belay station
- Girth Hitch—used to attach a rope or cord to a deadman, picket or ice axe
- Single Fisherman's—used to secure the tails after tying other knots such as Figure 8s and Bowlines

ALPINE OPERATIONS

- Double & Triple Fisherman's—used to join two ropes or both ends of a cord to make a prusik loop
- Prusik Hitch—a friction knot used as an ascending or self-belay device
- Water Knot (Tape Knot)—used to secure the two ends of 1" tubular nylon webbing (Spectra webbing should be sewn since it is too slick to hold a knot)
- Munter Hitch (Italian Hitch)—a friction knot used to belay or lower

Bowline on a bight backed up with a carabiner

Bowline

Bowline with Double Fisherman's

Figure 8 loop

Follow through Figure 8

Figure 8 with Single Fisherman's

Double Loop Figure 8

Figure 8 end knot

Overhand end knot

Girth Hitch

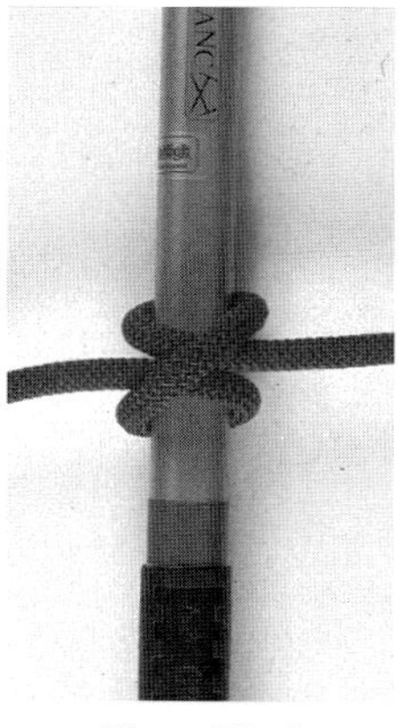

Clove Hitch

Clove Hitch

Water Knot

Double Fisherman's Knot

Triple Fisherman's Knot

Triple Fisherman's joining two ropes

Prusik Hitch

Munter Hitch

Figure 8 loops used to join two ropes

Klemheist friction knot

Bachmann friction knot

Other useful but not essential knots are:

- Double Loop Figure 8—is the most common end-loop used in rescue work
- Figure 8 on a Bight—to put a loop in the middle of a rope
- Bachmann Hitch—works like a prusik but is tied around a carabiner to make it easier to move
- Klemheist Hitch—another alternative to a prusik but allows the climber to use a cord or webbing
- French Prusik—an improvised prusik using a sling
- Butterfly Knot—used to put a loop in the bight of a rope
- Garda Hitch—an improvised ratchet for raising and lowering

"Red Hat" mountain warfare instructor supervises rappel training

RAPPEL TRAINING

Rappelling, known as abseiling by the British, is the technique of descending a rope in a controlled manner, usually by use of a descender or friction device such as a figure 8 rappel plate or some form of sticht plate (ATC). The military does not routinely carry descenders so they make a rappel by wrapping the rope(s) through a single locking carabiner two or three times. For short, easy rappels over low-angle terrain, it is also acceptable to simply run the rope around the body in the dulfersitz technique of rappelling.

In rescue work and on tactical operations static ropes are used for rappelling but in climbing and mountaineering, most rappels are done on dynamic climbing ropes unless semi-permanent fixed lines have been installed for a specific route.

Rappelling is an essential skill and one that can be taught on any cliff, or even more conveniently, on the side most buildings, fire training towers or dedicated rappel towers. It is also a skill that should be taught early in a program to give the trainees the opportunity to feel some exposure to height, and for the instructors to gauge the trainee's head for heights and ability to function under a little stress. Setting up rappels will also give the trainees their first opportunity to apply their newly learned knots, evaluate and install anchor systems, deploy ropes and begin to develop a trust in the equipment.

Alpine rappelling should be done in a relatively slow and controlled manner since fast rappels and hot-dogging generate too much heat between the rappel device and the rope and the rappeller is more apt to dislodge rocks—a danger to both the people below and himself.

The initial training rappels should be done with a belay man on the bottom of the rope who can arrest the downward speed of the student by pulling tension on the rope. The students should then be taught how to rig a safety/self-belay on the rope using a prusik loop.

During training, there is also a need for verbal communication between the trainee rappelling and the person on belay below. Essential commands used in rappelling are:

Command	Meaning
"Rope"	From rappeller—I am throwing the rope, beware below
"Clear"	From belayer—you are clear to throw the rope
"On Rappel"	I am on the rope and ready to begin my rappel
"Belay On"	I have the rope and am ready to belay you
"Rappelling"	I am beginning my rappel
"Rock!"	Rocks have been dislodged, take cover below
"Stop"	Stop everything, there is a problem

Royal Marines rappel on coastal cliffs. Note the use of a pack as a rope chaffing guard (Royal Marines)

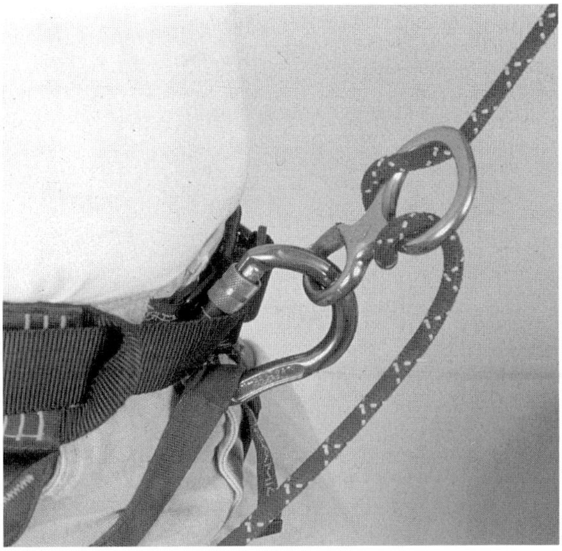

BD figure 8 rigged for a single rope rappel

Double rope rappel with a sport 8

In this case, all the control is in the right hand with the left used only for balance. However, without gloves rappels must be kept slow and care must be taken not to get the rappel line tangled with other hardware hanging from the harness loops

Opposite and opposed ovals used as a rappel device

RAPPEL ANCHORING

In basic rappel training, slings are placed around a "bombproof" anchor and then the rope is secured to the slings with a locking carabiner. This should be a strong simple system that leaves no margin for error.

Even when training in the mountains or on some local cliff, every effort should be made to secure the training ropes to solid objects such as large trees, rock outcrops, boulders or BFRs, usually by using two slings made from tubular webbing or heavier anchor straps. When using a large tree for an anchor, be sure that it is still alive and has a deep root system; similarly, even large boulders can be insecure if they are precariously placed. Living rock that is still part of the mountain is always preferable to both of these.

Once the trainees are proficient with strong but simple anchoring methods, they should be taught how to back-up all anchors and then introduced to multiple, self-equalizing anchors for situations where one solid anchor is not available. But even with multiple anchors, they will still be securing the rope with a locking carabiner.

Multiple anchor points could be a combination of natural and artificial protection including smaller rock nubs and cracks that require the use of various slings, nuts and cams. Creating these improvised anchor systems is an art all in itself as detailed in

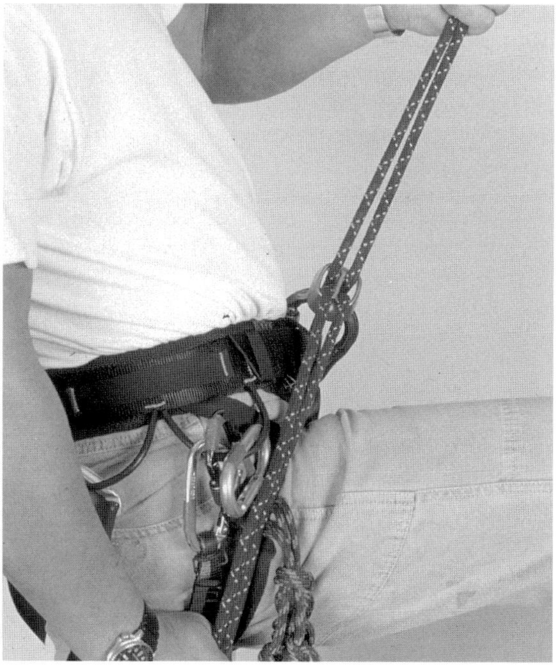

ATC used for a double rope rappel

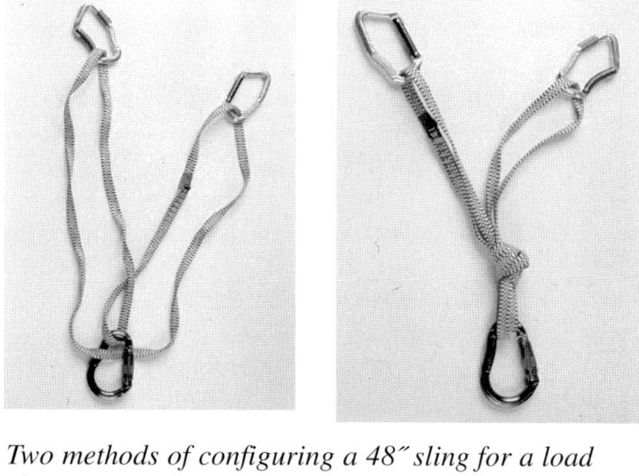

Two methods of configuring a 48″ sling for a load sharing anchor. The right is the preferred method

climber/writer John Long's numerous books "Climbing Anchors" and "More Climbing Anchors" (highly recommended). However, multiple sketchy anchors are still not as good as one solid, bombproof anchor point.

If working in a winter environment or summer alpine, trainees should be introduced to the various forms of rappel anchors that can be created in snow and ice. The most commonly employed protection in snow is a series of Deadman anchors using buried pickets, flukes or ice axes, but an anchor can also be created by digging out a large snow bollard 3-10 feet in diameter, depending on snow conditions and consistency. The drier and more powdery the snow the bigger the bollard should be. The bollard trench should be over a foot (30cm) deep and well packed, and the backside of the bollard may also need to be padded to prevent the rope cutting into the soft snow.

AAI guide shows students how to construct a snow bollard

On ice, large outcrops of living ice can be used to secure slings just like rock, but on waterfall ice there is always a possibility that the whole slab could detach from the rock wall if water is still running behind the ice.

A series of 17cm to 22cm ice screws screwed firmly into solid blue water ice is the most common form of protection, but natural protection can also be created by boring out a V thread hole to make an abalakov anchor. This is done with a long 22cm ice screw, a length of cord or webbing and the aid of a wire hook to pull the thread through the intersecting V channel in the ice.

If the conditions are right, a small bollard can also be carved out of solid ice with an ice axe or ice tool. An ice bollard should be tapered at the bottom like a reverse teardrop, about 24" long and 12"-18" in diameter at the top. The trench around the bollard should be at least 6" (15cm) deep and under-cut to form a mushroom so that the rope cannot accidentally slide up and over the bollard.

Both snow and ice bollards should be backed-up to pickets or ice screws for training. However, if a bollard is being used for a final rappel off of the mountain, then the back-up should be used while the first and heaviest climbers descend, but the last man down can remove the back-up slings and hardware if he feels confident that the bollard will hold and the rope can be recovered without hanging up.

This brings us to rappel situations where the rope needs to be pulled from the bottom to recover the rope, or after each pitch to be used for the next rappel. Obviously having it securely attached to a tree, rock or anchor point with slings and locking carabiners will make this impossible. The solution is to halve the rope and place it around the anchor, or thread the rope through the protection so that it can be pulled down after the rappel. With the middle of the rope secure, the climber rappels down with the two halves running through his rappel or belay device and then pulls on one side to recover the rope.

Using two ropes tied together at the top will allow for a longer 150-foot rappel whereas halving a single rope only allows for 75-foot descents. The

Rappel ring left and double ovals right. The overhand knot can be used to secure the bottom ends of the rope or to join two ropes.

use of a ring or carabiner as a link to the anchor webbing is also important since this will allow the rope to slide through smoothly when being retrieved. Running the rope directly through the nylon webbing can be done but is not recommended since it can cause the rope to twist and jam, but more importantly, can build up enough heat to weaken the nylon.

To make sure that the rope does not slide up through the rappel ring accidentally, the two ends of the rope should be also knotted at the bottom so that they cannot separate during the rappel. This knot will also help prevent the rappeller from inadvertently rappelling off the end of the ropes. If using two ropes tied at the top, it is important to check which side of the ring or sling the knot is positioned and which colored rope needs to be pulled to retrieve both ropes. With dissimilar sized ropes, the thicker rope should pass through the protection with the knot made on the smaller line's side, just incase one has to climb back up to free a jammed knot. Note: See Chapter 11 for more notes on Multi-pitch rappelling and Chapter 30 for Tactical Rappelling.

BELAYING

When rappelling, the climber is responsible for only his own safety when making a descent, but when belaying he is responsible for someone else's life and safety. Belaying is an important skill that must be mastered early in training since it is the foundation of all work as part of a climbing team. It is also a skill that can be learned and practiced in the convenience of a climbing gym or on a climbing wall.

Belaying is the process of feeding out or taking in rope in such a manner that the belayer can arrest the climber in the event of a fall. In most situations, the rope is being run through a two-direction friction device such as an ATC, Pyramid or belay plate attached to the belayer's harness with a locking carabiner. However, not all belays are done through a belay device. Climbers will often use a body belay, with the rope passed behind their back, if they are in a good position and the move is over moderate, low-angle terrain. In more precarious positions, the belayer must be anchored into the belay station so that he cannot be pulled off a ledge if or when the climber takes a fall.

Author does a body belay on climbers following up a relatively easy pitch

Climber has dug into the snow to make a solid belay position for crevasse rescue training

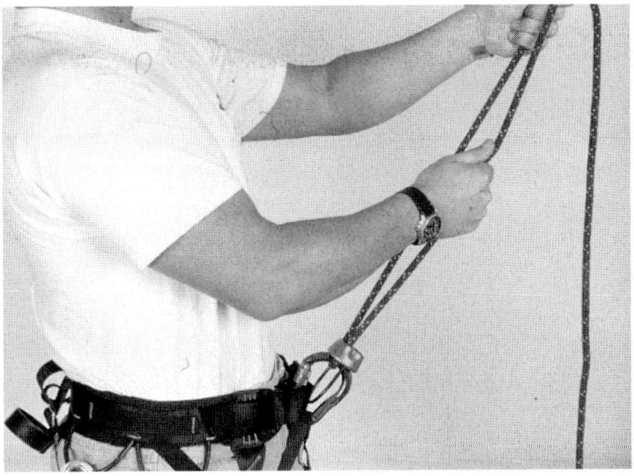

The belayer's brake hand must never lose contact with the rope while belaying a climber

Belaying is more difficult to master than rappelling since there are many more types of belays and belay positions requiring considerable experience and ingenuity to create the optimum belay. It is also necessary to be taking in or letting out rope in sync with the climber's moves, which are not always visible from the belay station.

The first belay learned is usually the top-rope belay used for basic rock and gym climbing. When belaying a climber on a top-rope, the belayer is taking in rope as the climber ascends the wall. Once the climber tops out the climb, the belayer then lowers the climber back to the ground by letting the rope slide through the belay device in a slow and controlled manner.

When belaying a lead climber from below, the belayer is letting out rope as the climber clips the rope into protection and continues to move upward. At the very top of the wall, the lead climber will clip into the last piece of protection and allow the belayer to lower him back to the ground cleaning the wall of protection as he descends.

On a multi-pitch climb, once the lead climber reaches the top of a pitch, he then becomes the belayer for the #2 climber, in which case he will be bringing rope in as the #2 climber climbs up to the lead climber's position.

Belayer using a Trango Belay 8

ALPINE OPERATIONS

Traversing steep or icy terrain with a lot of exposure is much like climbing in that the #2 climber will let out rope as the leader moves across the slope, and then brings in rope once across for those following.

When working with half-rope (double rope) or twin-rope climbing techniques which are more common in alpine climbing and technical ice climbing, the belayer will be required to deal with two ropes from both a bottom and top belay position. This requires a belay device that can accommodate two ropes. It also requires superior rope handling skills to not create a huge tangle of line at the belay station.

The key to good belaying is that the belayer, first and foremost, NEVER releases his "brake-hand" from the climbing rope, and that there be constant communication between the belayer and the climber. The belayer must remain focused and concentrate on anticipating the climber's moves so as to be an asset and not a hindrance to the climber.

The following are some of the voice communications commonly used between the climber and the belayer.

Once the skills of rope management, rappelling, basic anchoring and belaying have been mastered, it is time to get out of the classroom or controlled training environment and into the mountains to learn the fundamentals of mountaineering - a subject that can only be taught effectively on the mountain under realistic conditions.

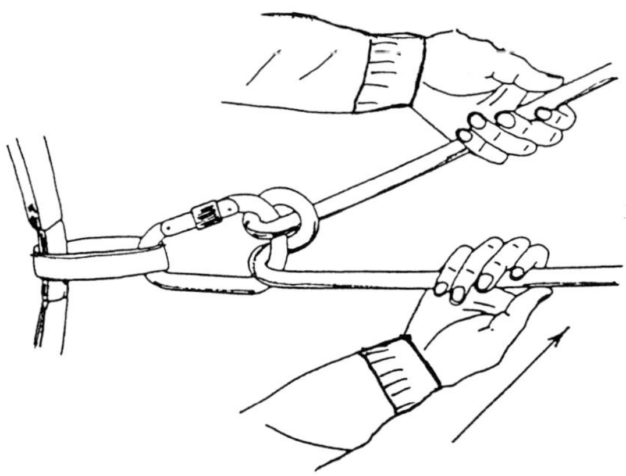

Belaying with the Munter Hitch

Command	Meaning
"On Rope"	The climber saying I am attached to the rope
"On Belay?"	The climber asking if the belayer is ready
"Belay On!"	The belayer telling the climber that he is secure and ready to belay
"Climbing"	The climber saying that he is beginning to climb
"Climb"	The belayer acknowledging that the climber is climbing
"Slack"	The climber asking for slack in the rope
"Tension"	The climber asking the belayer to tighten up and hold the rope
"Tension"	The belayer acknowledging that he has tension
"Falling"	The climber telling the belayer to brake and brace for a fall
"Rock!!"	Watch out below for falling rocks or hardware
"Ready to Lower"	Climber asking for the belayer to lower him (on a top rope)
"Lowering"	Belayer indicating that he is beginning to lower the climber

(See Chapter 11 on Climbing for additional climbing communications)

MOUNTAINEERING

There is considerable diversity in the types of recreational mountain climbing ranging from mountain walking to more technical mountaineering to extreme alpinism. The training and skills necessary to climb a particular mountain will be dictated by several factors—the height of the mountain, the season, selected route and its level of difficulty. To summit on mountains that have non-technical routes often requires no more than good boots, strong legs, suitable clothing and a little determination, where as other more challenging mountains will require considerably more equipment, technical climbing skills, a high degree of fitness and the right mental attitude.

The route that a team ultimately selects will be influenced by several factors including the mission, conditions, skill-levels of the weakest members, the time available and the overall strength of the team. Where experienced recreational mountain climbers may select a difficult route to challenge their abilities, the professional will usually select the easiest route to achieve the same objective with the least effort.

To better understand these different skill levels, they could be classified as basic mountaineering, technical mountaineering and then extreme technical climbing.

Basic mountaineering skills give the individual or team the ability to find and safely negotiate a route through the mountains, over glaciers and passes, along ridgelines and up to the summit, if that is the objective. Technical mountaineering skills take them to the next level with the ability to use protection, fixed ropes, ice tools and more advanced techniques to negotiate steeper and more challenging routes.

In contrast, the purpose of extreme technical climbing or alpinism is to develop the ability to negotiate near vertical faces of rock and ice—routes sought after by recreational climbers but usually avoided by mountain rescue teams and the military. So although technical climbing skills are an asset to the mountaineer, they are not always essential. In fact most basic mountaineering requires little in the way of additional equipment apart from the essential ropes, ice axe, harness, crampons, helmet and a few pieces of basic protection.

Military mountaineers moving up a ridgeline in the Austrian Alps (B Clearman)

Leading on vertical water ice

TERRAIN APPRECIATION

Before venturing into the mountains, it is important to be able to ascertain exactly what level of expertise and equipment will be required to reach a particular objective and how long the climb or movement would be expected to take. This begins days or even weeks before the operation with research into the nature of the terrain and expected conditions for that time of year. This is no substitute for on-site terrain assessment and route planning but it will better prepare the team for the planned operation.

Where a mountain guide's primary obligation is to the comfort and safety of his clients, he will still be looking for the most scenic routes to follow, or the most challenging climbs for a hardcore climbing team. The military or rescue team on the other hand will be looking for the easiest route between two points, taking into consideration suitable campsites and whatever time constraints the team may be working under. So, allowing for changes that will be made after the on-site terrain assessment, the planner will be interested in:

- The location of the objective in relation to the start point
- Suitable approaches and routes to the objective
- The type and difficulty of the terrain
- Potential hazards, obstacles, options and alternative routes
- Relative ease in route finding and following
- Availability of water and suitable campsites

In conjunction with a current topographical map, the best source of this information will be mountain guides and climbers who have first-hand knowledge of the mountain and experience on the same routes, at the same time of year and under similar conditions.

The next alternative is mountaineering guidebooks that give both narratives and photos or diagrams of the routes. Most books will also give a Grade or Class for each route that will give the planner some indication of the expected difficulty of a particular climbing route.

Each country and several climbing organizations have methods of grading and classifying routes and climbs so that they can be accurately listed in guidebooks, allowing the guide or climber to match the climb to their level of fitness and ability.

The British, French, Australians and Brazilians all have their own systems, as do the UIAA and Sierra Club. However the two systems most often quoted for climbing in the United States are the National Climbing Classification System (NCCS) and the Yosemite Decimal System (YDS).

The NCCS grades the overall expected difficulty of a multi-pitch alpine climb, taking into account: length of climb, number of pitches, difficulty, necessary commitment, route finding and ascent time.

The most popular classification system used in the US is a version of the Sierra Club system that was modified in the 1950s to become the Yosemite Decimal System (YDS). The primary modification was to expand the Class 5 rating into decimal points to better describe rock-climbing levels of difficulty.

Grade Ratings:

Grade I	Requires a few hours and could be of any difficulty
Grade II	Requires half a day and be of any technical difficulty
Grade III	Requires most of the day for the technical portion
Grade IV	A long hard day with pitches exceeding 5.7 level of difficulty
Grade V	Over a day and a half with pitches exceeding 5.8
Grade VI	A multi-day climb with difficult free climbing and/or aid climbing
Grade VII	Big walls in very remote locations that are climbed alpine style. Long climbs exposed to severe weather with complex approaches

Original Sierra Club System:

Class 1 Hiking
Class 2 Simple rock scrambling with occasional use of the hands
Class 3 Easy climbing where a rope might be carried
Class 4 Moderate climbing with some exposure, typically roped. A fall on Class 4 rock could be fatal
Class 5 Serious climbing where ropes, protection and belaying are required
Class 6 Aided climbing

Yosemite Decimal System (YDS):

The YDS was developed to sub-divide Class 5 climbs.

5.0-5.6	Typically suited to beginners and easy for experienced climbers
5.7-5.9	Intermediate difficulty for most recreational climbers
5.10	Very difficult and the level most dedicated climbers seek to attain
5.11-5.14	Experts only and even then requires natural ability, constant training and repeated working of a route.

For grading winter routes covered in snow and ice, the Scottish Winter Grade system is as good a guide as any.

- Grade I — Climbs requiring only one axe—45 degree snow gullies and easy ridgelines.
- Grade II — Axe and ice hammer are required because of steep snow and short difficult pitches or patches of ice.
- Grade III — Similar technical difficulty to Grade II but more sustained climb with possible rock and mixed climbing.
- Grade IV — Steep ice, short vertical steps, longer pitches of 60-70 degrees, requires more arm strength. More advanced techniques needed for mixed ascents.
- Grade V — Sustained steep ice of 70-80 degrees; mixed with linked hard moves.
- Grade VI — Vertical ice; long or sustained mixed routes; or short very technical routes.
- Grade VII — Long multi-pitch routes with thin vertical ice; mixed routes requiring a high degree of fitness and technical skill.
- Grade VIII — The very few hardest routes in Scotland.

Even without the benefit of these grading and classification systems, an experienced guide or mountain leader should be able to gauge the difficulty of a route and be able to make a GO or NO-GO decision for his group.

TERRAIN ASSESSMENT

Once at the start point of an operation an on-site assessment must be made of the terrain and planned route. In many cases the assessment will confirm the research done prior to the operation, but in others the terrain may be totally different because of freak weather patterns, heavy snow, recent avalanche activity, rivers rising or changing course, or simply the fact that the maps and planning materials were out of date.

Only by looking at the terrain will the experienced mountaineer or mountain guide be able to pick the route most suited to the team's equipment and experience levels, taking into account safety, economy of effort and ease of movement.

Terrain assessments will continue throughout the operation being performed every time the team crests a ridge and new terrain is opened up for evaluation. In addition to a good map and compass, a small pair of powerful binoculars (8X-10X) can be very useful for assessing more difficult routes over rock and ice before the team is committed to them. Once in rugged terrain it may be difficult to see even one rope length ahead.

BASIC MOUNTAINEER TRAINING

Although books are not a substitute for actual training, and mountaineering in particular is a hands-on type of activity, the first step towards becoming a competent alpine operator is to study as much as possible about the subject. So it is the purpose of ALPINE OPERATIONS to give the reader an understanding of the equipment and training necessary and to help eliminate any costly errors in either department.

However, it is beyond the scope of this text to go into great detail about individual mountaineering techniques, so we recommend "Mountaineering—The Freedom of the Hills" (6th Edition), which is considered the bible on this subject and is used as a basic text by many organizations. So assuming that every aspiring mountaineer will own a copy and take professional training, and rather than repeat what is covered in both, we have chosen to concentrate more on program development and the training progression for professional teams.

In addition to the mountain living and travel skills covered earlier in this text, the following are a list of essential skills that should be reviewed and/or taught in a basic mountaineering program.

- Knots & Rope Management
- Rappelling & Belaying
- Prusik Ascent Techniques
- Snow Travel, Ice Axe and Crampon Technique
- Self-Belay & Self-Arrest
- Glacier & Rope Team Travel
- Crevasse Rescue
- Balance Climbing
- Anchors & Placing Protection
- Basic Rock & Ice Climbing
- Risk Assessment & Accident Management

MOUNTAINEERING SCHOOL

Whether military or law enforcement, tactical operations or rescue, anyone anticipating work in the mountains must attend a basic mountaineering school since this forms the foundation for all future alpine training.

In doing research for this book I worked with several professional mountain guides, mountaineering schools and mountain warfare programs to observe the teaching techniques of those who do this for a living and to gain a better understanding of the problems experienced by the students in an outdoor classroom. One such course was Alpine Ascents basic mountaineering course taught in the Northern Cascades, designed for beginner and intermediate climbers wishing to develop strong fundamental skills in all aspects of mountaineering.

The published course objectives for this course are "to develop educated, self-reliant climbers with the ability to evaluate hazards and to increase technical knowledge and skills in all aspects of snow and alpine climbing", including: glacier travel, rope

& belay techniques, crevasse rescue, route finding, self arrest, rappelling, safe climbing, and cramponing.

Author's Note: Since the basic mountaineering program is the heart of this book and the foundation for all future mountain work, I felt that it would be more educational for the reader if I was to audit and profile an actual working school. This was done not with the intention of recommending any particular guiding operation but more to give a detailed insight into what to expect and what is expected during a basic course. After some careful research, I selected Alpine Ascent's North Cascades Mountaineering School as a prime example of a professionally run operation.

Basic 6 Day Mountaineering School—
An Overview

Preparation for a basic mountaineering course should begin weeks before the actual start date of the program for two very practical reasons:

1. To obtain the necessary equipment, or make changes to existing equipment
2. To allow time for essential physical conditioning

Both of these should be covered in detail in any reputable school's application package. To venture into the mountains with the inappropriate equipment and/or poor physical conditioning will be not only a miserable experience for all concerned, it can easily escalate into a safety issue.

AAI's registration package comes with a detailed Equipment Checklist of all the clothing and equipment required for their program and suited to the local conditions. This list in itself is extremely informative since it represents years of experience on the part of the mountaineering staff in terms of what works and what is absolutely essential. The sheet lists each item needed along with the weight, size, desirable features and recommended manufacturers. Applicants are strongly advised to follow this list as closely as possible.

Along with the equipment list comes a Rental Form checklist where applicants can check-off the equipment they already have and indicate those items that they will need to rent. Since applicants are expected to have backpacking and camping experience, they should already have most of the required equipment and clothing. However, for many this will be their introduction into mountaineering, so they may not have such items as plastic mountaineering boots, crampons, harnesses, ice-axes and helmets—the five items most commonly rented for the course.

AAI also supplies a very informative guideline for food selection and meal planning that encourages the reader to hit the supermarket in search of items that are lightweight, quick to prepare, nutritious and varied. The alternative is the pre-packaged freeze-dried meals available from stores like REI and Adventure 16, and although some of these are quite good, they are more expensive, offer less variety and are not as much fun to prepare. The Menu Planning sheet also gives some very useful advice on the importance of hydration and caloric intake to compensate for the anticipated energy output, temperature and altitude during the program.

Day 1 begins at 06:00 when participants who flew in to Seattle, Washington the night before are collected from the hotel and transported to AAI's office to meet the staff and for a thorough itemized gear check. This is an important part of the program since literally every item on the list is critical to the safety and comfort of the individual.

Each participant lays out all his or her gear and is check-listed by the guides—in this case Vern Tejas, Barbara Winkler and the school's director Willi Prittie.

Participants are also teamed up at this point into climbing pairs to eliminate duplication of items such as tents and stoves—making a welcome weight reduction of over 5 pounds each right there. To further reduce weight, the guides encourage everyone to leave behind any unnecessary items and in minutes each climber had built a small pile to be left behind—spare batteries, cameras, lenses, candles, non-essential toiletries, large tubes of toothpaste, soap, make-up, unneeded climbing hardware, books, journals, pocket knives, pens, pencils, excess first aid supplies, tent pegs, duct tape and a plethora of other gadgets sold in camping stores. (The only excess weight that I carry is a reliable but heavy F3 Nikon and a small waterproof notebook).

After the guides have supervised the packing and that packs are loaded correctly—with heavier items on top, no empty spaces and a minimum of equipment on the outside—fuel bottles are filled and the group loads into the van.

The drive time up into the Northern Cascades is a good opportunity for the trainees to get to know each other and for the guides to get a feel for the individual's experience. The guides also use this time for informal discussions about food selection, personal hygiene, hydration and then a brief orientation on topographical maps as to where the group will begin the approach climb into the mountains. There is also a quick stop at a supermarket for last minute additions to the food list, and one final chance to use a flushing toilet.

At the trail-head the guides distributed team equipment like ropes, shovels, pickets and hardware, and again encourage everyone to ditch any unnecessary items and excess weight. Looking up at the long, steep approach climb facing the group, virtually everyone found something that they "absolutely didn't need" to hump up the mountain.

The hike in began with a river crossing then a steep climb of over three thousand feet through forest and boulder fields to establish First Camp in the snow at about 5,700´. This area and the surrounding ridges would be the training site for the next three days.

The next hour was spent leveling the snow and pitching tents, all under the helpful supervision of the guides. This was followed by an orientation lecture on minimizing environmental impact with regard to trash and toilet requirements (Everything would be packed out).

Day 2 is a big day on the mountain where many basic skills must be learned or reviewed and practiced before the class can proceed to more advanced terrain. The day begins early on the steep snow above camp with lessons on walking on snow, kicking steps and axe handling and placement while walking. Emphasis is placed on the importance of sure, aggressive stepping to make steps and maintain balance.

Other exercises include: walking diagonally in a cross-step-up technique with the axe on the high side; executing switchbacks; downhill heel plunging and side stepping; ice axe belay (a self-belay executed by driving in the shaft into the snow); and steep ascent by planting the shaft of the axe and kicking in the feet. In all of these exercises, emphasis was placed on the importance of the first person making good steps with even spacing and that the following climbers work to improve these steps.

By late morning the class moved into glissading with and without the axe, sliding down snow slopes in a standing position while maintaining balance. (An entertaining exercise for the instructors). This was followed by self-arrest training—one of the single most important survival skills for the climber on an exposed ridge or slope, since lost footing could

Practicing ice axe arrest techniques

result in an accelerating out of control plummet of several hundred feet.

Self-arrests were practiced first without the ice axe and then with the axe carried in a pick-forward "walking position" (requiring the axe head be flipped in the hand to initiate the arrest). Self-arrests were then practiced with the axe carried in the pick-to-the-rear "arrest" position; and then self-arrest going back first and then face first down the slope turning to the right and left.

Throughout the day, trainees were being encouraged to stay hydrated and to continually pack their water bottles with clean snow that would then melt to make more water as the day warmed up.

After a steep climb, lunch was taken on one of the ridgelines overlooking the camp with the time being well used for an introduction to map and compass work. Students shot bearings on surrounding peaks and then applied those to the map to identify local features and their current location through triangulation (resection).

The afternoon of Day 2 was used to work on anchors and deadmen; the use of pickets, snow flukes and ice axes, planted vertically or buried horizontally. Equalizing anchors were created using cordelettes, stressing the importance of staggering the deadman anchors and not placing anchors on the same shear line of the snow.

Climbing on steep snow using the ice axe for a self belay

When the weather closed in, the group retreated back to camp and the instructor's tent to work on knots and rope terminology. The knots included the Overhand knot and Figure 8 knot with special attention given to Bowlines. The Bowline is AAI's preferred knot for tying into the harness so students worked on this, and then backed it up with a Double Fisherman's. A Bowline-on-a-Bight is used for situations where a climber is in the middle of a rope team, and this knot is backed up by clipping the end loop into a locking carabiner.

Clove hitches were practiced for use as an adjustable anchor in a belay position; the Munter hitch for belaying; and the Double Fisherman's knot for making Prusik loops.

After the weather cleared, belays were practiced using a basic hip belay by digging a pit in the snow to sit in; followed by Munter hitch and ATC belaying; then more work on equalized anchors; ending with rappels using two ropes and an ATC.

Day 3 began with a knot review, then the class made-up prusik cords in preparation for crevasse self rescue. Texas Prusiks are used to ascend the rope after a fall into a crevasse, so the class practiced prusik tie-in procedures in preparation for rope team travel and then configuring them for a rope ascent. Ascents were practiced near camp under controlled conditions on a rope strung between two trees.

The class then moved into a discussion of glacier travel and crevasse formation, followed by rope team travel and spacing. Students learned that a 3-man rope team is ideal, with 4 being the maximum and 2 the minimum.

For the 3-man rope team the climbers have 1/3 of the rope between each other and the remaining 1/6 coiled at each end. The 4-man team simply has 1/3 between each person, and the 2-man rope team has 1/3 of the rope between them and each carries just over 1/3 coiled.

Once roped up, the class was taught how to cross

horizontal crevasses at the perpendicular with the rope team in single file; traverse vertical crevasses with rope team line abreast; and to keep the rope extended and to not bunch up at stops. Staying well spaced is essential to arresting a fall if anyone of the team should break through into a crevasse.

To effect crevasse rescues, the class was taught about mechanical advantages and how to build Z rigs for a 3:1 advantage and a to add a C rig for a 6:1 advantage.

Next the class moved up to the ridgeline in rope teams practicing running belays using pickets in the snow. They practiced keeping their axes on uphill side and keeping the rope on the downhill side to avoid having their legs swept out in the event of a team member's fall. This rope travel exercise took the class up to 6,200´ where they again reviewed map and compass work while eating lunch.

Practicing 3:1 crevasse rescue / raising systems

A not wholly unexpected thunderstorm meant a retreat to the guide tent for knot review and Z rig review and practice. Later in the afternoon the weather cleared and it was onto the slopes for more ice axe arrest practice and glissading.

Day 4—At first light, the group broke camp and began a roped movement up to High Camp at the base of Eldorado Peak. This required climbing to an adjacent ridge and then rappelling down the backside to access the lower lip of the glacier. From there, the class roped up again and climbed onto the glacier, all the time being subjected to a variety of fast moving weather patterns that are not unusual for the Cascades.

Upon arrival at the High Camp area, it was necessary to form line-abreast rope teams and probe the area for hazards such as snow-covered crevasses. With the use of wands and a well-walked perimeter, it was possible to setup a "safe zone", then erect tents and establish plastic bag lined toilet pits within the perimeter. Nobody left the safe zone without being roped.

Once tents were pitched and secured, there was still time to climb a small but very steep peak, located to the east of camp, in rope teams using a running belay on pickets. This was further practiced in a cir-

Climbing in rope teams to a nearby ridge

cuit around camp to clean up some procedural difficulties since it can be very frustrating to have an individual on a team who is slow unclipping and reclipping as they pass pickets.

Late afternoon left just enough time for lectures on weather patterns and avalanche awareness.

High camp below Eldorado Peak

Day 5—As the rope teams left camp in the morning, they placed wands in the snow to mark the return trail in case there were whiteout conditions on their return. The three teams moved further up onto the glacier to find a crevasse that could be used to practice crevasse rescue techniques. With the shape of the terrain and the glare from the snow, these can be difficult to see until you are right on them.

Upon finding a large open crevasse, it was necessary to probe the training area and establish a "safe zone" back from the lip of the crevasse. Again, nobody would leave the safe zone or approach the lip of the crevasse unless they were roped.

After setting up anchors and belay stations, individuals practiced self-rescue techniques using prusiks to climb back up their own rope. Teams then built Z rigs to practice recoveries—first simulated with just a pack as the "victim", then actual rescues with a live victim. C rigs were then added to the Z rigs as a force multiplier as everyone was given the opportunity to be both victim and rescuer. To give an added sense of realism, the rope team leader would literally walk off the lip of the crevasse to give the others the opportunity to actually turn, drop and arrest the fall, and then stabilize the situation and effect the rescue.

Simulations were also done where a student had to first rappel down to stabilize a severely injured victim, then prusik back up to assist with the recovery effort above.

At the end of a very productive day, the class roped up again and followed the wands back to High Camp in actual whiteout conditions (no simulation there)—a situation that could have been quite uncomfortable without the benefit of the wands. Once safely in camp, everyone cooked dinner and prepared for an early morning assault on the summit of Eldorado Peak.

Practicing crevasse self rescue with both Petzl ascender and Texas prusiks

Rigging a 3:1 raising system and placing a pack on the edge of the crevasse to prevent the rope from digging in

AAI guide breaks trail on summit day and sets pickets for a running belay over more exposed areas

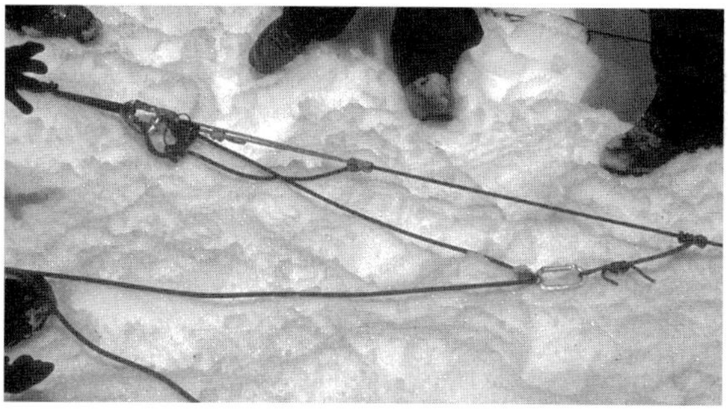

Rigging a 3:1 raising system using a prusik loop and small rescue pulley

Summiting at first light in near whiteout conditions

Day 6 required a very early start with the intention of summiting, then returning to break camp and get off the mountain the same day. Early morning starts are recommended in mountaineering to take advantage of the firmer and more stable snow conditions that will change when the sun hits them.

So at "zero dark early" (0400) hot chocolate was brewed and the group set off in three rope teams in near whiteout conditions caused by low clouds. The team placed wands every rope length all the way to the base of Eldorado to mark the return trail, then began the push for the summit, using a running belay with pickets for the more exposed areas.

By 0700 the team had summited, still in low

On the Summit at dawn

View of high camp illustrates how easy it would be to lose it in whiteout conditions

clouds, and returned to camp for breakfast. By 0900 high camp was broken down, holes were filled and the move was made back to low camp to recover gear that had been cached in a snow pit. By 1200 the team was on the grueling downhill hike through the boulder fields and trees to the waiting vehicle and a final critique by the instructors.

That concluded the 6 day program but was only half way for those signed up for the 13 day program, designed to increase technical knowledge and skill in all aspects of snow, ice and alpine rock climbing. This is a highly recommended package for those planning more challenging climbs or just wanting to be more confident in the mountains.
Additional topics that would be covered included:

- Top Rope and belay techniques
- Rappelling
- Protection placement
- Crack and face climbing techniques
- Route finding
- Lead and multi-pitch climbing
- Aid climbing
- Running belays and running protection

Day 7—Climbing techniques and top roping. Climbing techniques such as edging, smearing and proper balance. Top rope anchors to climb various routes. Much of the day is spent top roping several routes to give experience on both face and crack climbing. The day ends with a series of rappels.

Day 8—Additional advanced rock and varied climbing. Belay anchors and running protection for both lead and multi-pitch climbing. The goal being to teach the student how to move quickly and safely on moderately technical terrain. In alpine climbing it is important to be able to reach bivouac sites, etc. before dark.

Day 9—Aid climbing and advanced protection placement. This is the last day of formalized training before attempting the group's first alpine peak.

ALPINE OPERATIONS

Days 10-13—Ascents of alpine rock peaks. The last four days of the course are spent climbing some of the North Cascades finest peaks. The guides choose climbs that are in the best condition and which most aptly match the students' skill level. Potential peaks include: Mt Shuksan, Twin Sisters, Forbidden, Mt Sloan, Eldorado Peak, Torment, Buckner, Logan, Glacier Peak, Chair Peak, The Tooth and Guy Peak.

(Photo: S P Parker)

FEAR, TOUGHNESS & FAILURE

Fear is as much apart of climbing and alpine operations as is cold or fatigue. The climber is constantly exposed to some element of fear in the mountains, whether it is fear of heights, fear of exposure, fear of falling, fear of injury or fear of death, but just as we put on more clothing when cold or rest when tired, we must deal with fear in an equally positive manner. The individual that cannot overcome or at least control fear of heights for instance, should steer well clear of alpine operations.

Fear is one of the strongest check-valves in the human psyche and should be considered a healthy reaction to a dangerous situation. It is fear that keeps us from walking too close to the edge of a cliff, or makes us double check a piece of protection, but it is also fear that can cause a climber to freeze in a precarious situation. Left unchecked and undirected fear can spin out of control leading to panic—and from panic it is a very small step to tragedy.

No one is immune to fear but the professional should be able to first recognize it in himself and then turn that fear into a productive action. In climbing, fear occurs where the difficulty of a climb and individual ability come face to face—when one is suddenly confronted by one's own limitations and mortality.

At this juncture, attitude and emotional control is everything. The ability to stay calm, analyze the situation, harness the energy and extricate oneself from the predicament. And going down is not always the solution to the problem—many times continuing the ascent may be the best way out.

Fear is often the result of venturing into unknown territory, a factor that can be minimized through experience, but unfortunately we have to scare ourselves quite a few times to gain that experience. The trick to gaining experience without getting killed in the process is to know one's personal limitations and to bite off just a little at a time. To attempt progressively longer and more difficult routes but not to force the learning curve by taking on climbs way beyond one's ability.

The smart operator makes mistakes only once and then stores them away as part of his stock of valuable lessons learned. This is also known as "the burned hand theory". You only have to do one high-speed rappel with thin gloves; or have a belay rope get away from you; or get hit in the head by one small rock without a helmet to learn not to do that again.

Another important step to controlling fear in the mountains is not to allow oneself to be surprised by sudden changes in conditions. Nobody controls the situation in the mountains, particularly the weather, and it would require a massive ego to think one could. Instead, it is necessary to accept the often unpredictable nature of the elements, with our inherent lack of control over them, and compensate with careful planning and preparation, all the time considering the best bailout plan if conditions turn ugly.

When things go really bad and seem to be spinning out of control, as they inevitably do with enough time in the mountains, then it is only mental toughness that will save the day. A never-give-up dogged determination that allows some to overcome the difficulties that paralyze most other people.

This toughness is a bi-product of the mind and body's ability to adapt to extremes of either comfort or adversity, so individuals accustomed to a comfortable lifestyle will most often be the first to fold when conditions deteriorate, while those who have challenged themselves with difficult activities under harsh conditions will have a toughness and a higher probability of adapting to a nasty situation. In mountaineering this is most often manifested when

(Photo: S P Parker)

a storm catches a group high on the side of a mountain leaving the climbers cold, wet, tired and disoriented. The tough will accept the misery and work through it, while the weak will give up and lay down to die in the snow.

However, a "never give up" attitude is another of those proverbial double-edged swords that taken too far can get one killed. Although a high degree of determination is essential in mountaineering, it is equally important to learn how to fail and when to fail especially when trying to climb a mountain or make a dangerous rescue. When the difficulty is obviously beyond one's ability, the route turns sketchy or inclement weather suggests a hasty retreat, then one should rappel off satisfied in the knowledge that more lessons are learned from analyzing the mistakes and failures than celebrating the victories.

(Photo: S P Parker)

11

ROCK & ICE CLIMBING

As discussed earlier in this text, climbing and mountaineering are two different but related activities. The term "mountain climbing" is used for mountaineering but not all mountains require technical climbing ability. In many cases, even the highest mountains have routes that are little more than a very hard walk with the real challenge not being vertical rock and ice but the sheer determination and endurance required to overcome the cold, the weather and the altitude. More people retreat from high mountains because of fatigue, inclement weather and mountain sickness than because of some unclimbable obstacle.

However, the dedicated mountaineer with years of experience will seek out the more difficult routes on a mountain to challenge his ability. Many will also work to establish new routes so that they can claim the honor of a "first ascent". It is this type of mountaineering that does require considerable technical expertise and goes well beyond what would be required of the average mountain operations team.

Rock and ice climbing on the other hand have evolved into sports that have less to do with mountaineering, but are never-the-less good exercise and a definite confidence builder for those situations where technical climbing skills on the mountain are required.

A lead climber on any mountaineering expedition, and in particular professional mountain guides, should have an above average climbing ability so that they can establish fixed ropes or belay other less experienced members of the team. This does not mean that they have to be able to lead a 5.10 route on rock, a difficult task wearing heavy plastic boots and thick gloves, but at the same time, having strong climbing ability will make the mountaineer a more able, confident, versatile and valuable member of a team.

Sport climbing at both indoor gyms and local crags has become a popular recreational activity, but it also has many tactical applications. The European counter terrorist teams have long been advocates of free-climbing and building scaling as a means of accessing a target location, so it is no surprise to see climbing walls appearing in the training gyms and compounds of their US counter parts. In addition, Navy Seals, Delta Team, FBI-HRT, SOTG Marines and several major SWAT teams have all received training in both rock climbing and building scaling from nationally recognized climbers and mountain guides.

CLIMBING PROGRESSION

The training progression for technical climbing is a methodical building block process largely driven by the natural ability of the climber. It begins with basic rope management and culminates in the ability to lead difficult climbs, but to get to that point requires a lot of training and many hours of dedicated climbing. Even with natural ability, just to get to the point of being a competent lead climber on moderate routes could take many months or even years of training and considerable experience on a variety of rock.

Many of the skills learned in climbing are also covered in a mountaineering program but review and repetition can only improve the climber. The progression for a climbing course is as follows:

- Climbing Equipment & Terminology
- Knots & Rope Handling
- Rappelling & Belaying
- Balance Climbing on moderate Class 3 & 4 terrain
- Energy Conservation & Use of the Feet and Legs in Climbing
- Principles of Face Climbing—Mantels, Hand-holds, Under-clings, Stemming, Lie-backs, etc
- Top Rope Climbing beginning in the 5.5-5.7 range of difficulty
- Crack Climbing, Jamming, Chimneys, Traverses, Overhangs
- Down-climbing
- Placing Protection & Practice while on Top Rope
- Basic Lead Climbing on Bolted Routes (5.7-5.8)
- Lead Climbing & Placing Protection (5.8-5.9)
- Balancing Speed & Safety
- Belay Station protocol for Multi-Pitch climbs
- Multi-Pitch Climbs working on Economy of Effort & Movement
- Multi-Pitch Rappels & Anchors

ON ICE

- Introduction to Ice Climbing & Ice Tools
- Top Roped Ice Climbing
- Placing Protection on Ice
- Leading on Ice
- Mixed Rock & Ice Climbing
- Leading Mixed Routes
- Multi-Pitch Rappels on Mixed Routes

CLIMBING CONDITIONING

As with any sport, nothing conditions an athlete better than actually doing that specific activity—and this is especially true for climbing which requires a balance of strength, endurance and flexibility combined with minimal bodyweight. However, not everyone has daily access to a local crag or bouldering area so many climbers make use of indoor climbing gyms and a number of home gym techniques to strengthen the muscles and tendons used in climbing.

Rock gyms are an excellent form of training but tendons and ligaments must be given time to adapt to the specific demands of climbing

WARNING: It is not unusual for individuals when they first become involved in climbing, especially those with easy access to a rock gym, to over do it in the first week or two and find themselves suffering for several weeks from inflamed tendons in the hands, joints and shoulders. For this reason, climbing should be approached with a progressive build-up giving the body ample time to repair damage and adapt to the strains of a new activity.

Rock guide and former Seal, Scott Helvenston leads a rock pitch by taking advantage of a crack between two slabs

Mantle technique for moving up onto a ledge

A hand jam is a commonly used climbing technique

Lay-back technique

Climbing rock or ice is a whole-body workout that is both aerobic and anaerobic that requires not only upper body strength but also flexibility and a good power-to-weight ratio. Athletes aspiring to be good climbers, who already have exceptionally powerful physiques and considerable muscle mass, will find themselves striving to reduce that mass and transform themselves into a more sinewy climbing machine.

Powerful arms and upper body will allow a climber to muscle his way up through the first few moves but he will quickly tire as a result of not using his legs wisely and from moving more weight than necessary. Being a good climber is all about technique and economy of motion.

CLIMBING TECHNIQUE

The mark of an accomplished climber is a smooth, fluid technique with little or no wasted movement or energy expenditure. The route is studied and moves planned so that when the climber hits the rock he or she appears to be in a state of constant unhurried motion.

Maximum use must be made of the leg muscles for all upward movement with the hands used primarily for balance. A good climber will also try to maintain three points of contact while keeping the body away from the rock with weight centered over the feet so that minimum hang time is spent on the arms.

The climber strives to conserve energy on overhangs and long pitches using a straight-armed "monkey hang" to rest rather than straining to hold on with bent arms. Leaning back and hanging on straight arms is considerably less tiring than the muscle strain that goes with a "pull-up" type of technique.

Two aids that are essential to developing good technique are climbing shoes and a top rope. The first to better feel and grip the rock and the second to give the confidence necessary to experiment with new and challenging moves.

For the aspiring mountaineer who has mastered the basic moves wearing rock shoes, it is important to then try the same pitches with stiff mountaineering boots that do not allow the same smearing tech-

Climber uses an under-cling technique on an overhang

Straight-armed monkey-hang used to conserve energy

AMGA guide SP Parker demonstrates rock climbing techniques with double plastic boots

nique commonly used with soft climbing shoes. The climber must also learn to compensate for the additional weight and size of these boots.

The next step, while maintaining the safety of the top rope, is to try clipping into bolted practice routes and placing protection to simulate leading. Then, once a pitch has been mastered with the protection of a top rope, the climber should consider trying to lead the same pitch. Even a relatively simple climb can become a nerve-racking challenge without the security of a top rope, especially when the climber's safety depends on the location and positioning of each piece of his own protection.

To place protection while leading requires that the climber find a reasonably stable position where he can get his weight over his feet, hang by one hand, draw and place protection then snap into the rope. Not an easy task on a vertical wall. The first piece(s) of protection should also be placed immediately above the belay station so as to minimize the fall factor and reduce the probability of pulling the belayer off of a ledge.

Once mastered in the gym, these newfound skills should be taken to the mountains where the weather

Reaching for the next clip

and local conditions become additional unpredictable variables. Even though clothing should be kept as light and comfortable as possible, it must still supply adequate warmth, especially when not moving and belaying from an exposed ledge. The heavier clothing and mountaineering boots are significant restrictive factors for the climber compounded by the need for gloves in cold icy conditions. Pile pants for warmth and heavy Gortex shells simply do not give the range of motion that stretch climbing pants offer, so many climbers will go to a lighter Activent type of clothing with lightweight Windstopper vests to give a greater freedom of movement. But it is still necessary to carry something warmer for the long periods standing still belaying the other climber.

The foot and handholds found in nature are also a new variable. The carefully formed cracks, ledges and pockets in the gym now become wet rock covered with snow and ice. Handholds that first appear firm can disintegrate into crumbling rock and the falling debris becomes a very real hazard requiring the use of helmets.

Gloves must often be removed to make a difficult climbing move or place protection so warm hands become numb with cold and the wind chill can freeze any exposed skin. To lose a glove could be a fatal mistake so gloves should have idiot cords attached and spares should be carried—a lighter pair for climbing and a heavier pair for belaying.

Royal Marine Commando leads a pitch of hard ice

ICE CLIMBING

Climbing frozen waterfalls and man-made hard ice is somewhat of a contrived sport since it takes advantage of technical ice tools not usually carried in the mountains, but it is still an exhilarating form of climbing that has some application for the mountaineer.

Apart from just being a great workout, the techniques used in ice climbing will undoubtedly make the practitioner a more skilled and confident alpine climber. In addition, winter climbing in an icy canyon is a more realistic environment for the mountaineer than scampering around some sunny bouldering area in mid-summer. Similarly, having to climb in full Gortex and gloves with heavy mountaineering boots and crampons is also more realistic than training in shorts and rock shoes.

In addition to the boots and clothing, ice climbing is more equipment dependent than rock climbing requiring a pair of technical ice tools, suitable crampons and a rack of ice protection.

Technical ice tools are much shorted than conventional ice axes, usually about 50cm in length and come in pairs, one with a hammerhead and the other with an adze. Both have reverse-curved picks specially shaped to positively hook into the ice but designed to dislodge more easily than a conventional ice axe. The reverse-curve allows the pick to slide up and out of the ice, sliding on its upper edge when the handle is raised.

The more efficient tools are well balanced and have a comfortable swing weight that allows them to be planted in the ice with minimal effort. They have a bent shaft to better reach over ledges and to protect the knuckles when swinging into the ice, along with an ergonomic grip and wrist loops so that the climber can "rest" without having to grip the handles tightly.

Several companies make top-of-the-line ice climbing tools such as the Grivel Machine, Charlet Moser Quasars, Black Diamond's carbon fiber Black Prophet or lightweight Cobra, and the Trango Captain Hook. More economical tools with bent or straight shafts for a broader range of uses, are the

BD X-15, Grivel's Rambo and Geronimo, both used by the military, the CM Pulsar and the Trango Claw.

The heat-treated picks on these tools need to be kept sharp to minimize disruption of the ice and prevent shattering or dinner plating. It is also recommended to carry a spare pick and an Allen wrench to tighten bolts that vibrate loose with heavy usage.

As important as good tools are to efficient ice climbing, good crampons are even more important. Just as the rock climber maximizes the use of his legs for upward movement, the ice climber also strives to use the stronger muscles of the legs rather than depend on arm power.

To get maximum use out of the legs on wet slippery ice requires crampons with sharp toe spikes, and although standard mountaineering crampons can be used for ice climbing, there are others that are better suited to this task. Where S-12 type mountaineering crampons have considerable flex and relatively short toe-spikes to make them comfortable for walking, ice climbing crampons tend to be more rigid with longer and more aggressive toe-spikes to better grip the hard ice.

The crampons that have found popularity with ice climbers are models like the Grivel Rambos, Trango Harpoons and Black Diamond Switchblades—all available with either mono or dual points. Mono points are the favorite of lighter sport climbers but the dual points seem to grab more ice and are better suited to heavier climbers or those carrying equipment.

Another essential for the ice climber is good waterproof gloves, preferably with extra protection around the knuckles. The BD Ice gloves have proven to be excellent for this application.

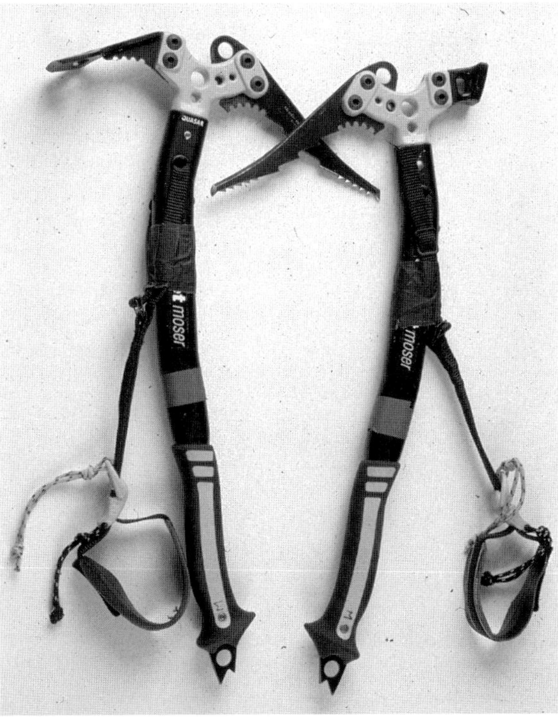

Matched pair of Quasar technical ice tools

Ice climbing guru Jeff Lowe coaches the author on the importance of foot work and crampon use at the Masters Seminar during the Ouray Ice Festival

ICE PROTECTION

The first rule of ice climbing is DON'T FALL—or at least climb with the attitude that you do not want to fall. Unlike rock climbing where falls and whippers are somewhat common, falling on ice can be extremely hazardous to your health—primarily because of all the sharp points involved.

A pair of crampons has a total of about 24 sharp points, and ice tools have spikes on both the picks

Trango harness ice screw clips

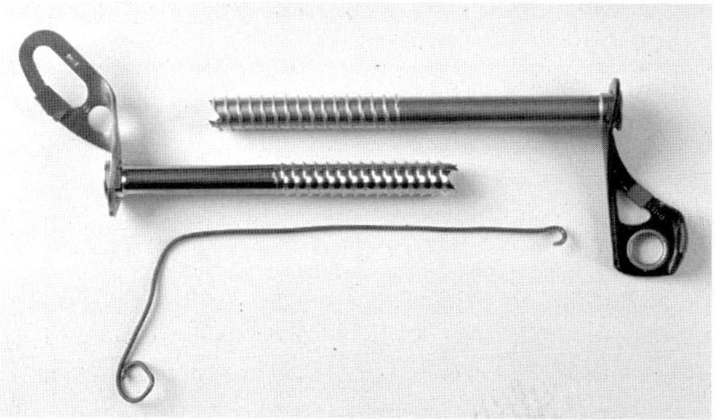

22 cm ice screws with wire hook used to pull through the cord or webbing on abalakov V thread anchor

Malcolm Daly uses two ice screws and an ice tool linked with spectra slings to create a secure belay station

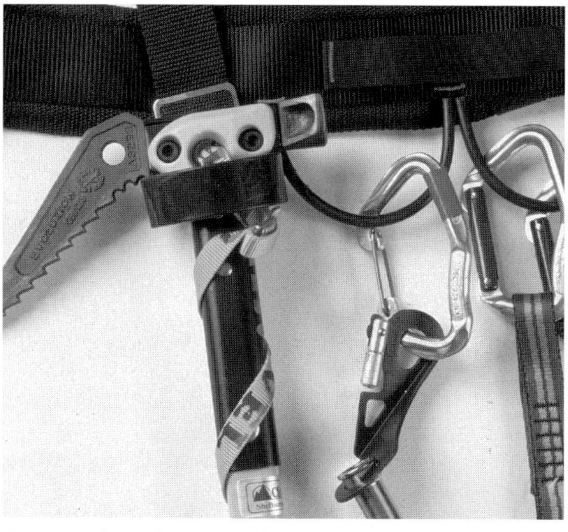

Harness loop for ice tools

SP Parker demonstrates screw placement at a 90–100 degree angle to the ice

and base of the handles. Add to that a rack of ice screws and that is all together too many sharp objects attached to your body that could cause serious harm in a fall.

The next rule is place protection wisely. Although slings can be placed around ice formations much the same way as they are used on rock, the primary form of protection used by ice climbers is ice screws. These tubular screws are made from aluminum or titanium and designed to be hand drilled into the ice to a depth of 10 to 22cm.

Several companies manufacture ice screws—Black Diamond, Charlet Moser, Grivel—but one of the most efficient is the BD Express Screw which has a small integrated handle that allows them to be quickly cranked into or out of the ice.

A well placed ice screw in good ice is good protection, but it is important to ensure that the ice does not have air spaces, will not come away in a big plate or that the screw will not bottom out on rock.

To plant an ice screw, first make a small hole with the pick, just deep enough to engage the threads of the screw, and then run the screw in at 90 degrees to the ice until it is flush. With Express screws, always flip the handle to the closed position after positioning the screw.

Another item carried by ice climbers is the Spectre, a drive-in device, much like a small pick, intended to supplement the rack of ice screws, not

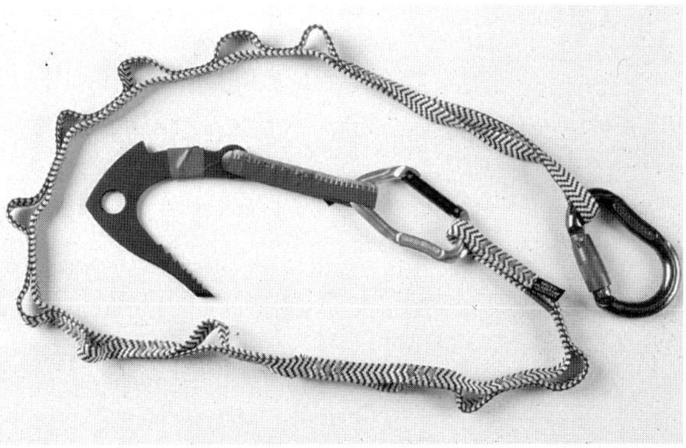

The Spectre hook and spectra daisy chain can be used for temporary anchoring and to rest while placing an ice screw

replace it. These are designed for fast placement in dicey conditions and can be easily placed in old pick holes.

Treated "Dry" climbing rope is the standard for most ice climbing with special care being taken not to accidentally spike the rope with crampons or a careless pick swing. For this reason, half and twin ropes are very popular with serious ice climbers giving the added security of two ropes should one get spiked.

Late in the day or toward the end of a hard climb when both of you are tired, someone is going to spike the rope and sure as hell neither of you will fess up to it. This makes it that much more important to be constantly inspecting the ropes for damage and running them through your hands feeling for irregularities.

ICE CLIMBING TECHNIQUE

As with rock climbing, the key to good ice climbing is to relax, climb smart and plan your route. This begins with an economy of movement, good footwork and crampon technique and a minimal use of arm-power. The climber uses a wider stance with the feet kept straight, level and perpendicular to the wall to prevent the toe-spikes tearing out. Upward movement is with the leg muscles using the arms and tools primarily for balance.

Tools that are correctly weighted and balanced are important in that they allow the climber to let the head of the tool swing home naturally with a flick of the wrist and minimum effort. Precise placement of the picks is also important, so the pick should be planted at 90 degrees to the ice. In addition, the climber should be constantly looking for the best ice to place the pick and natural depressions or old pick holes that will save the effort of having to make a new hold.

The bottom belay should be done from an off fall-line position to stay out of the line of falling ice, which can be considerable. A helmet is an essential and eye protection is recommended. Top belay positions should use multiple anchor points placed so that they are not on the same shear line of the ice.

Screw placement and removal is done at about shoulder level and only when the climber is confident with the three other points of contact with the ice—one pick and two crampons. A daisy chain attached to the climber's harness is also useful in that it can be clipped into the hole in the handle of a well-planted tool to give the climber an opportunity to rest his arms.

Note the left tool in a natural pocket in the ice and the feet 90 degrees to the ice

MIXED MULTI-PITCH CLIMBING

Technical mountaineering routes require a combination of alpine mountaineering, rock and ice climbing skills melded together into one long climb or a series of pitches—each pitch being one rope length or the distance to the next convenient ledge or belay station.

When climbing with significant exposure and risk of falling, climbers will generally work in 2- or 3-man rope teams. If there are four climbers then they will usually find working as two 2-man teams more efficient than one 4-man team. However, if a guide is leading a group, he will generally want to keep all his clients on one rope team as they move through more difficult areas unless he has a competent assistant.

San Juan Mountain Guide Mike O'Donnell stems a gap as he leads on ice. Lead climbers must take special care not to spike the rope with an inadvertent or careless crampon placement.

Since a climbing rope is only 50-60 meters long, and a days climbing could entail over several hundred feet of ascent, the rope team must establish an efficient method of climbing, belaying and moving upward on the single rope. This is known as multi-pitch climbing. If it is over a combination of rock, snow and ice then it is also mixed climbing requiring an array of climbing techniques and protection.

With a 2-man rope team, the climb begins with the belayer anchored into a ledge or belay station as the first climber leads off up the face, crack or ice. The climber has the rope attached to his harness as the belayer feeds it to him through a belay device. The belayer must be alert to the climber's movements, must never take his brake-hand off the rope and should avoid standing directly under the

Climber is able to brace her hip against a rock and use a chimney technique

When following a pitch it is important to avoid spiking the rope with an ice axe pick. This requires concentration especially when cold and tired.

The author stems a mixed pitch in the Uncompahgre Gorge

climber because of the risk of falling rocks and ice.

As he moves upwards, the lead climber places protection at appropriate intervals and clips in the rope. Ideally this could be a sling around natural protection or it may be cams, chocks or pitons in rock cracks, ice screws in water ice or pickets in snow. If the climber falls, he should not fall much further than the last piece of protection if it was placed correctly and provided the belayer is able to arrest the rope movement. The dynamic rope should help absorb the shock of the fall to both the climber and the last piece of protection.

When the lead climber reaches the end of the rope or arrives at a suitable ledge or belay station, he will set-up a series of anchors, attach himself to the anchors and prepare to bring up his partner. When he is safely anchored, he will communicate this to the belayer by shouting "Off Belay". The belayer will acknowledge with "Belay Off".

The #2 climber will now prepare himself to climb. When he is ready to begin he will call up to the lead climber, who is now the belayer, "On Belay?" which will be acknowledged with "Belay On". If it is too far or too windy, this may have to be done with rope pull signals.

As the top belayer feels or sees the #2 climber moving upwards, he will take in the slack rope through his belay device. As the climber passes the protection placed by the lead climber, he will carefully remove it and rack it on his harness or chest sling. This continues until the #2 climber reaches the lead climber on the belay station.

At this point, the #2 climber can become the lead climber and just continue climbing passed the belayer since he already has all the protection racked for the next pitch. This is known as swinging the lead and is the more efficient way to climb.

However, if he #2 climber is tired or does not feel confident in leading the next pitch, then he will first clip into the anchors and then pass the rack to the lead climber. If he then flips the stack of rope, it should be ready for the lead climber to lead off on the next pitch.

If climbing in a 3-man team it is best to use two ropes to maintain the 100 to 150-foot pitches. With a single rope, only short half rope-length pitches can be accomplished.

After the #1 lead climber has led the first pitch and clipped into the next belay station, he will signal for #2 to begin climbing. As the #2 climber begins climbing, belayed by #1, he will have the #3 climber's rope attached to his harness. Upon reaching the next belay station #2 will then clip into the same anchors as #1 and belay #3 as he climbs up. #2 will have already cleaned the route and pulled the protection placed by #1, so #3 should be able to move up quite quickly. Once all three climbers are secured they can decide who will lead the next pitch.

This type of climbing also requires some attention to clothing systems since it can be very hot while climbing but very cold sitting on an exposed ledge belaying the next climbers up. Difficult pitches could take twenty to thirty minutes to negotiate, often times longer for the lead climber. This means the belayer on a 3- or 4-man team could be sitting still for over an hour in icy conditions.

A secure belay station is a necessity on multi-pitch climbs

USEFUL VOICE COMMANDS FOR CLIMBERS

Communication between the climber and the belayer is an important aspect to efficient climbing. Climbers who climb together regularly come to know each other's style and are able to anticipate most moves. They will "feel" the climber on the rope much the same way that a hard-hat divers' tender is able to feel him working on the end of his air hose or umbilical. But when starting out, novices should get into a routine of using the following voice commands. Others can be made up to suit mission specific situations such as rescue raising and lowering systems. Military personnel will also be heard to use their names when rappelling or climbing, for example, "Corporal Stone, On Rappel," answered by, "Sergeant Granite, On Belay."

"On Belay?"	Are you ready to belay me?
"Belay on"	I am belaying you
"Climbing"	I am beginning or resuming to climb
"Climb"	I heard you, climb on
"Slack"	Give me some slack to move up or clip into protection
"Up rope"	Take up all the slack in the rope—usually to top belay
"That's me"	You have pulled up all the slack so don't pull any more
"Protection"	I am placing protection so not moving
"Cleaning"	I am removing protection so not moving
"Half way"	You have used half the available rope
"Watch me"	I am making a difficult move and may fall
"Falling!"	Assume your brake position, and be ready to catch my fall
"Rock!"	Falling objects, take cover
"Ice!"	Dislodged ice coming down, take cover
"Off belay"	I am secure and no longer need to be belayed
"Belay off"	I have removed the rope from my belay device, you can pull up excess

MULTI-PITCH RAPPELS

More climbers are killed descending a mountain than ascending, quite often because they are cold, exhausted, judgment impaired and either fighting bad weather or racing to beat an incoming storm front. Others have turned the climb too late in the day and find themselves descending unfamiliar terrain in the dark.

More climbers are also killed on rappel than down climbing, making rappelling one of the most dangerous aspects of mountaineering. This is in part due to the fact that they are trying to rappel off one piece of protection, which runs contrary to the practice of using multiple anchors in climbing. It is therefore critical that rappel techniques and procedures be learned and perfected before attempting long multi-pitch climbs that may require equally long multi-pitch rappels.

The multi-pitch rappel is considerably more complex than the relatively short rappels used in rescue and tactical operations. They are also more dangerous because the rope is not attached to a bombproof anchor with a locking carabiner and must be pulled down after each pitch. There is also a tendency to use a minimum of protection since the slings and hardware will be abandoned at each level when the rope is pulled. Climbers also know that they must conserve enough protection, hardware and slings to descend several thousand feet of mountain that may have taken days to climb.

If a climbing team has only one 60-meter rope, then it is necessary to halve the rope limiting each pitch to less than 30 meters. This is done by running the rope through a carabiner or rappel ring to the middle of the rope; then tying both loose ends together for safety; and finally tossing the knotted ends over the edge. When both climbers have rappelled down to the next ledge or rappel station and clipped into the newly established protection, they will untie the knot and pull the rope down.

Most climbing teams will use a single piece of protection for the rappel anchor even though this is not a practice that is recommended in climbing or rescue work. However, leaving multiple pieces of protection such as cams would be costly and could possibly not leave enough hardware to complete all the pitches required to rappel off of the mountain.

Long rappel off of Cerro Guilamet in Patagonia (SP Parker)

To add some safety to this practice, one alternative is to back-up the primary anchor with additional protection and have the heaviest climber rappel first, or have the first man down carry both packs. If the primary anchor holds, then the last climber to descend can remove the back-up slings and hardware and rappel down knowing that the one anchor will probably hold. However, removing the back-up

The same snow bridge from above and below

part of the mind-set that makes rope teams and climbing partners take responsibility for each other's safety and rescue.

It is good to have a healthy fear of crevasses but merely having a "rope up and go" attitude gives a false sense of security. It is only through training, coordinated teamwork and an elevated level of awareness that the rope team can move with any confidence over a snow covered glacier.

Obviously, following on the trail of those who have gone before adds a certain amount of surety but this is still no guarantee that one of your team will not be the proverbial straw that breaks the camel's back, or in this case the snow bridge.

The Khumbu Icefall behind Everest base camp moves at about 4 feet per day toppling huge Seracs that are a significant danger to the climbers who must negotiate this danger zone several times during their acclimatization and moves to higher camps.

ROPE TEAMS

The fundamental skill required for safe glacier travel is the ability to correctly form rope teams with the ideal being 3 on a rope but 4 can be used as a maximum and 2 as a minimum. For a group of 5 it is preferable to form a team of 3 and a team of 2 rather than deal with the awkwardness of a single team of 5, unless the lack of experience of team members dictates otherwise.

The way a 3-man rope team works is that the climbers would have 1/3 of the rope between each other and the remaining 1/6 coiled at each end. On

Rope team moves through snow covered crevasse field (SP Parker)

a 150 foot rope this would put 50 feet between each team member with the #2 man in the middle and #1 and #3 each coiling 25 feet around their chests.

The 4-man team simply has 1/3 between each person, and the 2-man rope team has 1/3 of the rope between them and they each carry just over 1/3 coiled.

When walking in a rope team, the pace is set by the #1 man who is constantly scanning the glacier for the best route and probing any potential problem areas. In general this would be the climber with the most glacier experience, but if the partner is a novice then the more experienced person may chose to be in the rear and better positioned to affect a rescue if #1 should break through into an unseen crevasse. When traveling roped to a client, a mountain guide must always assume that the client will not be able to rescue him so must be prepared to climb out of a crevasse under his own steam, as should any other team member.

When moving, it is important every team member work to keep the rope extended but not taut. The rope should drape naturally from one member's harness to the next with neither slack nor tension. Apart from the potential for tripping on the rope, if there is too much slack in the rope when a team member punches through into a crevasse he will sustain a longer drop and shock load the rope making it more difficult to arrest his fall. Conversely, if the rope is too tight, it will make travel and maneuvering awkward and irritating.

At stops and rest breaks, the team must stay extended and not bunch up since someone could be resting right over a snow-covered crevasse. Until a safe zone has been probed and marked out, the team should stay roped and correctly spaced. Multiple rope teams can pull up parallel to each other for rest stops if the first team simply steps off the trail and lets the next team move up into position. In this way, teams can also rotate their turn breaking trail on point.

Horizontal crevasses are usually crossed at the

perpendicular with rope team in line. Vertical crevasse formations can be traversed with the rope team line abreast or en echelon. Some crevasses are best negotiated by having the team stop as the lead man moves out around the end of the crack, and then holding position while others follow. But in all situations, everyone must be ready to go into an immediate arrest position to catch members either in front or behind.

For particularly hazardous areas like wide crevasses with narrow snow bridges or where the terrain forces the team to bunch up, it is best to stop, dig in and set up a proper belay position and actually belay each person as they make the crossing. This would also be recommended where it is necessary to climb down onto a snow bridge and then up the other side to make a crossing.

(Photo: SP Parker)

Rigged for glacier travel with Texas prusik cords already attached

EQUIPMENT

Everyone on a rope team should be equipped to handle both a self-rescue and the rescue of a team member. In addition to a rope and harness, this will require that each individual carry:

- 2 Locking Carabiners
- 1 ATC belay device
- 4 Non-locking Carabiners
- 1 Small Rescue Pulley
- 2 Prusiks—one short and one long with leg loops
- 1 Pack Rope
- 1 Picket or Ice screw

In a self-rescue situation, the climber will slip out of his pack allowing it to hang on the pre-rigged pack rope, and then prusik up out of the crevasse using Texas prusik loops.

In a rescue situation, one team member will arrest and hold the victim while an anchor is rigged with pickets and/or ice axes. Then a recovery raising system can be rigged with the pulleys, prusiks and extra carabiners.

If towing a sled on a glacier, then the back of the sled should also be secured to the rope with a prusik to prevent it from falling on the #1 man as he goes into the crevasse.

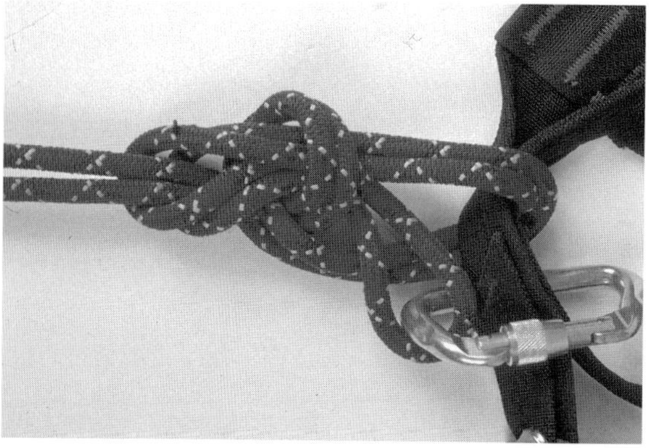

The bowline on a bight clipped into a locking carabiner is used to tie into the harness

Small Petzl pulley used to improve efficiency in a rescue raising system

Denali expedition team moves out onto the Kahiltna Glacier. Note the sleds tied into the main rope with prusik loops (Steve Dundorf)

CREVASSE RESCUE

If for no other reason than the remoteness of most locations, the assumption must be made when traveling on glaciers that your choices are either self-rescue or no rescue. If moving in a two-man rope team, it may take all your partner's strength and effort to simply hold your fall. It will then be up to you alone to climb out using prusiks or ascenders.

In the event of your partner punching through into a crevasse, if he is injured or unconscious, it will be again up to you alone to secure an anchor and then rig and operate the raising system. This can only be done by being organized and having recently rehearsed arrest and rescue procedures.

To successfully arrest a fall, it is necessary to be constantly alert and prepared to immediately turn and drop into an arrest position, digging in first the pick of the ice axe and then kicking in crampons or boots. For a mild fall with little or no slack in the rope, this is not difficult, but for a more dramatic break through it can be more challenging.

Assuming a three man rope team is traveling across a glacier when #1 punches through into a crevasse, then:

1. #2 and #3 immediately turn and drop into arrest position
2. Once the fall is arrested, then #2 will hold the arrest position while #3, after verbally ensuring that #2 has a secure position, rigs the anchor. This would normally be done with each of their pickets but could also be done with ice axes or other items used as deadmen anchors
3. The load can now be transferred from #2's harness to the anchor system
4. At this point one team member can approach the lip to check on #1 while the other belays him for safety
5. Check on #1 by asking if he is injured or can do a self rescue and simply climb out

6. If not, then have #1 describe his situation and what the lip of the crevasse looks like from below
7. Tell #1 that you are rigging a Z rig raising system
8. Secure an ice axe or pack under the rope at the lip of the crevasse to stop the rope from digging in too deeply
9. Rig the 3:1 raising system using a Z rig, and add a C rig for a 6:1 if necessary
10. Check with victim before raising
11. Raise #1 in a series of small smooth lifts
12. Assess and attend to #1's injuries
13. Re-organize gear and configure the rope team for continued glacier travel

Digging a T slot in the ice to deadman a picket

If traveling as a two-man rope team with no other assistance, then arresting a fall and setting an anchor can be quite a demanding task. This requires that both climbers be well rehearsed in this procedure and carry all the necessary equipment.

At a minimum, each man should carry two suitable anchors in a location where they can be reached from the arrest position. This is usually on the harness or side of the pack.

Once an ice axe arrest has been made, it is necessary to try and transfer the weight of the rope from the arms to the legs by kicking your boots deeper into the snow. By still gripping the ice axe with one hand, it is then necessary to roll onto one side, draw out the anchors and plant then firmly in the snow. This is best done by digging out T trenches with the fluke or pickets themselves, or even your hand if necessary, burying the anchors and then transferring the load from your harness to the anchors. Once satisfied that the anchors will hold, you can then check on your partner and rig a raising system if necessary.

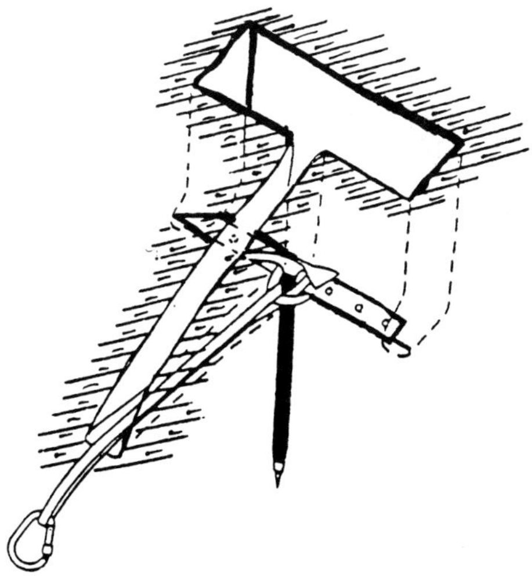

T-slot in snow using a picket and ice axe deadman

As the climber proned out on the right holds an arrest position, other members of the rope team rig an anchor to support the weight of the victim

One useful technique that can be used to minimize the fall in a two-man team is to tie a series of overhand knots in the rope. In this manner, when one of the rope team takes a fall, the knots should dig into the lip of the crevasse and jam the rope, minimizing the pull on the arresting partner.

In some instances, if the victim is badly injured, bleeding, unconscious or wedged in the bottom of a crevasse, it may be necessary to first rappel down to stabilize him or her and render first aid. This is a decision that should not be taken lightly as it now endangers another person who may or may not be able to render any useful assistance. There is also the extra time required to rappel or lower a rescuer down, versus the time it would take just to pull the victim.

If it is deemed necessary, then a completely separate rope should be used if available. This will leave the victim's rope free for the raising system and will allow the rescuer to climb up along side the victim to render further assistance, for example in maintaining the airway or prevent bleeding.

The rescuer should dress warmly since he may be in the cold crevasse for some time. He should also take prusiks or ascenders to climb back out and any tools or rigging material that may be necessary to dig the victim out and rig a more stable lifting system to cradle the victim's chest, head and neck.

Climber performing a self-rescue with Texas prusik loops

It is recommended for all glacier travelers to practice crevasse rescue techniques before moving out onto the ice (SP Parker)

Placing figure 8 loops in the rope every 2 meters will help arrest the fall as the knots jam into the lip of the crevasse

13

AVALANCHE TRAINING

Also known as "White Death," avalanches can be lethal when triggered, especially in remote locations where assistance is not readily available.

Since so much is known about avalanches—how they form, where they form and what triggers them—one would think that avalanche accidents should be a non-problem. It does not take a rocket scientist to know to avoid the steep leeward side of smooth slopes where avalanches form, but there will always be the uneducated and the foolhardy that venture into avalanche country oblivious to this danger. Or the snowmobiler, encapsulated in his helmet, engrossed in the speed and noise from a high performance machine that will not hear or see the avalanche that sweeps him away and buries him.

However, at the risk of contradicting myself, there are also situations where it is difficult to avoid traveling through avalanche prone areas to reach an objective. But in these situations, the climber or back country skier should at least be cognizant of the risks involved, take all necessary precautions, carry the right gear and proceeding quickly through the danger zone.

Most avalanches are triggered by the victim, with an average of about one in ten of those caught in an avalanche being killed. These are most often young males in their 20s and 30s—usually back country skiers, snowmobilers and climbers. Most die in 20 to 30 minutes from suffocation, or are killed by blunt trauma as they are smashed against trees or rocks. Hypothermia ranks in a distant third place as the cause of death since even if one survives the avalanche, the snow sets up hard and fast making it impossible to move and ice is quickly formed by the exhaled breath causing suffocation.

Every wilderness professional should know how to search for and rescue a victim of an avalanche, but even more importantly, be able to recognize the terrain and conditions that could produce one.

Education, understanding and a healthy awareness are everything—but avalanche prediction is still not an exact science. However, with training, hands-on experience and careful evaluation, the backcountry traveler can develop a fairly accurate picture of the local snow conditions and related potential for avalanches.

AVALANCHE STATISTICS

- About 120 are killed each year in the Alps and over 20 in the U.S. (26 for 97/98)
- 95% of the people who get caught in an avalanche triggered the avalanche
- Half the people caught in avalanches will be buried
- Asphyxiation is the primary cause of death, usually within 30 minutes
- After 3 hours there is only a 10% chance of survival

Atlas photo ©1999 John Claussen

- Colorado leads the US figures for fatalities, followed by Washington, Alaska, California, Utah and Wyoming
- France leads the world followed by Austria, Italy, Switzerland, the USA and Canada
- February has the most deaths followed by January and March.
- Snowmobilers are now leading the pack in fatalities, followed by climbers, back country skiers and snow boarders
- See **www.avalanche.org** or **www.csac.org** for additional stats and trends

Tracking statistics can give an accurate insight into the trends of avalanche fatalities. Snowmobiler deaths have increased as machines become more powerful and riders access more remote areas. In the 98/99 season, 8 of the 13 avalanche deaths in Alaska were snowmobilers, and 6 of those in a single avalanche (see report). One reason for this is a game called "high marking" where snowmobilers compete to see who can get highest on a snow slope. Unfortunately, this game is very conducive to triggering avalanches. Climbing steep slopes, the noise and vibration from the machines, and then turning and cutting across the slope at the top, all contribute to the inevitable avalanche if snow conditions are appropriate.

AVALANCHE FORMATION

Avalanches form on the leeward sides of mountains where the slope is shallow enough to hold snow but steep enough for it to break loose when conditions are right—usually about 30-45 degrees.

Avalanches are gravity powered but it is the wind that moves snow from the windward side of a mountain to the leeward side where it builds up or forms heavy overhanging cornices. However, avalanches are also dependent on weathcr, temperature, snowfall, wind speed, solar energy and snow conditions.

At the risk of over simplifying the subject, under the right conditions, snow stratifies into weaker and stronger snow layers or shear planes, indicated by the constantly changing crystalline formations deep inside the snow pack. These unstable shear planes allow huge avalanches to be initiated by a small disturbance when the crystals of the weaker snow layer below can no longer hold the heavy, stronger layer above. They then detach and the top layer slides down hill taking with it tons of snow and debris. In addition, the vibrations from the first slide can trigger other sympathetic slides in the immediate area where conditions are similar.

A dry avalanche can have a slab weight of 1500 tons and move from 50 to 150 mph. A wet avalanche may only move at 2-3 mph but is heavy, unstoppable and extremely destructive. Hard slab avalanches are made up of churning blocks as big as cars but even soft snow can bury a person in snow that compacts and becomes so hard that they cannot move or dig themselves out.

The factors contributing to avalanches are:
- Snow—the amount that has fallen
- Terrain—relatively steep open areas
- Weather—the cycle of warm sunny days and colder nights

Conditions suitable for avalanches:
1. Mountains that are high enough to form snow
2. A slope angle of about 30-45 degrees. If it is too steep it won't hold snow; too shallow and it won't break loose
3. Regular avalanche trails devoid of trees and marked by scattered debris

Three Types:
1. Powder
2. Wet—late in snow season
3. Slab—a real killer. Hard top layer breaks loose, turns to chunks and accelerates up to 80-100 mph

A group stops to evaluate snow stability on a slope (SP Parker)

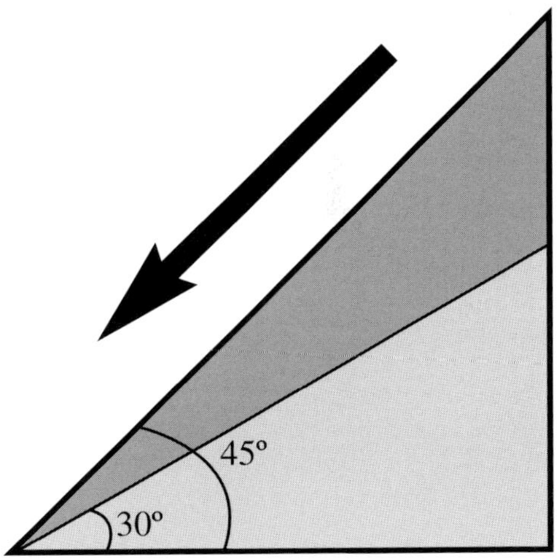

Caution: The Avalanche prone slope angle of 30°-45°

Avalanche debris is an obvious warning sign for these climbers on the Ruth Gorge crossing in Alaska (SP Parker)

An important part of avalanche evaluation is to study a cross section of the snow pack and try to identify layers that may be too weak to support the weight of the slope (SP Parker)

SLOPE & SNOW EVALUATION

There are several non-scientific methods for evaluating snow conditions in the field but if the slope and weather conditions discussed above are in evidence then the risk factor is already significant. Especially if the area has a history of avalanches or recent signs such as broken trees and debris.

The fundamental principle behind evaluating the snows propensity to break loose and avalanche is based on an examination or test of the various layers of snow in the snow pack. By digging a pit and examining the wall of exposed snow, the trained observer will be able to distinguish the various layers of snow that have been formed by the preceding weather conditions, snow fall and temperature changes.

Referring back to the information in the first part of this chapter, the evaluator is looking for a thick slab of dense snow resting on a weaker layer below. This could vary from a hard slippery icy layer to a powdery loose layer.

With training and a magnifying glass, the various crystalline structures can be studied and interpreted, but a cruder and more field expedient method is to physically test a block of snow and see how easily it breaks loose and slides down hill. One commonly taught method of doing this is the Rutschblock test (German for "glide block or shear block test").

The Rutschblock test consists of loading a large block of snow in a potential avalanche area with the weight of a skier in specific increasing increments until the block breaks loose. To do this, one first selects a suitable slope of at least 30 degrees and then digs a trench on three sides of a block of snow measuring a ski length (2 meters) across the slope by a ski pole length (1.5 meters) down the slope. The trench walls must be even and plumb (vertical) and dug to a depth passed where weaker layers are suspected (usually 1-1.5 meters). The fourth or uphill side of the block is then carefully cut with the tail of a ski, a snow saw or cord to the same depth.

The test is to then see how much force it takes to

Skier jumps on the cutout for the Rutschblock test (SP Parker)

dislodge the now freestanding block of snow. The increments are as follows:

1. Failure under the blocks own weight
2. One skier with skis steps carefully onto the block from above
3. Skier weights the skis by making rapid knee bends
4. The skier jumps once
5. The skier jumps a second time
6. A person jumps onto the block without skis
7. No failure observed in the block

The reaction of the Rutschblock test can be interprcted as follows:

- Failure with load levels 1, 2, or 3 = Extremely unstable
- Failure with load level 4 = Unstable
- Failure with load level 5 = Potentially unstable
- Failure with load level 6 = Relatively stable
- Level 7 = Stable or Very stable snow conditions

Although this test can be performed in about 15 to 20 minutes, the difficulty is in finding a slope steep enough and safe that is representative of the start zone of the avalanche area. Obviously this could be dangerous if performed in the middle of a significant avalanche zone.

A similar test can be done on a much smaller scale with a shovel or tail of a ski on an 11″ (28cm) wide block where the ski is inserted in the cut behind the block and increased pressure is applied until the block releases. But then an assumption of conditions for the entire slope must be made on a very small sampling.

Another test is the ski test where a watcher is assigned to observe as one of the team skis high and fast directly across the top of the suspect slope. This test should be approached with caution and limited to short slopes and areas where no serious consequences could result.

After the skier has skied across, he or she can look back to see how much movement is visible on the surface of the slope. This technique is also used in conjunction with ski stabilization of ski cutting work to release small avalanches that will break up the weaker layers on a slope.

Ruth Gorge below Moose's Tooth (SP Parker)

MOVING IN AVALANCHE COUNTRY

The following are a few recommendations for traveling in avalanche prone areas:

1. Do not travel alone
2. Carry avalanche beacons and test them before setting out.
3. Discuss the possibility and have a group plan in case of an avalanche.
4. Stay to the sides of avalanche prone slopes, gullies and couloirs.
5. Always have an escape route planned.
6. If it is necessary to traverse a suspect slope, cross as high as possible, removing any ski pole straps and leashes so that these items can be discarded in the event of an avalanche.
7. Cross one at a time so that no more than one person is exposed to the risk.
8. If caught in an avalanche fight for your life and yell out.

9. Try to stay on your feet and move towards the sides of the avalanche.
10. Look for trees or rocks to grab onto.
11. Try to avoid being driven into trees and rocks by the avalanche.
12. Discard any loose equipment that may act as an anchor or drag you down.
13. Fight to "swim" to the surface and stay high in the flow.
14. As the avalanche slows, bring your hands to your face and try to make a breathing space.
15. If you feel you are near the surface try to get a hand or leg above the snow.
16. Work to enlarge the breathing space before the snow sets up or forms ice.
17. Never give up.

PATROL & RESCUE

In resort ski areas, ski patrollers perform a valuable service by monitoring local conditions and triggering any potential avalanches with explosives before they develop into hazards for the recreational skiers. For the ski patrol, the art of predicting avalanches is more intuition and experience than science.

However, there will always be extreme skiers and snowboarders who insist on skiing out of bounds and in areas that have not been cleared of avalanche hazards. The same is true for the ignorance and arrogance shown by many snowmobilers who play hill climb games at the base of avalanche prone slopes.

In the event of an avalanche and a victim being buried, the other members of the group must immediately try to locate and rescue the victim. Since the victim may have less than 30 minutes before suffocating to death, there is precious little time to go for help.

For a professional rescue team responding to an avalanche situation, the first requirement will be to evaluate and stabilize the slopes above the accident rescue site so that another avalanche does not endanger the rescuers out on the slope. This may require blasting or bombing from a helicopter.

After interviewing witnesses, evaluating the debris field and looking for clues such as skis, gloves and hats, the search party will begin a slow and methodical line search with probes—but this may not help the trapped survivor for whom the first 30 minutes to 3 hours is critical.

If the victim is wearing an avalanche beacon, this will greatly help the searchers, but it will not guarantee that they can mobilize and get to the accident site in time to save a life. Search dogs are also excellent at finding victims but like their masters, will seldom be on site fast enough to make a rescue.

This leaves the only realistic expectation for the rescue of victims before they suffocate to their own party, if they can immediately locate and dig them out. This can only be done if the right equipment is on hand—beacons, probes and shovels—and the group has had Rapid Rescue training. That first 30 minutes is critical so the #1 priority is to search, not leave for help.

ESSENTIAL EQUIPMENT

AVALANCHE BEACONS - Avalanche beacons are rugged, light-weight transceivers that emit a radial homing signal that can be picked up by other units set on receive mode. Modern avalanche transceivers transmit and receive on 457kHz, which is now the accepted frequency for the United States, Europe and Canada. The older units that operate solely on the less powerful frequency of 2.275kHz should be retired since this signal cannot receive or be received by newer 457kHz units.

Avalanche transceivers are carried turned on and in the Transmit mode when moving in avalanche country and then turned to Receive mode by searchers if someone is buried. The group must then move quickly to locate and rescue their teammate before he or she suffocates.

But just wearing an avalanche beacon is not enough—the user must also be trained and practiced in its use, especially the search procedure for a buried victim. It is also important to check everyone's avalanche beacon at the trailhead before setting out to ensure all of the units work in both transmit and receive mode.

There are several excellent beacons on the mar-

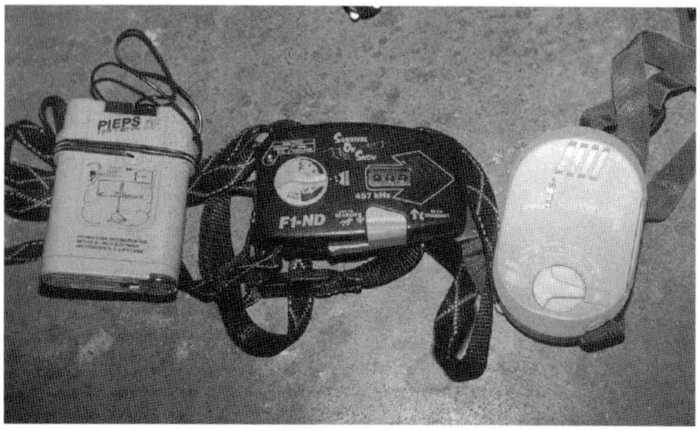

Assorted beacons used at MWTC—Pieps, SOS F1-ND and Ortovox F1

ket but the units that we use at STTU and have found significant acceptance with both the military and search & rescue teams are the SOS F1-ND, the Ortovox F1 and now the newer M1.

The Ortovox F1 operates solely on 457kHz (the F2 operates on both frequencies) and has an operational range of up to 70 meters. The F1 operates for approximately 250 hours with two AA batteries, weighs 9.8 ozs, has a rugged case and harness system, a distinct loudspeaker and three-light visual indicator.

Survival on Snow's SOS F1-ND avalanche beacon also operates on 457kHz, has a range of 90 meters with a 7 distance scale selection range and an extremely directionalñprecise locating system. Other features are recessed buttons, auto forced on switch (no forgetting to turn on), a 2 point body harness with separate chest and neck straps, and a new tapered body contoured design to make it easier to wear and store.

The F1-ND operates on two AA batteries for a minimum 300hr transmit time, has a simple battery check, built in safety checks in receive and transmit modes, audible signal in receive mode, earphone jack, light graphics bar for more accurate directional pinpointing and a weight of 230g (8 oz).

The Ortovox M1 Transceiver—The simpler and more foolproof an avalanche transceiver, the better the odds a victim will survive being buried in an avalanche. So ease of use was the priority of the people at Ortovox when they designed the M1 avalanche beacon. The clear graphics lead the rescuer through the search process, complete with green and red warning lights to indicate "proceed" or "stop and check the screen for visual clues." One very useful feature of the display is that it also indicates the distance to the victim.

Thc 475kHz M1 weighs in at 8 ounces in a tough water resistant case with a well designed harness system. It also takes two AA batteries, has a visual display for battery capacity, a transmit time of 250 - 300 hours depending on temperature, and a range of 70 meters. Other features include a piercing loud-

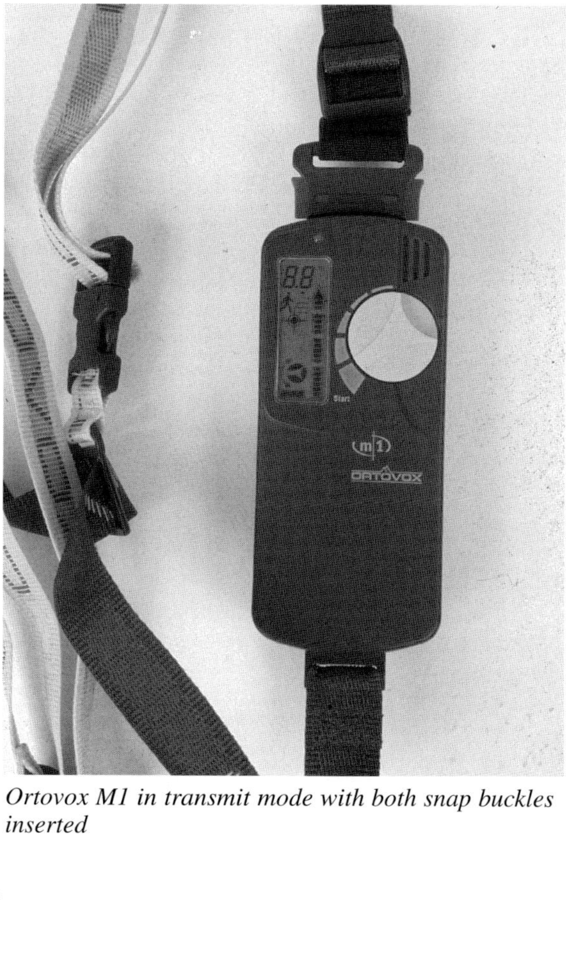

Ortovox M1 in transmit mode with both snap buckles inserted

speaker, a volume control, a jack for an earplug, digital display for a menu-controlled search, and a search arrow with analog signal.

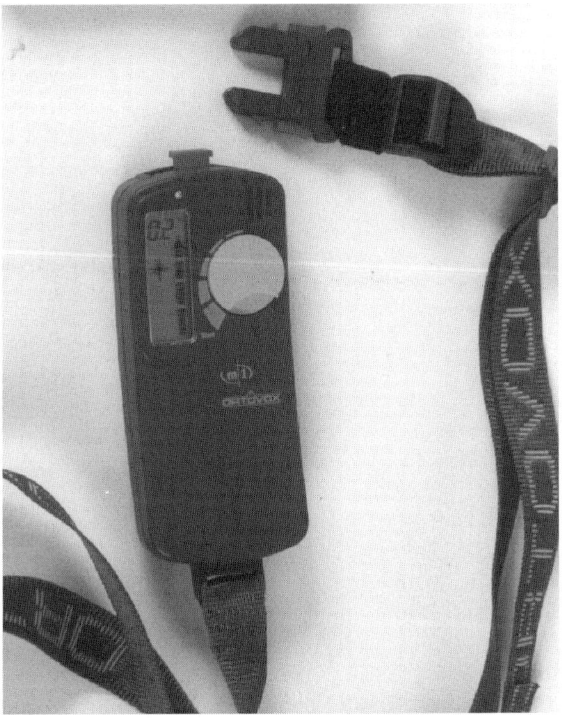

The Ortovox M1 automatically goes to search mode when the large snap buckle is removed

AVALANCHE PROBES—The purpose and use of the probe is quite obvious. Searchers are able to line-up across the avalanche debris field where the victim is suspected to be buried and then run their probes down into the snow at predetermined intervals to try and feel for a hit.

An avalanche probe is a collapsible pole with a total length of at least six to nine feet (2-3 meters) that must be both lightweight and convenient to carry.

A good example is the SOS probe, which is 2.8 meters long and made up of seven 400mm sections, joined with a new quick-release ratchet locking mechanism. It is constructed from 12mm aluminum tubing with an oversized top for better extraction from snow and comes complete with nylon bag for a total weight of 0.8 pounds (370g).

A field expedient alternative to an avalanche probe is the Leki or Atlas ski poles that come with a special joint section that enables the two ski poles to be joined at the tops to make one long probe.

SOS probe, shovel and F1-ND beacon

EMERGENCY SHOVEL—A snow shovel is not only a very useful tool for alpine operations and camping above the snowline, it is essential for both snow profiling and avalanche rescue. Without shovels, the probability of rescuing a buried victim in less than 30 minutes, under several feet of snow, is greatly reduced.

Since there is no predicting which unfortunate soul will trigger the avalanche, every member of the party should be equipped with a probe, beacon and shovel. One shovel for a group may work when camping in the snow but it is not acceptable when traveling in avalanche country.

Most snow shovels suited to avalanche rescue work are lightweight with a wide mouth to move a lot of snow efficiently and a folding handle for travel.

SOS out of Alberta, Canada has three models available: Basic with a detachable handle; Regular with a telescopic handle; and Deluxe which has a telescopic handle with wood or snow-saw in the handle. All their shovels are constructed of high quality aluminum with reinforced blades that are approximately 24mm x 28mm. Extended they are 72cm long and collapsed 48cm with a weight of 870 grams (30 oz).

The Ascension shovel comes with a particularly comfortable D shaped handle and the BD Shovel is also an excellent tool that is popular with professionals for its compact size and rugged construction.

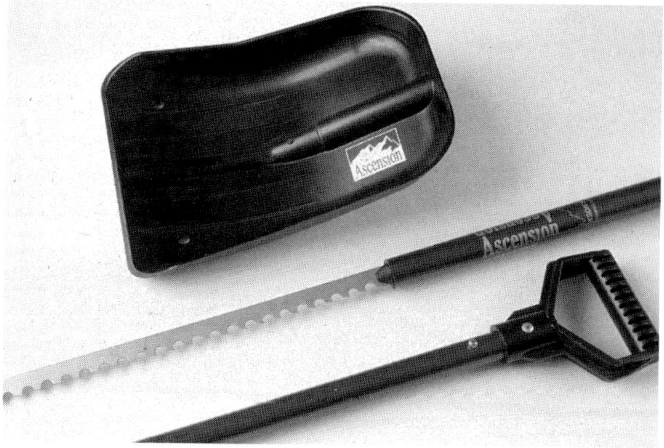

Ascension snow shovel and saw combination

Leki ski poles that convert to an avalanche probe

SNOW STUDY TOOLS—There are several tools that are useful for examining and evaluating the snow pack for its propensity to break loose and develop into an avalanche.

The Life-Link Snow Study Kit (Professional) is recommended for snow controllers, backcountry skiers, snowmobilers, mountaineers and avalanche forecasters throughout the world. It includes two Celsius Dial Stem Thermometers, Inclinometer (Slopemeter), Snow Crystal Card, Snow Pit Card, 20x Magnifier and a Snow Study Organizer Pouch.

The Economy Snow Study Kit designed for a quick evaluation of the snow pack. includes two F Dial Stem Thermometers, a Slopemeter, Snow Crystal Card, Snow Pit Card, 5x Magnifier and a pouch.

Life-Link snow evaluation kit

A team can only respond well to an avalanche accident if each member is properly equipped with a beacon, probe and shovel

CONCLUSIONS

Rapid Rescue training, probes, beacons and shovels are essential but avoidance is still the wisest solution to what should be a non-existent problem.
Recommendations:

- Know the planned route, terrain and weather
- Travel early in the morning
- Avoid open slide areas
- Avoid moving under or on cornices

WEB RESEARCH SITES

The following websites offer a wealth of useful information about avalanches, research, accident statistics, training, schools, programs, professional links and current avalanche conditions: www.avalanche.org & www.CSAC.org

Note: The following report has been included because of the magnitude of the incident and the value of the detailed information compiled from the investigation. It is only through studying accident reports and post-operational debriefs that a rescuer will be able to share in and benefit from the experience of others.

TURNAGAIN PASS, ALASKA AVALANCHE ACCIDENT

Prepared by Doug Fesler and Jill Fredston
Alaska Mountain Safety Center, Inc.
May 4, 1999

Synopsis: On Sunday, March 21, 1999, a slab avalanche ½ mile wide and up to 7½ feet deep was triggered by snowmachiners in Turnagain Pass, in the Kenai Mountains of Chugach National Forest. In all, 6 snowmobilers were killed, 3 injured, 1 was partly buried but uninjured, at least 5 others were dusted as they tried to escape, and 9 snowmachines were destroyed or damaged. An unknown number of other snowmachiners (6-12, estimated) narrowly escaped. Immediately following the avalanche, witnesses in the area responded with shovels and probes to search for survivors. Organized rescue arrived within an hour but the scene was chaotic, with a large search area and an unknown number of victims. Two victims were dug out Sunday night while three others were found as part of a large search effort that took place March 22-24 (Monday-Wednesday). One victim has not been located to date. Special Note: Because of the size and complexity of this accident, and the difficulty in obtaining consistent information from a variety of sources, certain facts may be in error or missing completely.
Report: Sunday, March 21, 1999 was a clear, warm sunny day following a succession of storms involving heavy snowfall and a winter of cooler-than-usual temperatures. By noon, the snowmobile parking area at Turnagain Pass was jammed to capacity (100-150 snowmachines) with overflow spilling onto the sides of the Seward Highway. Turnagain Pass is a mile wide valley divided by the Seward

Highway which runs in a north/south direction. Snowmachiners are restricted to the west side, while the east is set aside for backcountry skiers and snowboarders. Midway on the western side of the valley is a natural ramp, called the Knob, that snowmobilers use to access the western ridge overlooking Turnagain Pass and Seattle Creek, a parallel valley to the west.

Around 2:30 (times vary slightly) P.M., a large avalanche, 4-6 feet deep by ½ mile wide released naturally to the north of the Knob, in an area out of view to most snowmobilers, but witnessed by several skiers (Bocarde, Havens, Wunnicke, et al) on the opposite side of the valley. Two snowmobilers, Brent Snow and his wife were at the base of the path when it released. Unable to get his machine started, Snow jumped on his wife's machine. As they sped out of the runout zone, they were dusted by powderblast (a fast-moving, billowing wave of displaced air and snow particles preceding the bulk of the descending snow) but escaped unharmed. When they returned for Snow's machine, they found it partly buried by debris at the toe of the slide, as were all of the tracks from previous snowmobilers that day. Another couple snowmachining nearby were not caught. Even earlier, another slide (the east face of Pyramid Pk) had released naturally 2 miles to the north on the same aspect. Unfortunately, many of the other snowmachiners in the area either never heard about these two slides or failed to recognize the importance of the message.

At approximately 3:45 to 3:50 P.M., another release occurred ½ mile to the south of the Knob. Only minutes earlier several snowmachiners had been in the area hill-climbing, but none were caught. It is not known whether the slide was natural or human-triggered. Moments later, those gathered near the base of the slide talked about the hazard and decided they better "pack it in" but first they wanted to make one more climb. Earlier in the day, Aaron Arthur and friends had been involved in making an extreme snowmachining film.

At around 4 P.M., a fourth slide released, most likely triggered by highmarkers trying to reach the top of the mountain. This fracture, interconnecting several avalanche paths, extended roughly ½ mile south from the southern side of the Knob, almost linking with the avalanche that had run 15 or 20 minutes earlier. Breaking at approximately 2800 feet elevation, the fracture depth measured 6-7½ feet thick for most of the distance across (see fracture profile information). The debris and powderblast reached the maximum normal run-out of the paths, stopping at approximately 1,000 feet in the northern portion of the path, near the powerline alignment. Horizontal distance from the crown face to the toe of debris was .9 miles and 1,750 vertical feet. When the slide occurred, numerous people where spread out on, above, and below the slope. Four men (Ken Seagle, Dan Demers, Aaron Arthur, and an unknown man riding a "Power Special") were in the process of hill climbing.

Ken Seagle was just within the southern flank of the fourth slide in the process of hill-climbing (about ½ way up), when he looked over his shoulder and saw the whole mountain to the north avalanching. He turned his machine and rocketed toward the debris of the earlier slide in a successful attempt to escape. He said he was traveling at speeds of approximately 50 mph over debris blocks 4-5 feet in size, causing him to become airborne approximately 75% of the time.

To his right (north), approximately ½ way up the mountain, was Dan Demers, who had gotten his RXL short track stuck while hill-climbing and was digging it out when the avalanche ripped out above him. Demers was unable to free his machine and never had a chance to escape. He and his machine were completely overcome, tumbled, and carried approximately ⅓ of a mile (horizontal distance) and 500 vertical feet down-slope. His machine, though totally destroyed, was partly visible on the snow surface (facing down-slope, cocked to the right) and found within minutes by Seagle and others. The first person to arrive with a probe found Demers after 1-2 minutes of probing. Demers was upslope from the machine 5½ feet and 2 feet to the left (south). His feet were under 2-3 feet of snow and his head was approximately 5½ feet under. His body was horizontal to the slope. According to Seagle, Demers was found at 4:46 in the afternoon. CPR was performed by a nurse without result.

To the right (north) of Demers was an unknown individual riding a Power Special who was last seen out-running the avalanche. It is believed he escaped, since no one was reported missing, but to have escaped from this location required exceptional luck.

Aaron Arthur was in the process of making a long high traverse from south to north and was approximately ¾ of the way up the northern portion of the path when it released. According to witnesses, Arthur passed several bald areas (devoid of snow) at mid-elevation until he reached a point just below a steep slope (measured at 48°-50°) near a patch of willows. When he was at his highest mark on the traverse, the avalanche released, causing Arthur to be swept off his fast-moving machine. Carried down-slope an undetermined distance, Arthur remains buried as of the time of this report. The fracture line was estimated to be 300-400 vertical feet above him (or 500-600 linear feet).

His machine, a borrowed 700 RMK (Rocky Mountain King), was found approximately 600 vertical feet below his last seen location (approximately ±1600 linear feet down-slope of the top of the Knob). Arthur most likely took a parallel trajectory slightly to the south of his machine's descent. Of the four men watching Arthur from their vantage point at the top of the Knob, three managed to get their machines started and speed off as the slide headed their way. Estes, unable to start his machine, opted instead to run north toward some nearby mountain hemlocks. Almost immediately he was hit by the powder-blast, but managed to hold onto a tree as the turbulent powder-blast hit him and debris passed 20 feet away. His machine was only dusted. The other three, blitzed over the southern edge of the Knob and were almost immediately overcome by the slide which, by then, was traveling an estimated 80-90 mph. They were caught, violently tumbled, and carried toward the lateral edge of the debris approximately 600 vertical feet and 1600-1700 lateral feet below the top of the Knob. When the snow came to rest, Ray lay dazed with a possible concussion and a punctured lower lip, 4 feet down-slope from his broken machine (broken headlight and cowling) which was lying on its side, skis pointing uphill.

Ray remembers seeing a snow machine hurtling through the air just before being hit by the slide. He is unsure of whose machine it was. Fred, approximately 30-40 feet above Ray, came to rest buried beneath his overturned machine, which lay on the surface. Despite a badly bruised leg, he managed to crawl from his location and establish contact with the others. Both Rex and his machine ended up about 50-60 below Ray, out of view over a steeper roll-over of debris. Rex, stunned by a broken nose and lacerated face, started running toward the edge of the slide, 20-30 feet away. His machine landed upside down with only its ski tips showing. Broken glass, anti-freeze and blood could be seen at the site the next day. These men were extremely lucky not to have been killed.

The avalanche continued down-slope at 80-100 mph, hitting several groups of people in the process. Victor Jones, who had been sitting on a bench below the Knob, north of the center of the path, was just starting downhill when he was hit by the slide. The force carried him and his machine 1,700 to 1,800 linear feet (and ±100 vertical feet down-slope), burying both. Jones, who was not found until Wednesday, was uncovered from 7 feet of snow 111 feet diagonally upslope from his Polaris 440 machine.

Jones' partner, Ray Debor, was approximately 150 feet down-slope from Jones, heading downhill when he was hit from behind. Debor never knew the slide was coming. He and his machine were carried to within 70 feet of the toe of the slide. His machine was buried to the windshield while he was still sitting on the seat, buried to his knees but uninjured. The other two members of this group, Shane Brown and Brian Kirk were at the bottom of the slope, roughly 150 feet below Debor, waving Jones down. When they saw the slide coming, Kirk drove downhill towards the power-line and Brown drove south, parallel to the alignment. Both were dusted by the powder-blast but uninjured.

Two other victims, Chris Scott and Jodi Combs were riding together somewhere in the run-out of the path (location and direction of travel unknown) when they were hit, tumbled, and buried not far from Jones resting position. Scott was found dead

Sunday evening buried under 7 feet of snow, approximately 18 feet diagonally upslope from his machine (Summit) which had its seat and handlebars above the snow. Combs' machine (600 Rocky Mt. King) was found Sunday night with its handlebars sticking out of the snow. Combs was found Monday afternoon, approximately 132 feet diagonally upslope from his machine. He was buried roughly 7 feet deep, with his head uphill, supine on his back, with his feet higher than his head. Scott's machine was located almost directly uphill of Combs' machine.

The snow machine (Arctic Cat) of the sixth victim, Jeff Saunders, was found Sunday night crossways underneath Scott's yellow Summit. Saunders' position at the time of the slide is unknown. He was found Monday afternoon, approximately 50 feet upslope diagonally from his machine and buried roughly 7 feet deep.

Along with the three machines (belonging to Scott, Combs, and Saunders) found Sunday night in the lower search area, two helmets were also located nearby, on the surface, roughly 6-8 feet apart. One was positively identified as Victor Jones'. The other (black, older model, gold lettering) must have belonged to either Chris Scott or Jodi Combs. Jeff Saunders had his helmet on when found.

Evidence indicates that the avalanche was quite turbulent right to the end of the run-out zone. A video of the avalanche, shot from Tincan Ridge on the other side of the highway shows 6 to 8 other snowmobilers fleeing the slide. The total number of victims caught and close escapes may never be known. Immediately after the dust settled, witnesses responded from all directions to several areas of the avalanche and 911 calls were made to alert authorities. Debris covered an area greater than ½ mile by ½ mile, typically deeper than 10 feet. For the first couple of hours searchers worked independently in three search areas—"South Bench" where Dan Demers was buried, "Lower North" where Scott, Combs, Saunders, and Jones were buried, and "Upper North" where Aaron Arthur is likely to be buried.

The first organized rescuers to arrive on the scene were the Alaska State Troopers and the Girdwood Fire Department, followed shortly afterward by Chugach Powder Guides, the Alaska Mountain Rescue Group, the Alaska Search and Rescue Dogs, the National Ski Patrol, Mt. Alyeska Pro-Patrol, the U.S. Forest Service, the Alaska Mountain Safety Center, and others. Several helicopters were on-site and used to transport rescuers to the site. An aerial, scene safety assessment was done by the Dept. of Transportation and Public Facilities avalanche forecaster in AST Helo 1. By nightfall, roughly 7 P.M., the decision was made by on-scene Troopers to clear the area and call the search off until daybreak. They felt this was the only way they could gain control of the scene and determine how many people were unaccounted for. Mission organization took place Sunday night at the Girdwood Fire Station.

On Monday, roughly ±200 people were involved in the search effort for more than 12 hours. On Tuesday, approximately ±350 volunteers searched another 12 hours, including 98 snowmobilers to transport food, equipment, and searchers to the site. Roughly 450 people responded Wednesday, including National Guard personnel. The weather during the operation was generally poor to bad. On Wednesday, there were 30-40 mph winds, heavy wet snow, and poor visibility. Due to storm conditions and increasing hazard, the official search for the last person was terminated early Thursday morning, March 25. On Friday and Saturday, approximately 40-50 friends and relatives of the remaining victim (Aaron Arthur) probed "unofficially". Saturday night, the family of Aaron Arthur asked all searchers to stop searching because they did not want searchers placed at risk. That afternoon, another snowmobiler was killed by an avalanche while playing highmark near Eureka, in the Chugach Mts. 100 miles to the northwest of Turnagain Pass. In all, 12 avalanche fatalities occurred in Alaska from March 21 to April 29, 1999.

Note: The Turnagain Pass avalanche accident report of March 21 contains some errors. The actual number of fatalities is 6, plus 3 additional injured and 1 partly buried with 2 others dusted and numerous (about a dozen) very close calls. In all, 9 snowmachines were damaged or destroyed. The body of the sixth victim was recovered on May 15, 1999, located 553 ft downslope of the machine.

Probe line (Charlie Shimanski)

The first 30 minutes is critical in any avalanche search (SP Parker)

Atlas Photo: ©1999 John Claussen

14

SKI TRAVEL AND SKI MOUNTAINEERING

There is no question that a team can cover more terrain with less effort on skis than on snow shows or by breaking trail on foot. However, skiing is a specialized skill requiring significant initial training time and the opportunity to maintain those skills with regular practice.

Unlike snow shoeing that most can pick-up the first time out, grasping even the basics of backcountry skiing can take several training sessions just to gain minimal competency and tone-up the necessary muscle groups. This may be less of a problem for individuals who have a significant amount of alpine (down hill) ski experience, however alpine ski gear is not suited to long distance backcountry travel.

Conventional alpine ski gear used for downhill skiing is not designed for extensive travel over varying terrain since the boots are locked rigidly into the skis not allowing for the more natural ski-glide/walking motion that is necessary. The boots are also designed to be very high and very stiff to give more control over the skis at speed and the extra weight of these skis and bindings is not an issue since they are never carried any distance.

Ski Patollers at ski resorts are able to make good use of conventional alpine gear for rescue work since they will take a gondola or chairlift to the top of the mountain and then work their way down to the victim from the top, usually over or near groomed slopes. They will then package the victim in a specially designed stretcher and ski them down to the bottom of the slope where an ambulance will be waiting.

For the mountain rescue team working out-of-bounds or the military unit operating in remote regions, there is ski gear that is lighter and better suited to backcountry travel—Nordic/telemark cross-country equipment or Randonnee/Alpine Touring (AT) ski mountaineering gear.

Free-heeled telemark gear (tele gear) is lighter but requires more training to become proficient on the steep downhill grades. AT gear is more like conventional alpine ski gear except that the heel can be released to allow for free-heeled movement over flat terrain or climbing hills.

There is also a significant cost difference in ski equipment versus a pair of snowshoes. Where a good pair of snowshoes would cost a couple of hundred dollars, a ski package including skis, bindings, boots, poles and skins could easily run twelve to fifteen hundred dollars. This additional equipment, apart from being expensive, can also be quite burdensome in terrain where skiing is not practical and everything must be carried.

However, the cost and weight of ski equipment are offset by the distances that can be covered and the ease with which the skier can move over snow covered terrain. Like snowshoes, skis spread the weight of the skier over a greater surface area allowing them to "float" on the snow. This is an advantage on glaciers since the skier is gliding across

Military skiers on the move

snow-covered crevasses as opposed to exerting direct downward force by walking or standing on them. There is also less likelihood of punching through a snow bridge because of the broader weight distribution.

Any time the snow is deeper than about 20″, snowshoes or skis will make for easier movement and eliminate the tiring process of wading through deep snow, posting and climbing out. However, if planning for a group where some are proficient on skis and others are not, then snowshoes may be the most prudent choice. To put individuals on skis who are not comfortable or proficient will only stress them and frustrate those who can ski.

MOVEMENT ON SKIS

There are two basic approaches to training novice skiers for movement on skis based on the expected level of proficiency at the end of training. One is the Marine Corps approach of "stand up and shuffle forward," designed for training a large number of Marines in a short period of time to be basically ski mobile over relatively easy terrain.

The other is the military special operations approach where operators are expected to operate for extended periods of time behind enemy lines in varied terrain and possess a much higher level of ski proficiency.

The Marines are able to achieve their goal with a battalion of young Marines in a few weeks during the MWTC winter warfare package. For Special Forces, the training is longer and more intense but they have the bigger budgets and fewer operators to train. They are also able to go outside of the military system to seek out the best professional civilian training they can to meet their mission goals.

The Marine Corps winter warfare instructors teach what is known as "Fundamental Skiing". This meets their mission requirements since they are not looking for speed, performance or grace, just a more efficient forward movement than would be expected on foot. They want to be able to move into an area, dominate ground and prepare to engage the enemy with as little energy expenditure as possible. This also allows them to train and travel with their NATO counterparts when participating in cold weather exercises in Northern Norway.

MWTC ski instructors work on the principle that once a student has developed some feel for the balance required just to stand-up and move forward on skis, it is not difficult to teach them to side-step up a hill and glide or side-slip down a hill. If it is a gentle rolling terrain, they will not have much problem just pushing on their poles for forward movement or gliding down the slopes provided there is an open run-out area at the bottom.

Where things become more challenging for the student, and entertaining for the instructor, is in steeper, tighter terrain with less room to maneuver such as going through trees and wooded areas. The NATO type tele skis and rubber VB boots used by trainees simply do not give sufficient control over the skis for aggressive turning, so at these times the Marines will simply trade their skis for snowshoes and walk it.

The difference in philosophy between conventional warfare and special operations is also reflected in the equipment. Where a Marine may be equipped with a set of old tele skis (not so affectionately known as "white rockets"), inexpensive tele bindings and rubber VB boots; the special forces operator will have a high-dollar pair of skis built to their specifications, with Silvretta 400 or 404 bindings and double plastic mountaineering or ski boots.

There is also a significant budget consideration in that the Army trains smaller numbers of soldiers at a time (20-30) while the Marines train at a battalion level. This means that in addition to all the ski gear at MWTC required for training, the Marines must also have sufficient equipment for war stocks and to go with a MEU for overseas deployment. However, it does not mean that one group is any better than the other in achieving their objectives, just that they have different mission-driven equipment needs.

SKI EQUIPMENT

Once the decision is made, based on terrain, mission and suitable personnel that a unit is to have a ski capability there are several factors apart from just ski equipment that need to be considered. The first of these is based on the planned area of operation.

If a team is to be active in the backcountry then there is a high probability that they are going to be skiing in either avalanche prone areas or on glaciers or both. So the first steps in preparation will be avalanche training and crevasse rescue (See Chapters 12 & 13), and in addition to essential ski gear, avalanche beacons (457 kHz), snow shovels and probes will become required pieces of equipment for movement in avalanche country.

Some consideration should also be given to clothing systems, packs and climbing or rescue gear. Skiing is an aerobic activity that generates considerable body heat and requires freedom of movement. As such, clothing systems need to both breathe well and be of either stretch material or cut for a full range of movement. Close fitting gaiters are also an essential to keep snow out of the boots and socks dry.

In selecting a pack, consideration must be given to its width so that the elbows can move freely when poling. It should also have a secure harness system so that the pack does not move around while skiing throwing the skier off balance. Heavier loads may be best pulled behind the skier in a sled.

Climbing, rescue and tactical gear should be carried inside the pack so that there is a minimum of loose items flapping around on the pack or interfering with efficient ski technique.

This brings us to the actual ski equipment. It is difficult to make specific recommendations in this area since personal preference plays such a big role and ski gear seems to be in a constant state of evolution, with each season bringing new products, designs, improvements and changes. But whichever system is most suited to the team's needs, telemark or AT, the essential pieces of equipment will be skis, bindings, poles, boots and climbing skins, but to get optimum performance and comfort, virtually every piece must be fitted to the individual skier.

Ski length is proportionate to height, weight and skill level; ski boots must be comfortable for prolonged wear and yet snug for control; ski poles must be adjusted to the height of the skier; bindings must be adjusted to fit the boots and release at the right loading; and ski skins cut and trimmed to match the profile of the skis.

SKIS

Although there are dozens of brands and models suited to backcountry skiing, there are only two basic directions to go with ski selection. The first is the traditional free-heeled telemark ski which comes in two basic forms—either the lightweight, edged, Nordic-camber design or the wider alpine-camber telemark skis.

The second type is the wider and slightly heavier alpine touring (AT) / randonnee gear more popular for ski mountaineering. However, some wide tele skis and even alpine skis can also be mounted with an AT binding.

The traditional Nordic skier, much like the long distance runner, enjoys the freedom of movement that comes with lightweight gear so telemark ski equipment tends to be lighter and designed more for sustained travel in ideal conditions. The randonnee skier is looking for more extreme conditions so AT gear is built more rugged and, with the option of locking down the heel, will work in steep terrain where alpine gear would normally be more suited. Both telemark and AT skis used for professional purposes should be relatively wide and have metal edges to better turn and hold in a variety of snow conditions including ice and crud.

One advantage that the AT set-up has over the telemark is in the boot department. AT skis with 400 or 404 bindings can be run with conventional plastic mountaineering boots where as the tele bindings only work with tele boots that have a specially constructed toe block. Even though there is a significant loss in ski performance when using plastic mountaineering boots as ski boots, if the skis are only used for an approach to an area where the bulk of the climbing will be on foot, then the trade-off is justified.

There are several companies that manufacture telemark and AT skis including TUA, Atomic, K2, Karhu and Black Diamond, all with desirable characteristics such as lightweight, strong, torsionally rigid, medium flex and sufficient width for flotation. Skis with a softer flex are favored for skiing out of bounds in deep virgin snow (off-piste), while stiffer skis are preferred for skiing on groomed slopes (on-piste), so a medium flex ski that is torsionally rigid is ideal for a wide variety of mountain and snow conditions.

Ski professional Ed Cereda mounts Pitbull 2 bindings on K2's Heli Stinx (left) and Silvretta 404s on the Tua military skis (right) for STTU evaluation

A ski's float characteristics are derived from a combination of its length and width, but longer skis are more difficult to maneuver in tight terrain so one can compensate with a fatter ski. This is also a growing trend with newer skis. Where traditionally x-country skis were long and narrow with a virtually parallel side-cut, the current trend is toward wider skis that are narrower in the middle but wider at the tips and tails. In the last few years skis have gone from a 80-60-70 profile all the way up to a 108-72-100. (These numbers are the width in millimeters of the tip, center and tail of the ski respectively). This width obviously gives a ski more float in deep snow and allows the skier to select a ski that is 10-20 cm shorter than traditional lengths.

The three skis that we have been evaluating and using at STTU are models from TUA, K2 and Atomic, all suited to backcountry operations.

The ski that TUA built for the military is a stiffer version of their Excalibur Plus supplied with a white top in 170, 180 & 190 cm lengths, with a profile of 94-70-84. This wider, stiffer ski is ideally suited to the heavy loads carried by both soldiers and Marines and weighs in at about 6.2 pounds per pair in a 180 length. The people at TUA have also shown a willingness to work with military units to customize a ski to their needs if the quantities are sufficient.

Atomic Tourcaps, TUA Excalibur Plus, the civilian version of TUA's white military ski, and K2's Heli Stinx

The K2 skis were recommended to us by an instructor at the army's mountain warfare center with the K2 Heli Stinx being the one we are currently evaluating. This ski is available in lengths from 170-195 cm, with a profile of 99-69-88 and weighing in at 6.5 pounds without bindings in a 190 length. The Heli is a lighter softer version of the Piste Stinx, well suited to powder and long slogs with good downhill performance.

Atomic had two skis that interested us, the Tourcap Light and the Nomex-cored Tourcap Guide that was used for a descent on Everest. The Tourcap Light is a light all around touring ski with laminated wood core construction and steel edges for positive hold on icy slopes. It is available in lengths from 160-190 cm, with a profile of 91-69-81 and the 180s weighing in at 6.3 pounds a pair without bindings and 5.35 pounds each with the 404 binding.

These are not the only skis out there, in fact the military also has a large inventory of Elan skis which we have not had the opportunity to evaluate along with other models from Black Diamond, Karhu and Dynafit. With this in mind, individual test and evaluation is recommended to find the ski most suited to local conditions and the required mission.

Just looking at the latest editions of Back-Country and Couloir magazines shows dozens of new models for the year 2000, however, keep in mind that "functional skiing" with strong reliable equipment is more important to the professional than speed and performance.

BINDINGS

Like skis there are several brands and models of bindings available, but rugged construction and reliability under tough conditions should be the primary criteria for selection. The binding must be strong since it is the interface between the ski and the skier so any failure at a critical time could mean disaster.

Bindings fall into three basic categories: telemark, randonnee/AT and alpine, but only the first two apply to the backcountry skier.

The BD Pitbull 2s, recommended to us by SP

BD Pitbull 2 binding with Garmont Veloce tele boot

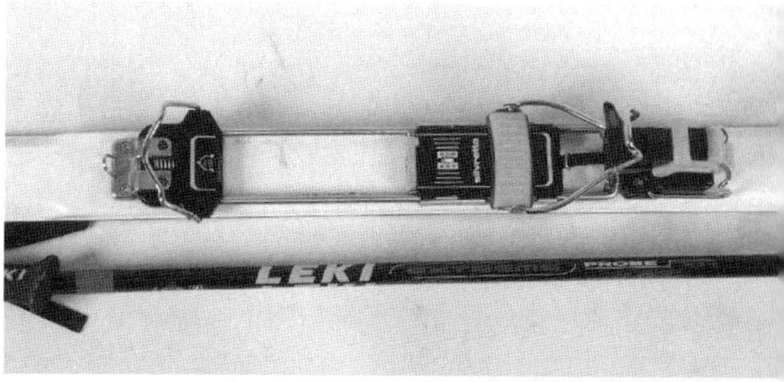

Silvretta 404 randonnee / AT binding

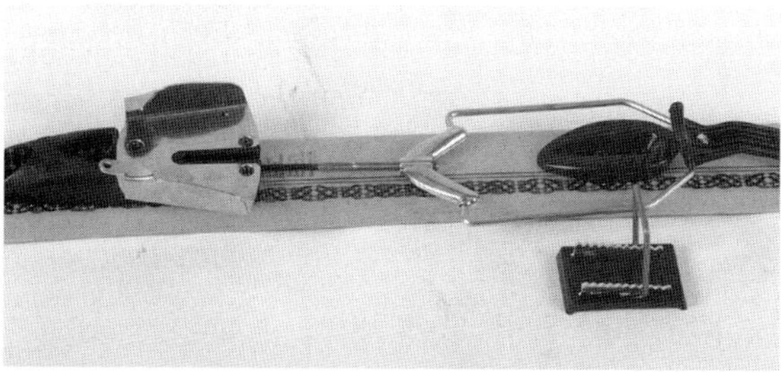

Heel lift to be installed for easier up-hill climbing

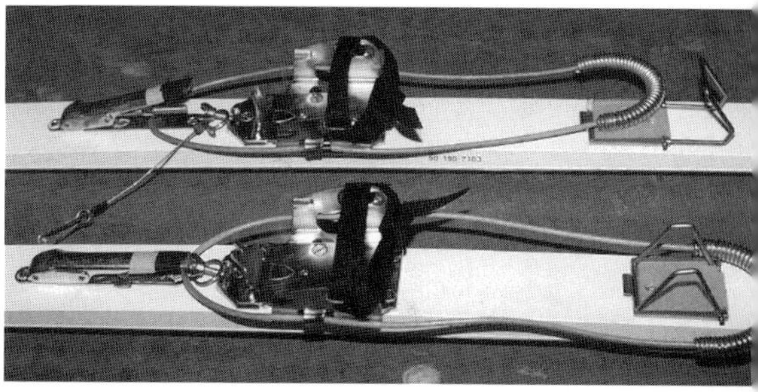

Military adjustable tele binding with heel lifts for hill climbing

Parker, a civilian mountain guide and ski instructor who has worked with the military, are about the strongest tele binding on the market and although more expensive and heavier than most, they will stand up to the beating that big men with heavy loads will give them. They weigh in at 1.35 pounds each, as opposed to the lighter Rainey at 1 pound which is also worth looking at. Many of the other tele bindings are just too light for the daily use and abuse encountered with heavy loads and rough terrain.

The US marines use a Voile binding with some success and the British Royal Marines prefer the Riva 1 for its simplicity and field maintainability. All are relatively inexpensive compared to a randonnee binding.

The US Army uses the Silvretta 400 and 404 so that they can take advantage of the added features of a randonnee binding.

The Silvretta 404 binding from Climb High is a DIN rated binding with forward and lateral release capabilities, easy adjustment for varying boot sizes and a pole activated snap down lock to switch from free-heel touring mode to downhill turning. They are heavier and more substantially built than tele bindings to handle the loading that goes with more aggressive downhill skiing. Most tele bindings are secured to the ski by only three short screws in the toe area, while the AT binding is also secured in the rear.

Other advantages of the 404 binding include their compatibility with both AT and plastic mountaineering boots, along with a ski-crampon for movement over ice traverses. The Silvretta crampon can be dropped into the binding and is locked in place by the boot (see page 178).

The newest version of the Silvretta, designed and built at the request of the German and Austrian military, is the Silvretta 500 which takes the best features of the 404 and combines them with the carbon fiber Easy-Go.

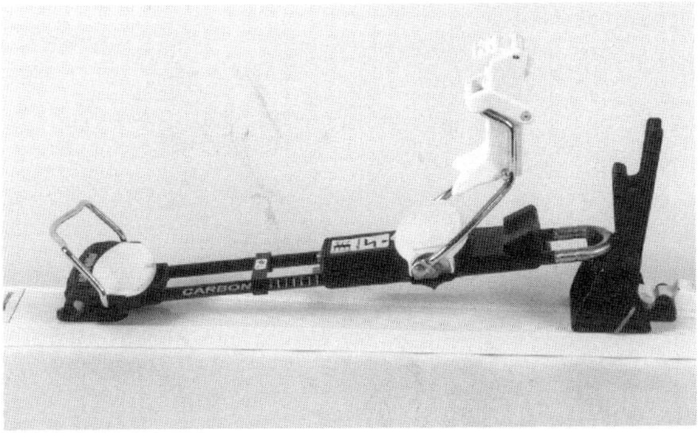

New Silvretta 500 military binding with variable heel lifts

BOOTS

Boot selection will be based on which type of skis and bindings are to be used. Telemark skiers have long favored low-top leather ski boots for their convenience and comfort, and even the military still has leather tele boots in their inventory. However, the two deficiencies of leather ski boots are that they are relatively soft so do not edge well and when wet are very difficult to dry. This may not be a problem for single-day outings on rolling terrain, but they are not suitable for prolonged multi-day operations in cold, wet snow.

Another problem that the military has with leather boots is that they mold to the skier's foot so cannot be returned to the ski locker for reissue to future trainees. There is also a health problem with re-issuing leather boots that have absorbed sweat and fungal bacteria from previous users. This has led the Marine Corps to go to Scarpa T3 plastic boots with a separate liner.

Stiffer plastic tele boots, like plastic mountaineering boots, are one hundred percent waterproof if used in conjunction with a good gaiter. They improve skiing by being more responsive and giving the skier more edge-holding capability on steep ice and crud. Having a double plastic boot also allows the military to discard the liners for health reasons and reissue the boots with a new liner.

The boot that we use at STTU is the Garmont Veloce which is similar to the T3 but with a higher top and maybe a little warmer. The Veloce is an excellent boot for skiing with heavier packs and where downhill control is a high priority, including alpine style turns. If slightly lighter weight, lower price, better marchability and telemark style turns are required in a touring boot, then the Libero is a good alternative.

In selecting a tele boot, comfort, performance and warmth are a direct product of fit. A boot needs to fit snugly for performance and even though a loose boot may be warmer, this will be transmitted to the ski in the form of little or no edging. But if a boot is too tight, the resulting circulatory compromises will negate any insulation built into the boot. So it is partially a personal issue which brand and model one selects and which will be warmer. The better a boot fits, the warmer the feet and the better the performance.

Knowing that fit and comfort is everything in selecting boots, ignore the size chart and try the boots on. Wear them around the shop for an hour to find any hot spots and have the boot doctor make adjustments where necessary. The Mondo Point size system is however useful since it corresponds with

Garmont Libro, a low top version of the Veloce

the length of the foot, i.e. a Size 30 is a foot that is 30 cm long.

Although one advantage of the 404 binding is that they can be used with mountaineering boots, even a double plastic boot is a poor substitute for a dedicated, high-topped AT ski boot like the Scarpa Denali or Tambo. Mountaineering boots are built more flexible for walking and are usually purchased a size or two larger than the user's foot so that extra sock layers can be added for warmth and comfort. Both of these features degrade the ski mountaineers control over the ski. An AT ski boot on the other hand is designed to be more rigid and fit the foot snugly for more positive control over the ski. This allows the skier to descend steeper terrain with the added security of quicker and more positive turns.

404 binding with Kolflach mountaineering boot

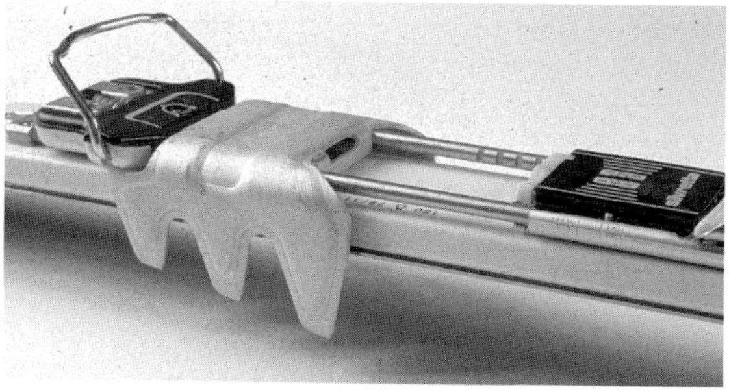

Ski crampon for the 404 binding

BD Tambo AT Boot in the 404 with binding released for free-heeled skiing

WAX & CLIMBING SKINS

The wax used on skis primarily protects the bottom surface of the ski but also affects the glide characteristics of the ski over snow. Alpine skiers will generally use one type of wax for cold days and another for warmer days, but for the avid cross country skier there are many more choices and options. Apart from numerous temperature sensitive choices, the tele skier uses two distinctly different waxes—a glide wax and a stick wax.

The glide wax is applied to the front and rear thirds of the ski to allow unweighted glide across the snow, but the concave middle of the ski under the binding has a stick wax to grip the snow and give the skier some forward drive. This wax is designed to stick when the skier weights that ski and then rise up off of the snow when weight is moved to the other ski. However, this sticky wax is not sufficient for direct movement up step hills so the skier must stop and "skin" his skis.

Climbing skins allow the backcountry skier to go up hill without having to sidestep or use an awkward herringbone technique. This is achieved by hair like bristles on the nylon or mohair skins that fold down when gliding forward and stick out and grip when pushed down and back. The army mountain warfare school has been using Pomoca Mohair skins but will be changing to Coll-tex, available through Climb High in Vermont. The USMC-MWTC uses Ascension Climbing Skins which are now available through Black Diamond.

Ascension Climbing Skins use a nylon construction and are 80mm wide. They adhere to the ski but are also held in place at the tail with a flat metal clip and at the tip with a rubber stretch/ wire loop device.

To install this type of skin, the skier starts by placing the hook on the tail of the ski, then sticking the skin down working towards the tip of the ski. At the tip, the skier pulls the stretcher up and over the tip, securing it with the wire loop.

Ascension climbing skins with tail clips and stretch-rubber wire toe attachments

Skin Care & Usage:

Skins are stored by folding the adhesive side against itself—best done by folding the ends into the middle. The skins are stored folded like this to avoid contamination of the adhesive and should not be stored stuck on skis. It is also important to ensure that the skins are dry before prolonged storage.

Although the adhesive is designed for use in temperatures as low as zero degrees Fahrenheit, it works better when warm. On very cold days, it is best to keep skins inside your coat until ready to use.

Adhesive skins are only as good as the adhesive on them, so do not hesitate to add more if they are not sticking properly. In addition, regular treatment with Ascension Glop Stopper will help prevent snow build up and improve glide.

SKI POLES

Poles for backcountry skiing are used for balance and timing on the downhill runs and for propulsion on the flat and uphill areas. To be really effective, poles should be different lengths for going uphill (longer) versus downhill (shorter). This makes adjustable tempered aluminum poles that can also be set for different lengths on long traverses, the hot set-up.

To calculate the personal length of the shorter downhill pole, flip the pole upside-down with the grip on the deck and your hand under the basket wearing ski boots. With your elbow against your side, your forearm should be parallel to the ground. At the longer end of the scale, again held upside down, the basket of the classic cross-country kick and glide pole should be at about armpit level. This is the range of adjustment needed in a backcountry ski pole.

The hand-grips need to be easy to grip and comfortable with both light and heavy gloves and should have breakaway straps to avoid shoulder injuries caused by a basket getting snagged in the trees. If the poles do not have breakaway straps, then do not put your hands in the loops when moving through trees.

Medium sized baskets are recommended, work well and have less swing-weight than the large heavy baskets that used to be popular with x-country skiers.

SKI TRAINING

The purpose of backcountry skiing, or what the military calls Nordic Expeditionary Skiing, is to get a group of operators from one point to another in an efficient and timely manner. It is also important that everyone who begins the movement completes the movement in good condition and ready to work or fight. To do this, everyone must be functionally competent on skis and in good **physical condition.** However, in addition to the strains of just skiing, changing weather and snow conditions along with having to carry heavy loads will tax the strength and stamina of even the most proficient skier. This makes physical fitness a critical factor in backcountry skiing.

Fortunately, anyone involved in alpine operations, patrol or rescue, should already possess a high degree of physical fitness and leg strength. However, skiing also requires some specificity in training since it works different muscle groups to those used for walking and climbing. Once those muscles have been conditioned by basic ski training, **endurance** must be developed so that the skier can move more efficiently while maintaining a high level of awareness and coordination over uneven terrain.

Flexibility also plays an important role in ski conditioning in that it helps prevent injuries and gives the skier a wider and more comfortable range of movement. Once trainees have warmed up, they should be run through a range of stretching exercises wearing skis to improve their flexibility and better prepare them for the ski drills.

To be able to respond smoothly and quickly to constant changes in terrain, the skier must be **relaxed and alert.** This requires the instructor to maintain a positive learning environment and always consider the comfort and safety of his students. By keeping the students active, interested and in good humor, he will also be keeping them relaxed and eager to learn. This also requires that the students come to training rested, fed, well hydrated, and at thermal equilibrium. They should have comfortable clothing suitable for ski training and warmer layers to put on when waiting their turn or riding the chairlift.

The instructor must also consider the **training environment** and if the temperature and snow conditions are conducive to the training process. Pushing beyond the limits will frustrate both the students and the instructors and can ultimately result in injuries when people become too cold or tired to care.

Balance is the foundation of all skiing and should be the primary focus of both the student and the instructor during training. Fortunately, if the student is already fit, flexible and relaxed, and the training progression is logical, good balance on skis should be a natural by-product of the program.

Skiing through the trees can be a real challenge for new skiers and pulling a sled only adds to the level of difficulty

INSTRUCTIONAL CONSIDERATIONS

The single most difficult part of ski training for the new skier is being taken onto steep, **intimidating terrain** for basic lessons. To have a successful learning experience, the instructor must select a site that is both safe and appropriate for the skill level of the students and the drills being taught.

In developing a teaching methodology, instructors are encouraged to use a process of **guided discovery** where the students are presented training exercises in a logical progression and learn primarily through doing. Skiing is a "watch me and try it" kind of activity and military and law enforcement alpine operators are the "can do" type of people. However, if they are presented with problems beyond their capability and terrain beyond their skill level, they will become tired and frustrated. As a result, the learning process will diminish quickly and may eventually cease altogether. Either that or they will aggressively bomb down the mountain with youthful exuberance trying to impress the instructors, damaging both themselves and the equipment in the process.

So it is important to first explain the drill and its finer points and answer any questions the students may have. But no more than 10% of training time should be spent on talking.

The next step is **demonstration** since this is even more important to pragmatists like lawmen and soldiers who have a "show me" attitude. This means that the instructor must possess **demonstration quality** skiing ability, and if he does not, he should have someone more proficient demonstrate. Students must see all drills demonstrated correctly the first time since the first mental image of an exercise is the most vivid and the one that will be recalled for self-visualization. But again, no more than 20% of training time should be spent on demonstrations.

Practical application should make up over 70% of the ski training process since only through actually skiing will they develop the balance and muscle memory necessary to progress to the next level.

Adequate time should be allowed for rest and constructive critiques, but it is equally important that the skier not be permitted to dwell on failure. He must be encouraged to get right back up after a fall, as many times as it takes, and continue until he gets the feel for the exercise. With success comes the confidence that is necessary to becoming a good skier and to progress to the next level.

Training progression is a building block process that makes the assumption that the trainee has built a solid foundation with previous exercises and is ready to build on that knowledge and skill level. If it is found that a trainee is not ready to progress to the next level, it is better to keep him working on more basic skills until he shows proficiency and is ready to progress at his own rate.

This is somewhat contrary to the usual military training methodology where groups are moved through the system with little consideration for the individual, but skiing is one area of training where more individualized attention is necessary and a higher instructor to student ratio is recommended. There should also be one or two instructors specifically assigned to work with the slow learners.

At the end of each training evolution or specified skill level there should be some form of formal testing to quantify the student's progress. This does three things: it gives the students a goal to strive for; it gives the instructors a formalized method of evaluating the students' readiness to move to the next level; and it gives the trainees the added confidence of having achieved their goal.

Eagle Becker Patrol pack suited to carrying skis

Todd Burleson with a client and two Sherpas on the summit of Everest, 29,028 feet—1992 (Alpine Ascents)

15

HIGH ALTITUDE OPERATIONS

Not everyone wants nor can afford to climb Everest (29,028´) but every season several thousand adventurers attempt climbs like Denali (20,320´), Aconcagua (22,829´) and any of the numerous more accessible peaks above fourteen thousand feet. Many also die on these mountains every year—some to be left frozen in place and others to be evacuated by friends or local rescue teams.

Although sudden deceleration (falls) is the number one killer of climbers, altitude also claims its share, directly or indirectly. Statistics show that for every four or five who summit Everest one dies on that mountain. From 1922 to 1999 there were 161 deaths for 808 summits—36 of those deaths occurred after the individual had summited and was descending. In 1996 alone, 15 climbers lost their lives on Everest; in 1970, 4 summited but 8 died; and in 1983, 18 summited but 11 died.

Another mountain that has claimed its share of victims is K2. In 1986, 27 summited but 7 died on the descent for a total or 13 fatalities for that year. In 1994, for 17 successful summits, 5 died trying.

The sheer weight of these numbers tells us something about the dangers of high altitude mountain climbing, but it should also be remembered that it is not the mountains that kill people. The mountain merely sits there and allows climbers to kill themselves through a combination of poor judgment and a "summit at all costs" attitude.

OXYGEN AND THE CLIMBER

There are several peaks higher than 20,000' that are quite accessible and popular with recreational climbers, some of which have routes that can even be classified as non-technical. But even though a route can be walked with little more than crampons and poles, people are still seriously injured and die on these mountains each year because they underestimate their own ability to handle the fatigue, time commitment, cold, difficult conditions and the effects of altitude.

Apart from storms, avalanches and plain bad luck, there are several contributing factors to these deaths such as dehydration, snow blindness, loss of equipment (gloves) and frostbite. But most deaths at altitude are the result of poor judgment, brought about or compounded by the effects of altitude and its corresponding lack of oxygen.

The oxygen that we draw from the air with each breath is one of the most important life supporting elements to our existence. Without food we can last weeks and without water days, but without oxygen our life expectancy is only a matter of minutes.

Filtered from air and transferred into our blood stream through our lungs, oxygen is fed to every cell in our bodies including our brains. When we go underwater we must carry compressed air in tanks on our backs, without which we are rendered unconscious and within minutes begin to die. Similarly, when we climb above sea level, the air becomes thinner and we again begin to deny our bodies this essential element.

At sea level, air is approximately 21% oxygen and if that level is dropped below 16% we risk hypoxia and unconsciousness, and at 12% anoxia and death. The exception to this rule is that if the level is dropped gradually over a period of time, the body will adapt to a certain extent.

Even though this adaptation begins as soon as we leave sea level, for most climbers the real altitude acclimation begins at around 9,000 feet and anything above 14,000´ must be taken very seriously.

At 16,000 feet, because of the lower partial pressure, the oxygen in the atmosphere is almost equivalent to one half that of sea level and already below the level that can cause anoxia. If an individual were taken from sea level directly to 20,000 feet the average person would have a few minutes of consciousness and be dead in about 15 minutes.

To more graphically illustrate this point, I recall one incident where I was sitting in an un-pressurized aircraft at 24,000´ preparing to make a HALO (High Altitude Low Opening) parachute jump. We were all plugged into and breathing off the plane's on-board oxygen supply but carried additional bail-out bottles for the freefall. The jumpmaster, standing behind the cockpit talking to the pilots, was also plugged into the on-board oxygen, but as we turned to make a pass over the target, the jumpmaster disconnected his oxygen line from the main panel and walked back to the door with the intention of plugging into the panel in the rear of the aircraft. He never made it. He collapsed from hypoxia.

Over 26,000 feet the available oxygen is approaching one third that of sea level and yet hundreds have successfully summited at these elevations without the aid of supplemental oxygen. However, thousands more have been turned back by their inability to function in this rarified atmosphere.

So if lack of oxygen is the big problem, the casual observer might ask, "why don't you just carry bottled oxygen with you?" Not a stupid question since we have all seen climbers summiting on Everest with oxygen masks; and jet pilots routinely use oxygen when flying at altitude. In addition, the hundreds of empty cylinders scattered around the high camps have received more than a little negative press for the climbing community's past lack of environmental consideration.

For the informed climber, the answer is obvious. The body does seem to adapt well enough to altitudes up to 24,000´ and one simply cannot carry sufficient oxygen for the days or weeks spent adapting to altitude. For extreme high altitude climbs it is burdensome and costly enough just to have oxygen available for the push on summit day and a little extra for sleeping at high camp. Each bottle of oxygen weighs 7 pounds and can cost about $300, plus the cost of the mask and having it packed in to base camp. At a meager 2-liter per minute flow rate a bottle of gas lasts about 5-6 hours, so to make the 12 hour run to the summit and 6 hour descent would require 3 bottles. Expeditions allow several additional bottles for use in high camp while resting up for the summit push but seldom use it below 25,000´.

The more traditional answer from the purists and the "travel light, go fast" alpinists is that it is better to adapt to altitude without the aid of oxygen, since if a climber were at altitude and dependent on supplemental oxygen and were to suddenly lose it because of a frozen valve or leaky hose, he would be in a very bad way in a very short time. A climber suddenly losing oxygen at 28,000´ will feel as if he is suffocating, his vision will darken and blur and he could easily collapse into unconsciousness without warning, or even worse, act irrationally and panic endangering the whole climbing team.

Seasoned alpinists who climb without oxygen also believe that they can move more easily and expend less energy if they do not have to carry the added weight of the heavy cylinders, but for most, supplemental oxygen is still a necessity in the "death zone" above 26,000 feet.

GOING TO ALTITIUDE

Natural hazards like avalanches, storms and lightning are an accepted risk in high altitude mountaineering, and many of the 8,000-meter peaks reach up into the jet stream where 100+ mph winds, plunging sub-zero temperatures and extreme wind-chill factors are not uncommon. So in light of all the dangers that mother-nature can throw at us, to survive a high altitude climb, personal mistakes and human error must become an unacceptable part of the equation. But then again, we all accept the reality that climbing is hazardous to our health and climbers are habitual risk takers.

Lonsdale prior to a high altitude jump to evaluate a military freefall rig. Note oxygen mask and bail-out bottle on left hip

Sitting in the relatively oxygen rich environment at base camp, it is easy to plan the final push to the summit and rationalize all the possible contingencies and bail-out plans. But once at altitude, climbers become "oxygen dumb" and even simple tasks like melting snow for drinking water or securely fastening crampons become a major chore. Their oxygen starved brains have the mental agility of a retarded child, so any chance of making rational decisions based on a complex array of problems and pressures becomes quite challenging.

The author on HALO jump from 30,000′ on oxygen with an outside air temperature of minus 40 degrees

However, lack of oxygen is not the only problem. Higher altitude means drier air and considerable moisture lost through respiration. Exercise burns calories and produces increased fluid loss through sweat, and the cold is a natural diuretic. All these contribute to another major problem at altitude—Dehydration.

Diarrhea is another major problem in high altitude mountaineering that contributes to dehydration. The body's ability to absorb fluids and nutrients is impaired so becomes weakened, and does not recover at altitude. Individuals suffering from severe diarrhea should be assisted to a lower altitude—below 14,000 feet—until their condition improves and their strength returns.

In struggling to maintain the thermal equilibrium necessary at altitude; where one bakes on the snow during day and freezes at sunset; where every hundred meters higher can mean a few degrees colder; the climber must constantly add and subtract clothing layers. This results in additional perspiration and fluid loss. To compensate, for this fluid loss, a climber should plan on carrying additional fuel to melt snow and make sufficient water to stay hydrated—at least three to four liters of liquids per day. At higher altitudes it is necessary to carry at least a weeks supply of fuel since unexpected storms can

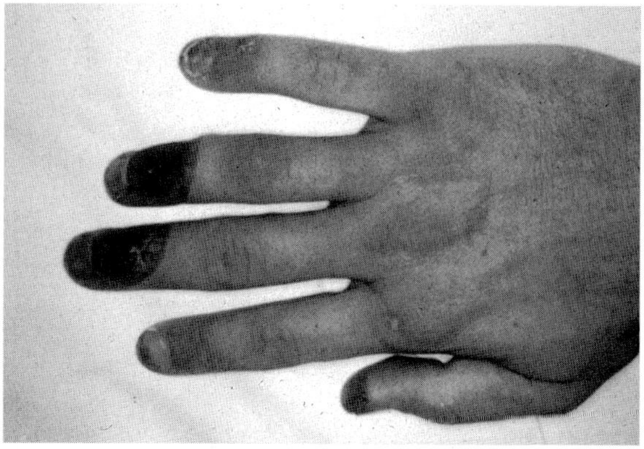

Frostbite

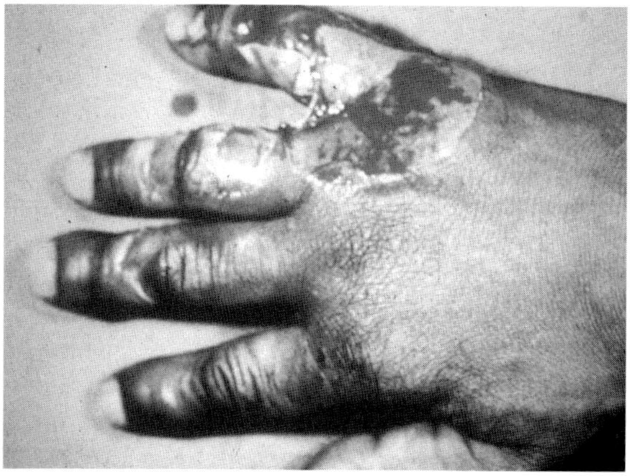

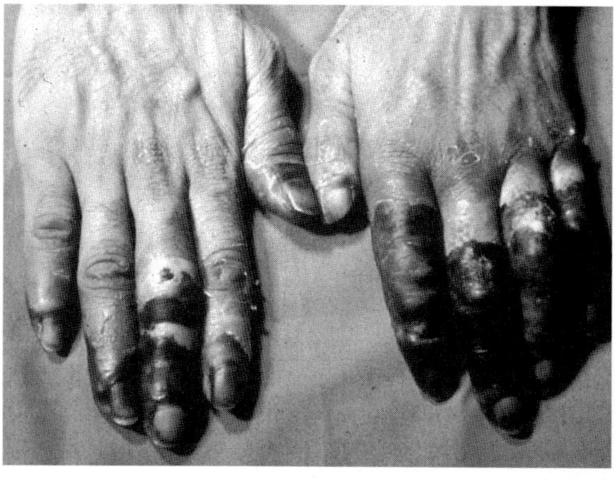

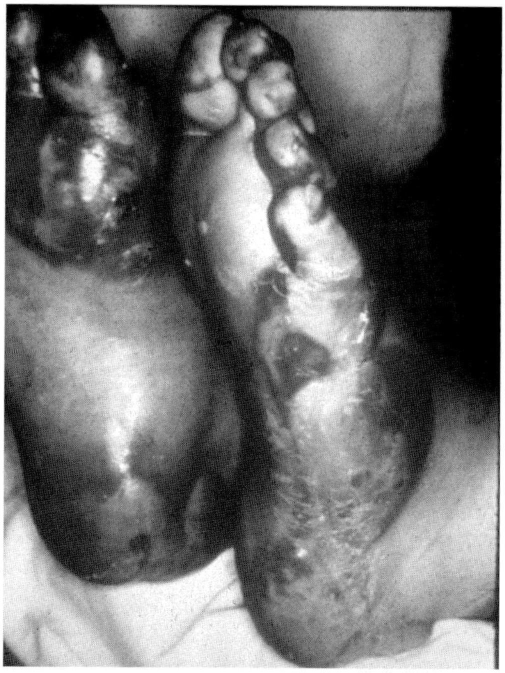

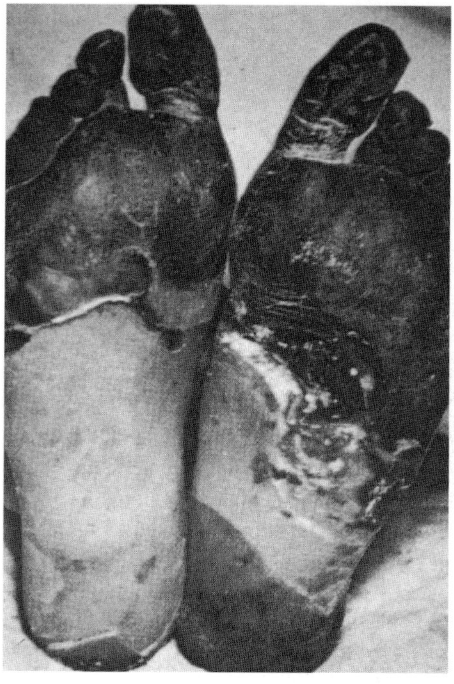

stall a team for several days in howling winds and zero visibility.

Even when fuel and snow are in abundance, climbers still become dehydrated because they simply do not have the energy or motivation to make water. It is important to fight this lassitude and get in the habit of filling water bottles as often as possible and keeping them in the sleeping bag at night to prevent freezing. "Hydrate or Die" is more than just a slogan.

With all this energy expenditure and effort to generate warmth, calorific intake requirements can rise to several thousand calories per day, and yet most climbers lose their appetite and eat even less than if they were resting at lower altitudes. One reason for this is the blood flow that is so essential to feed the muscles and maintain circulation to the extremities, cannot be spared for digestion. So it is not unusual to lose over ten pounds in less than two weeks at altitude, but this is a dangerous cycle that must be broken by forcing oneself to eat and drink more to compensate for the dehydration and calories burned just trying to stay warm.

The first night above 16,000′ this young porter accompanying the STTU researchers came down with quite severe altitude sickness. He had come up from a lower elevation and was not from the Sherpa region.

Another problem exacerbated by altitude and dehydration is frostbite. Extremities such as the fingers, toes, nose and cheeks, already starved of blood flow, oxygen and warmth, quickly become frozen when exposed to the elements.

Where most people fall victim to frostbite, is on the final push to or retreat from the summit when they are severely fatigued and their judgment is blunted by lack of oxygen. The initial pain and discomfort that should be a warning signal of impending frostbite is ignored, allowing it to progress to the point where the extremities become numb, hard and white, then through absence of blood flow, turn black and die.

The best safe guard against frostbite is prevention through adequate clothing, along with early recognition of the signs followed immediately by aggressive treatment. So as we move higher on the mountain, adjustments must be made to the equipment list.

- Clothing with greater insulation to maintain core temperature is needed
- Extremities such as face and hands must be protected from cold, wind and sun damage
- Spare gloves, socks, goggles and/or glacier glasses are essential
- Sleeping bags must be of the highest quality
- Tents must be able to withstand violent storms and high winds
- Stoves must be suited to high altitude (MSR-XGK Expedition)
- Additional fuel is needed to melt snow and heat food

In doing the research for this book, I spent time in the Himalayas and Andes, both regions frequented by a large number of climbers and trekkers. These two areas were selected for the sheer number of people that can be observed passing through the fourteen thousand foot mark where many begin to experience some of the more serious problems associated with altitude adaptation.

The HRA clinic at Pheriche on the approach to Everest (14,000´)

On the Everest approach, the HRA clinic, a small stone building in Pheriche, offers daily programs on altitude sickness so attracts many trekkers, plus they are equipped to test blood oxygen levels. My partner and I both checked in at 14,000´ with a blood-oxygen of about 85% and were feeling good considering the altitude and usual difficulty in sleeping. A trekker who came through at the same time had a blood-oxygen of 50% and looked like the walking dead. He was actually quite blue and had to be put on oxygen until he could be moved to a lower elevation.

A young man from Japan who came through was feeling good at 14,000´ but the next day arriving in Lobuje (16,200´), he collapsed and had to be evacuated back down the mountain while he was still ambulatory and before his condition worsened. As he descended, he was still protesting that he just needed to rest the night there, but that is not a gamble anyone should take. Others, also feeling mild effects of altitude wisely turned at this point and descended to a warmer and more comfortable elevation.

Very few of the individuals that we studied pushed on to Gorak Shep (17,000´) and those that did were not happy campers. Similarly, as we reached Everest Base Camp (<18,000´), the manager of the Iranian climbing team, suffering from pneumonia like symptoms and looking quite ill, had to be walked down to Pheriche for subsequent helicopter evacuation.

Working in the Andes between Argentina and Chile was equally edifying. In the course of just two weeks on Aconcagua (22,829´), many climbers were laid up at base camp for days with headaches and stomach cramps and several Asians had to be pulled off the mountain either unconscious or dead.

In one instance, at about 17,000 feet, we came across a Korean gurgling and unconscious in his tent, who had to be literally dragged down to base camp by mountain guide Marty Schmitd and a group of volunteers. Without this immediate action he would have been dead in a matter of hours. The victim was poorly equipped with inferior clothing, no sleeping bag, a cheap tent and was out of fuel and water. He had probably lain down to rest because he didn't feel well, only to pass into unconsciousness as his lungs filled with fluid.

Two others froze to death on the infamous Caneletta because they tried to summit too late in the day and became trapped on the mountain. One more died near the summit, and even one of the local mountain rescue team had to be evacuated with pneumonia like symptoms brought on by inferior sleeping gear.

One climber took a fall on a relatively easy trail at 20,000´ but still managed to break an arm, a leg and a hip and had to endure a painful trip down the mountain to a waiting rescue helicopter. At the same time we observed a Japanese woman who arrived at high camp (19,400´) with no tent or stove hoping to sleep in one of the small wooden refuges already packed with European climbers. A day later a French climber became incapacitated with HAPE and had to be carried down when his partner left him at Berlin Camp to recuperate and climbed on alone leaving him. (Nice guy!)

Then there was the man who lost his glove but made the decision to continue—a decision that cost him all the fingers on one hand to frostbite and almost his life as he realized that he could not hold or handle a rope.

Guides and team leaders working at altitude should be aware that individuals climbing or trekking in larger groups have a higher potential for problems than those traveling alone or in pairs. An

Injured climber being carried to the helicopter at Aconcagua base camp. Rescue teams had brought him down from 20,000´ the night before.

Victim being loaded onto the rescue helicopter at 14,000´

individual, let's say a strong healthy male, will be less likely to acknowledge that he is having problems with altitude, especially when there are women and older people in the group who are having no problems at all. This is largely due to ego and peer pressure. He simply does not want to admit to the others or himself that he "cannot hack it". This is a more significant problem with military types who are competitive by nature and conditioned to "suck

Another fatality being brought down from high camp by the Mendoza police mountain rescue team

it up" and just keep going, and to not slow the group down.

Solo climbers and pairs on the other hand, are much less affected by peer pressure and group dynamics. They will be more in tune with their own condition, more dependent on their individual health and strength, and will simply slow down, stop or descend at the onset of symptoms of altitude sickness. The key to moving to altitude is to have extra acclimatization days built into the schedule from the start and to not hesitate in using them.

ACCLIMATIZATION TO ALTITUDE

Climbing high and the process of gradual adaptation is not rocket science or some secret shared only amongst elite alpinists, evidenced by the fact that virtually every book on mountaineering, mountain medicine or even high altitude trekking expounds much the same advice on the subject. Obviously it helps to be in good cardio-vascular condition and use to vigorous exercise, but it is not essential since a broad cross-section of society, from the very young to the very old, has successfully hiked or climbed to over 20,000 feet.

First and foremost, it is important to listen to your body, to not push too hard and to not force yourself to conform to someone else's acclimatization schedule. The next thing is to stay well hydrated, well fed and to rest when needed. Even though the going can be slow and steep at altitude, the actual distances covered each day between camps are relatively short for a reason.

The rule of thumb for altitude gain above 10,000 feet, camps and topography permitting, is to climb no more than 1000´ (300m) per day and every 3000´ (900m) rest for two or three days. It is wise to always allow a few extra days in any expedition for rest and altitude adjustment. Extra time should also be built into the schedule in anticipation of team members simply being fatigued and moving slower as they climb higher.

Climbers are also heard to say, "Go slow—Climb high, sleep low." This refers to the practice of taking a hike or climb above camp during the day to better acclimate, but then returning to camp to sleep at night. This is especially beneficial during the rest days when a two or three hour climb up to the nearest ridgeline can be made without over exerting.

When ferrying loads on expeditions, climbers will carry loads up to the next cache and then return to camp the same day. On the second day they will move to the new camp, then on the third day ferry a load again and return to camp to sleep. This procedure of multiple carries continues all the way to within striking distance of the objective and further aids in acclimatization.

Captain Matt Marcenelle and the author in front of the base camp medical tent. At 14,000´, it does a brisk business in the summer months

The approach to Aconcagua base camp is 38 kilometers with a rise in elevation from 9,000 to 14,000 feet.

However, even following these schedules, there is still shortness of breath, mild headaches, difficulty sleeping, and a lack of appetite, but the breathtaking majesty of the high alpine regions and sticking to predetermined schedules often inspire a climber to continue ascending. If the headaches and discomfort continues or worsens then the insidious onset of Acute Mountain Sickness must be considered.

ALTITUDE SICKNESS

Acute Mountain Sickness (AMS) is always a concern with rapid ascents above 10,000 feet, and especially to or above 14,000 feet, so it is important to know the difference between mild discomfort and the onset of AMS. The most common complaint is headaches ranging from mildly annoying to "a nail driven between my eyes", and if accompanied by any of the following is probably indicative of AMS.

The common signs of AMS:

• HEADACHE Insomnia
• Loss of appetite Nausea Vomiting
• Fatigue Weakness Loss of coordination

AMS rarely has a dramatic presentation, coming on slowly, often within a day of arriving at altitude so may not be in evidence immediately. With only mild symptoms, AMS can pass in a day or two with rest and hydration, but it can also progressively worsen and become more severe. The best remedy is to descend immediately to a more comfortable altitude, usually about two to three thousand feet down mountain. If the symptoms diminish upon descent then this is a confirmation that it was AMS.

After an extra day or two acclimating and hydrating at the lower elevation, most climbers can then begin to re-ascend, but all the time continuing to listen to their own bodies.

When the headaches worsen to the point they become debilitating and/or breathing becomes more difficult and a wheezing sound is heard in the chest, this may now be HACE—High Altitude Cerebral Edema, or HAPE—High Altitude Pulmonary Edema.

With HAPE, body fluids seep into the lungs and interfere with respiratory function causing weakness, coughing and loss of sleep. If left untreated this is a potentially fatal condition. Unfortunately, the early signs of HAPE are not unlike other more benign problems such as the coughing caused by bronchial irritation brought on by the dry air encountered at altitude.

If left untreated and allowed to advance, there will be crackling noises heard from the lungs and the lips and nail beds may appear blue, a sign that the oxygen is being blocked from transferring into the arterial blood by the fluid in the lungs. This is a life threatening condition and immediate descent is required. Oxygen is also very beneficial, administered at 4 liters/minute for 4-6 hours or as needed to maintain a Sa02 of 99-100%*.

Another condition encountered at altitude is HACE—High Altitude Cerebral Edema. HACE can strike as low as 10,000 feet and is caused when vessels in the brain do not respond well to high altitude and begin leaking, ultimately causing brain swelling within the cranial cavity. HACE is avoidable since it is invariably the result of continuing to ascend even though having symptoms of AMS, and then becoming trapped at altitude by weather or sickness.

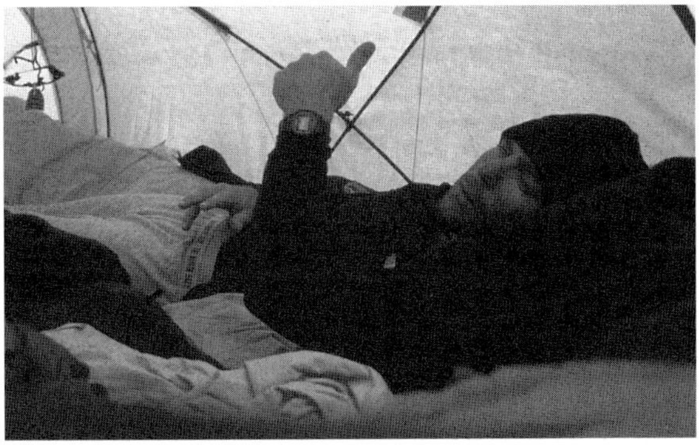

At 17,600´ the thumb says "okay" but the face says "I'm hurting"

Common signs of HACE are headaches, loss of coordination, stumbling and a change in mental state, but victims may also deteriorate quickly to where they are unable to walk or even become comatose. If left untreated, HACE can advance rapidly so immediate descent is critical to survival. The victim should be given dexamethasone and moved rapidly and without delay to the previous lower elevation where he or she was symptom-free, and even waiting for a rescue helicopter could further danger the patient. Oxygen should also be administered if available at 4 liters/minute for 4-6 hours*.

There are a couple of drugs that can be of some use at altitude but should only be considered after consultation with a physician and even then used judiciously. They are not a substitute for a slow and sensible approach to higher altitudes.

Diamox (acetazolamide) is used to treat altitude sickness but also aids in acclimatization and helps prevent altitude sickness. It is more commonly used when one has to make a fast ascent to an altitude above where one has acclimatized, however many climbers avoid its use because of some annoying side effects. One of the more significant ones being that Diamox is a diuretic that further contributes to fluid loss in addition to breaking up sleep patterns when one is forced to take a pee in the middle of the night. This can be a real pain when good quality sleep is already hard to come by at altitude. Used prophylactically 125 mg can be taken twice a day 24 hours prior to ascent*.

Many high altitude climbers also carry dexamethasone, a steroid that reduces cranial swelling and temporarily negates the symptoms of cerebral edema. This can give the victim of HACE a twenty-four to thirty-six hour safety window to get to a lower elevation, weather and conditions permitting.

Dexamethasone can be injected intra-muscularly or taken orally—the recommended dosage for AMS or HACE being 8mg initially followed by 4mg every 6 hours*.The injection takes effect more quickly and should be used on coma cases but most climbers still carry both since it could mean the difference between life and death. Pre-loaded syringes of dexamethasone can be protected in empty toothbrush holders and carried inside the climber's jacket to keep them from freezing.

*See High Altitude Medicine Guide at www.gorge.net/hamg for additional information on altitude sickness and treatment

"ACCIDENTS IN NORTH AMERICAN MOUNTAINEERING"

Edited by Jed Williams, and published by the American Alpine Club and Alpine Club of Canada.

HAPE & HACE—Washington, Mount Rainier—1998

While spending the night on the summit of Mount Rainier, off duty climbing rangers Gauthier and Patterson were contacted by the leader of a scientific research team, who informed them that one of his team members was suffering from mountain sickness. The rangers found Haley in his tent, disoriented, suffering from slurred speech, and unable to support himself. Team members indicated that his condition had deteriorated significantly in the previous 15 minutes and that his behavior was abnormal. A helicopter was requested, but nightfall and altitude prevented a safe response. Within 15 minutes of the initial assessment, Haley became unconscious and unresponsive. Assisted rescue breathing was provided throughout the night while a climbing team from Camp Muir attempted to reach the summit with oxygen, but newly formed crevasses and whiteout conditions prevented their efforts and Haley was evacuated shortly after sunrise by an Army Chinook helicopter and flown to Madigan Hospital. He was unconscious and in critical condition.

Analysis:

Haley suffered from High Altitude Cerebral and Pulmonary Edema (HACE and HAPE.) Although Haley was part of a team that had spent four days and three nights transporting gear to the summit, he still reported headaches and a feeling of nausea. Haley also commented that he did not drink much water on his summit day. By the time his condition had deteriorated seriously enough to alert his team members, Haley was immobile, so a safe, rapid descent was out of the question.

Climbers on Mount Rainier frequently feel the effects of altitude sickness, but most do not get seriously sick because they descend to sea level before many of the medical complications arise. Acclimatization on Rainier is difficult because most climbers go from sea level to 14,411 feet in less than 24 hours. Parties that expect to stay at altitude should have a systematic plan of acclimatization. Anyone exhibiting or complaining of the signs and symptoms of acute mountain sickness should be taken seriously. A rapid descent is the best treatment. It is imperative that team members watch each other. It is not uncommon for climbers to dismiss their symptoms as other maladies and push on. (Source: Mike Gauthier, SAR Ranger, Mount Rainier National Park)

CLIMBING INTO THE DEATH ZONE

Before moving to extremely high altitude, climbers must accept full, personal responsibility for the potential ramifications of their actions. They are about to enter an environment from which many have failed to return, and many more have lost hands, feet and noses to frostbite.

The body will adapt up to about 18,000 feet, but

after that, even though it continues to adapt it ceases to recuperate since the oxygen equivalent is below half the 21% that the body is used to. Even rest, food and re-hydration do little to physically improve the situation. The oxygen and blood flow that is so essential to life is just not sufficient for digestion, warmth and recuperation.

Climbing even higher toward 25,000 feet, one moves into the "death zone". The place where few can survive without bottled oxygen and all things begin a slow death. For some, that "death zone" begins at much lower altitude but they are driven on by zombie like determination to make the summit and get that photo.

Whatever discomfort was felt at lower altitudes becomes more noticeable as the climber moves into a colder and thinner atmosphere. First there is the constant headaches, shortness of breath, dry hacking cough and mild diarrhea. Then the headaches become more nagging, the wind howls, you have no appetite, you are always cold, your partner tosses and turns and sleep is impossible. Just putting your boots on to go outside is an hour long process that leaves you exhausted and breathless. The cough becomes worse and pneumonia like symptoms set in, but hey, you can see the summit and others have made it, so why can't you? The fire of life begins to go out as fingers and toes become cold and painful, then the pain stops, not because of re-warming but because the tissue has frozen. You find it harder to concentrate, simple tasks become difficult challenges, conversations are down to a grunt or a nod and you find yourself stumbling more. Well time out!!

The most difficult decision to make on any climb is to turn around, especially after having invested so much time and effort to get to that point—and many do not, with fatal consequences. The climber must never lose sight of the fact that reaching the summit is only half the climb, with maintaining sufficient energy reserves to get back down being the most important consideration. Whether it is grim determination or "summit fever" that keeps the climber painfully putting one foot in front of the other, this drive must be tempered with good judgment—and yet the lack of oxygen blunts the ability to make sound decisions.

The climber must be listening to his own body and instincts. Determination says to go on; fear says to be careful, but only cold logic can say go back—but that thought process is also warped by lack of oxygen and the burning desire to summit "at all cost".

The higher you go, the more mental and less physical the effort becomes. The mountain and Darwinian selection filter out most but a few with the right determination get a shot at the summit. Without determination one cannot reach the summit—but too much determination and someone is going to die. Ask anyone who has summited and then lost hands and feet to frostbite if it was worth it?

As the climber moves higher and the morning that began at midnight the night before becomes afternoon, the oxygen-starved brain must make several critical calculations: Is the team strong and moving well? Are they on schedule? What time is it? Is there time to make the summit? Is it passed turn-around time? What is the water situation? How does the weather look? Is there sufficient oxygen for the retreat? Are there sufficient physical reserves to make it back down? Who is going to turn back and who is going to push on? All good questions, but ones ignored by climbers every year who pay the ultimate price.

The plans and solemn pacts that were made in the comfort of base camp are now forgotten and ignored in striking distance of the summit, even though the last few hundred meters may take hours to negotiate putting the team well passed their agreed upon turn-around time and exhausting all physical reserves for the retreat. Like drunks with a loaded gun, the climbers are now playing the dangerous game of Russian roulette, a game played by many climbers before them.

Apart from the importance of personal decision-making, there are other outside influences that can impact on high altitude climbing safety. One of the most dangerous situations on a mountain is created by inexperienced, ego driven teams or individuals who push upward way beyond their abilities and in the process endanger other teams on the mountain who must now come to their assistance.

A good example of this is the Taiwanese team involved in the tragedy on Everest in 1996. They had failed on other mountains; seven had to be rescued off McKinley in 1995; and then they turned up on Everest where they endangered the other climbing expeditions with their lack of experience, incompetence and disregard for climbing protocol. Teams like this will intentionally follow in the tracks of stronger teams on their summit bid, knowing that they will come to their assistance in an emergency. This is only further indicative of their lack of confidence in their own abilities and disregard for others safety.

BLOOD OXYGEN SATURATION

As climbers move higher on the mountain, some seem relatively unaffected by the altitude while others become weaker, slower and less alert. One contributing factor to this loss of performance is poorer acclimatization to the altitude and correspondingly less blood oxygen being transported to the brain, muscles and vital organs. Technology has now given us a convenient way to track and compare this adaptation process.

At higher altitudes, blood oxygen saturation (%SpO2) decreases making physical exertion more difficult. This decrease can now be tracked and monitored by guides and climbers with the **"SportStat"**, a very small, lightweight, pulse oximeter—a device that measures blood oxygen saturation (%SpO2) and pulse rate.

%SpO2 is the percentage of hemoglobin molecules carrying oxygen. Under normal sea-level conditions, an individual would be expected to have a %SpO2 reading of about 97, which indicates that 97% of the hemoglobin molecules are carrying oxygen. Ascending through 4,000 meters a climber would be expected to have a %SpO2 of about 85%; at 5,000—80%; 6,000—75% and then above 6,500 meters it tends to flatten out at 60-65%. Readings less than these standards could be indicative of poor acclimatization and a need to stop or slow the ascent.

To get these readings, the **SportStat** shines red and infrared light through a finger and detects the fluctuating signals caused by blood flow. Pulse rate is determined from the signals received by a light detector. The ratio of the fluctuation of the red and infrared light signals is used to calculate the blood oxygen saturation (%SpO2) of hemoglobin.

The **SportStat** weighs less than 2 ounces (60 grams) and its advanced digital circuitry requires no adjustment (calibration) or periodic maintenance other than battery replacement. It is built to withstand extreme conditions of +32 to +122° F (0 to +50° C) with an operating range up to 30,000 feet (9,144 meters) of altitude and 3 atmospheres of pressure.

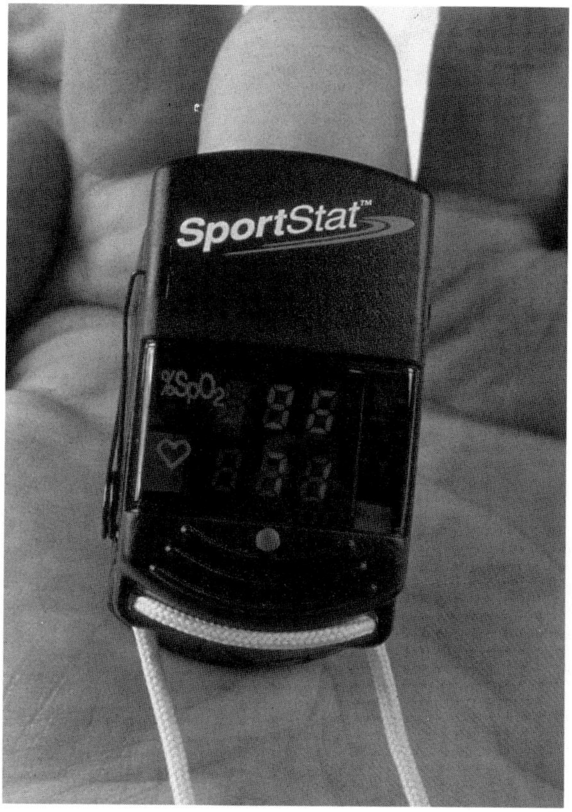

The compact and affordable SportStat by Nonin Medical showing percentage blood oxygen and pulse rate

HIGH ALTITUDE RESCUE

It is not unusual for high altitude climbers to leave their dead at altitude for no other reason than it is too arduous a task to bring them down. There are approximately 120 bodies to date that have been left on Everest, some of whom were lost in storms, a few who fell to their deaths and others who were simply left where they died from lack of oxygen and hypothermia. Those that were left, were left because the conditions that caused their deaths probably also endangered the rest of the team—lack of supplemental oxygen, turning too late in the day, storms, winds, fatigue, plunging temperatures and hypothermia. Any delay or expended energy to recover the body would only mean more unnecessary deaths.

At an altitude where one can barely carry a light pack and put one foot in front of the other, any possibility of actually carrying or even dragging a dead or unconscious climber off the mountain is slim to none. Even the rescue of Beck Weathers off of Everest in '96 was an astounding feat of skill and determination on the part of those involved—and only possible because his rescuers were superb alpinists who had the benefit of being acclimatized and well rested at lower camps.

Rescuers recover the body of a climber at 17,000′ on Denali who unclipped from the rope and did not have his ice axe attached to him. The cause of the fall was most likely a gust of wind on a rather stormy day when temps were around 10 and winds were in the 50 mph range. The probable cause of death was impact with rocks on the 1000-1500 ft. fall. The body was not found on the day of the accident. (Steve Dundorf)

This goes back to the concept of taking personal responsibility for your actions and fate when going to altitude since very few of the big mountains have professional rescue rangers such as those found on Denali (See Chapter 19 for more information of High Altitude Mountain Rescue)

For the fortunate few who can be evacuated to a lower altitude, there is always the possibility of hyperbaric treatment in some form of portable recompression chamber. With the increased traffic on some mountains such as Everest, many major expeditions and a few high altitude rescue organizations have invested in relatively lightweight portable hyperbaric treatment bags such as the Gamow ($2500), purchased in the US through Chinook Medical.

Gamow bag on the Rongbuk Glacier at 18,000′, Tibet, 1992 (Chinook Medical

Note: The following section, concerning the hyperbaric treatment of climbers, has been downloaded from the High Altitude Medicine Guide (http://www.gorge.net/hamg/) with the permission of the author, Thomas Dietz, MD.

HYPERBARIC TREATMENT

By Thomas Dietz, MD

Several portable hyperbaric chambers are now available, and are very helpful in treating severe forms of altitude illness. They are all similar to the extent that they are air-impermeable bags that completely enclose the patient, and are inflated to a significant pressure above ambient atmospheric. This effects a physiological "descent." This can be demonstrated with an altimeter inside the bag, and marked improvements in oxygen saturation are measurable with a pulse oximeter. The extent of the descent depends on the altitude at which the bag is used; as an example, at 4250m (14,000 ft), the inside of the bag is equivalent to 2100m (7,000 ft).

Typical treatment protocols are to put the patient into the bag, pump it up until the pop-off valves hiss, and keep the patient at pressure for one hour. Unless your bag has a CO_2 scrubber system, you need to continue pumping several times per minute to flush fresh air through the system (consult the specific manufacturers' information to determine how frequently). At the end of the hour, the patient is removed from the bag and reassessed. Additional cycles of "descent" and reassessment are continued as needed until either the patient is clinically improved enough to not need further hyperbaric treatment, or is able to actually descend. Experience has shown that HAPE typically requires 2-4 hours of hyperbaric treatment, and HACE typically requires 4-6 hours of hyperbaric treatment.

Compressing and decompressing the bag should be done slowly, while talking to the patient. Slow down if she or he experiences ear pain. Despite the concerns about tympanic membrane barotrauma, however, I have not personally seen this happen nor heard any reports of it happening.

Patients with severe HAPE may not tolerate lying flat; this problem is readily addressed by putting the bag on a rigid surface (such as a bed) and propping one end up 30-40 cm (12-16 inches). However, it is so much easier for these patients to breathe inside the bag once it is pressurized, that this maneuver is usually not necessary.

For maximum therapy, it is possible to put patients inside the bag with oxygen on (I suggest 4-6 l/min); obviously the bottle must be inside the bag with them. Do not hook oxygen up to the pump intake.

Remember to put a sleeping bag in with the patient - it can get very cold lying motionless for an hour at high altitude! Conversely, if you are outside in the sun, remember to shade the hyperbaric bag, as the sun is intense at altitude and will "cook" the patient.

Absolute contraindications to using the bag include lack of spontaneous respirations, as you can't ventilate the patient from outside the bag. There is a "Gamow Tent" which is about twice the diameter of the standard bag, and accommodates two patients simultaneously. This is useful for treating a critically ill patient with a medical person present, or treating a child with a parent present.

Relative contraindications to using the bag are middle ear congestion (small risk of barotrauma), inability to protect the airway in a deeply comatose patient (consider intubation), and claustrophobia.

GAMOW BAG—The Gamow Bag is the original portable hyperbaric bag, invented by Dr. Igor Gamow at the University of Colorado. It is a bright red-coated nylon fabric cylinder with a long zipper to allow entry. Small clear windows allow viewing in/out of the chamber. Conversation is slightly muffled by the fabric, but it is easy to talk to patients. Inflation is with a raft foot-pump, to 2 psi (105 mmHg) above ambient pressure. At the HRA clinic in Pheriche (4250m/14000 ft) the equivalent "elevation" inside the bag is roughly 2100m/7000 ft.

The patient is placed completely within the bag, the zipper sealed shut, and the bag is inflated with the foot pump. Two pop-off pressure valves set to 2 psi (not adjustable) start to hiss when maximum pressure is reached, and prevent over-pressurization of the bag.

All of the currently available bags must be continuously pumped (about once every 5 seconds) while someone is inside, not to maintain pressure, but to flush fresh air through the bag to prevent CO_2 buildup - this is a significant effort at 4000-5000 m.

Advantages: Well tested. Relatively lightweight (the bag, pump, and daypack carrying case is roughly 7 kg (15 lbs.). Fast and easy, you can have someone in the bag and "descended" in one to two minutes.

Disadvantages: Cost is US$2400. The zipper takes a lot of wear, and is the weak point—this is where leaks eventually show up. On the other hand, Gamow bags receive daily use in the HRA clinic and last several years. Due to the location of the reinforcing straps, it can be difficult to get a patient in and out of the bag.

Gamow bag in use in Nepal (Photo by Chinook Medical)

PROFESSIONAL GUIDING TO ALTITUDE

When first attempting new peaks, there is no substitute for having a professional certified guide who has experience on that particular mountain and at the target altitude. But not all guides are equally qualified, and as a result of recent accidents and fatalities at higher altitudes, concerned professionals created a set of guidelines for guides and expdition operators working the 8,000m peaks (26,000′+).

The following Code was approved by the General Assembly of the UIAA in 1997 and adopted in 1998.

THE INTERNATIONAL MOUNTAINEERING AND CLIMBING FEDERATION

Recommended Code of Practice for High Altitude Guided Commercial Expeditions

1. Definition: This Code applies to commercial expeditions attempting 8000m peaks, which offer to guide or accompany climbers above Base Camp. It is not concerned with the many expeditions that employ trekking agencies to supply transport etc. to Base Camp, and may also supply Base Camp services and H.A. porters.

2. Rationale. A variety of organizations offer to take clients on 8000m peaks. They vary from those that provide a full service to the summit or nearly to the summit, to those where there is minimal support for clients above Base Camp. However at the present moment it is difficult for clients to deduce from brochures exactly what is offered in terms of guiding and support, and whether it corresponds to their needs. This Code supplies clients with pointers to assist them to make an informed choice.

3. High Altitude Warning. Mountaineers climbing at very high altitude, especially above 8000m are at the limit of their mental and physical powers and may not be capable of assisting others as has always been traditional in mountaineering. This fact is of particular importance to mountaineers of limited experience who rely on professional guides to bring them safely up and down 8000m peaks.

THE CODE

The leader or chief guide and as many as possible of the guides should have *experience* at least to the altitude of the peak to be climbed. [There is no qualification appropriate to high altitude guiding, so the term "guide" does not imply that the person holds a professional qualification. Clients can only judge from the *previous experience* of the guides, who may be westerners or Sherpas or other local mountaineers].

The staff on the mountain must be adequate for the aims of the party and the services offered. A doctor in the party is very desirable but at the very least advance arrangements must be made for medical help. Advance arrangements must also be made for evacuation assistance in case of emergency.

The minimum safety equipment available must be walkie-talkie radios, a satellite phone, medical oxygen, and recommended First Aid supplies.

Advertising must give a true picture of all the difficulties and dangers involved, and avoid promising the impossible. Biographical information about the team should be included.

Clients should not usually be accepted for 8000m peaks unless they have previous altitude experience to 6000m or 7000m.

Information supplied in advance should include a clear statement of the guiding and support offered.

The expedition must take account of the UIAA Environmental Objectives and Guidelines and follow the UIAA Expeditions Code of Ethics.

Climbers move through the Khumbu icefall (Alpine Ascents)

ANG-MWS military mountaineering instructor, SFC Gordon (right) tops out on a summit in the Austrian Alps

16

MOUNTAINEERING EXPEDITIONS

Mountaineering expeditions are truly the highlight of being involved in alpine training and operations and an event that most mountaineers look forward to on an annual or at least semi-annual basis. For those in the peacetime military, or former military, they give a sense of mission and a challenge not unlike preparing for a combat operation. They also stimulate the close-knit camaraderie that only comes with total dependence on one another.

For anyone involved in alpine rescue or mountain warfare, expeditions offer a tremendous amount of experience in a relatively short time and are an excellent form of training.

The ability to plan and execute a successful expedition is not unlike planning an actual operation but on a larger scale since the average expedition can run several weeks; and that's not counting the months of pre-planning and preparation. Expeditions are also serious endeavors in that they require meticulous attention to detail and the coordination of a number of assets to bring everyone home alive and in one piece.

For all the time, money and energy that go into an actual expedition, most expeditions have very humble beginnings and start with no more than a spark of an idea. One individual will see a documentary, read a book or hear others talking about a particular mountain in some exotic far away place, then before he has even realized it, the thought of climbing this mountain has seized his imagination and the cerebral juices of the right brain begin to flow.

In voicing his interest to friends or colleagues, many will scoff at the notion but others soon become equally inspired and the daydream begins to take on an element of reality with words of encouragement such as, "Others have done it so why can't we?"

But with reality comes the first two hurdles to overcome—money and time. Everyone begins checking bank accounts, discussing it with loved ones and seeing if they can get the time off work. This is where the true adventurers are separated from the idle dreamers. Those that really want to go will make it happen, even if it means taking a few credit cards to the limit to do it. The advantage of credits cards being that if you don't survive the expedition, you won't have to deal with the bills when they come in.

For professional rescue teams and military personnel this process may be facilitated by the fact that expeditions can be considered work related training and valuable experience so with Uncle Sam picking up the tab, the only hurdle will be to qualify for a place on the team.

PRE-PLANNING

An expedition begins many months, and sometimes over a year, before the actual date of departure. This is to allow sufficient time for the all important planning which will be a pivotal factor in the success or failure of the whole effort. But before the real planning begins, a feasibility study must be made, along with a period of basic fact-finding to ascertain if the final objective is within the realm of reality. This is made easier by the number of books available on most climbing destinations and the wealth of information put on the World Wide Web by professional guiding companies.

Initial consideration must be given to such things as the difficulty of the mountain, selected route, climbing season, availability of permits and if the mountain is within the current capabilities of the team.

The next question is climbing style—Alpine or Expedition. In some cases this may be dictated by the accessibility and nature of the mountain, but in most it will be the climbers' choice.

Alpine style infers going as light as possible and taking one shot at the summit without significant logistical support. Although usually cheaper and less time intensive, alpine style climbing leaves very little margin for having to wait on bad weather, adapt to altitude or negotiate other obstacles that may slow progress. This is not the ideal for most high altitude climbs except for the final strike at the summit.

Expedition style climbing involves packing or ferrying in sufficient food and equipment to lay siege to the mountain. An expedition could take anywhere from 2 weeks to 2 months depending on the approach routes, necessary acclimation, weather, and number of camps required. The recommended time for mountains like Denali and Aconcagua is 3 to 4 weeks, with Everest expeditions taking 7 to 8 weeks.

The time of year for the planned expedition is also important since the clothing and equipment list will be directly affected by anticipated temperature and conditions. Winter expeditions obviously requiring warmer and more expensive clothing, along with additional fuel and food to compensate for colder temperatures and slower progress.

The final destination will have a significant influence on the cost of an expedition. Apart from the initial cost of international airfare to the destination country, there will also be additional costs for the excess luggage that goes along with most expeditions, plus domestic flights to get closer to the final objective.

Additional costs will be incurred with ground transportation. Once the flying is finished, and you are left standing beside a mountain of bags, there will still be many hours or days of overland travel by bus, truck or four wheel drive, followed by several days on the final approach packing in to base camp. Then you can add to that the cost of specialized equipment, hotels, food, permits, guides, porters and pack animals.

Military climbing team arriving in Mendoza and being met by their liaison officer on left

PLANNING

Once the destination has been selected, one that is not beyond the capability of the team, and some quick numbers have been crunched, the expedition planning can begin in earnest.

Again, this is made considerably easier these days by the masses of information available from books and the internet, but there is still no substitute for talking with those who have been before you. Preferably individuals who have been on the mountain in the last couple of years and at the same time of year since conditions and routes can change.

The next order of business is to select and assemble a team that has the training and experience to give the expedition a high probability for success—and this means reaching the summit.

Barring having prior experience on the selected mountain (a major plus), and assuming everyone already has the necessary equipment, the following would be considered desirable attributes for members of the team:

- Training & Experience based on the Difficulty of the Selected Route
- Demonstrated Fitness & Determination
- Compatibility, Attitude and Good Humor under adverse conditions
- Willingness to Share the Workload
- Team Spirit & Commitment to the Endeavor

To handle all the planning and logistics that go along with expedition planning, the team can be broken into smaller working groups with individual assignments. This division of work is an essential part of the team building process that makes everyone feel like they are contributing. These assignments should be allocated based on the individual's expertise, knowledge or experience.

After selecting a team leader to serve as an overall planning coordinator and a treasurer to organize the financial side and collect individual contributions, other tasks that need to be handled include:

1. Collection of maps and topographical information for route planning
2. Pricing and booking air and ground travel
3. Organizing travel visas, park permits and inoculations
4. Liaison with local authorities, guides, expedition support groups and hotels
5. Procurement of all the necessary team equipment and standardizing individual equipment
6. Planning menus and obtaining all the necessary food and provisions
7. Arranging sponsorship from organizations or manufacturers to help offset the costs
8. Organizing communications and contingency planning for emergencies

EXPEDITION COSTS

Since the cost of mounting an expedition is an important consideration, the following should be researched and shopped around to get the best deals:

- Individual Equipment Requirements
- Team Equipment Requirements
- Visas and Inoculations
- International Air Travel & Excess Luggage
- Domestic Flights & Luggage Restrictions
- Local hotels preferred by climbers
- Ground Transportation
- Park & Climbing Permits
- Guides, Porters & Mules
- Communications
- Rescue costs, helicopter evacuation, hospitals & rescue insurance
- Contingency funds for the unexpected

A high altitude helicopter rescue can be a huge unexpected expense, so membership in the American Alpine Club (AAC) is highly recommended since it includes rescue insurance.

EXPEDITION EQUIPMENT

Again, assuming that each individual on the team already has all his personal mountaineering equipment, there may still need to be some additions based on the anticipated climbing conditions. Particular attention should be given to having the appropriate clothing, thermal protection, gloves, boots, gaiters and sleeping bag. An expedition size pack is also an essential, and sleds, snowshoes or skis may be needed for the approach.

Team equipment will also have to be assembled with special attention given to tents, stoves and fuel suited to the planned altitude, wind and snow conditions. Menus have to be planned and sufficient food packed to sustain the whole team for weeks on the

The questionable soviet surplus Mil-17 helicopters used to ferry climbers up to Lukla, 9,350´ in the Himalayas. They were grounded soon after for lack of air-worthiness

mountain, and this can amount to a considerable load. Fresh food for the approach and base camp can be bought locally in country but pre-packaged, lightweight, freeze dried, alpine foods will need to be purchased and securely packed along with everything else. The total weight of food to sustain a 6-man team for 15 days would be about 180-225 pounds (2-2½ lbs per man per day), then add to that another 45 pounds of stove fuel (8 ounces per man per day)

Since everything has to be packed to fly international, and then repacked for the trip into base camp, weight will be an important factor when selecting and packing gear. It is not unusual to pack and re-pack everything several times trying to pare down the load and identify only that which is essential.

In packing to fly, all equipment should be stowed in large duffel bags such as the XL Gregory Long Haul Duffel. When food and gear has to be packed into base camp by porters or mules, we have found the rugged North Face Base Camp Duffel bags to be ideal.

THE TIME FACTOR

Even though the length of an expedition has a direct effect on cost, extra time should always be allowed for any of the many problems that can arise when traveling in third world countries and dealing with mother nature.

Colonel Eduardo Garate on the summit of Aconcagua, 22,829´—January 1999

The first snafu that can rear its ugly head is airline delays and this is compounded when team members may be flying in from different points of origin. The next one, every traveler's nightmare, is when you arrive but your luggage does not. Not a big problem on most business trips, but when the missing bag contains essential equipment like your climbing boots, tent, Gore-tex, etc, you are essentially disabled until it turns up. If insufficient time is left for airline connections, especially from international to domestic flights, then you may be getting further away from your luggage with every flight.

Then there are all the bureaucratic delays that can arise when trying to obtain visas and permits—unless the right palms have been greased in advance. This alone can often justify the use of a reputable local facilitator or guide service.

Ground transportation is also notoriously unreliable in countries where vehicles are held together with baling wire and the proverbial chewing gum, tires are bald and gasoline is of the lowest grade.

Once away from the filth and corruption of the cities, there are still all the delays that mother nature can throw in your path—rain, storms, flooded rivers, washed out roads and bridges, rock slides, white-outs, deep snow, howling winds, steep climbs, poor trails, icy conditions and avalanches.

On a more personal level there are several medical conditions apart from physical injury that can cause delays: diarrhea, fever, disease, dry hacking coughs, fatigue, dehydration, headaches, acute mountain sickness, sunburn, heat stroke, hypothermia and frostbite, to name a few.

With the above in mind during your planning, ensure that there is adequate time built into every step of the expedition to allow for these inevitable delays.

"ACCIDENTS IN NORTH AMERCIAN MOUNTAINEERING"-
Edited by Jed Williams, and published by the American Alpine Club and Alpine Club of Canada.

FALL ON SNOW, EXCEEDING ABILITIES, ASCENDING TOO FAST
Alaska—Mount McKinley, West Rib

A ten member British Army expedition "Summit to the Sea" checked into the Talkeetna Ranger Station for their briefing on June 2, 1998. Their original plan was to ascend the West Buttress, which they had changed to doing the upper West Rib, assessed as an Alaska Grade 4. The Park Service recommended they stay to the West Buttress (Grade 2), since several members had very little glacier or ice climbing experience. They would make their decision once they reached 14,200 feet on the West Buttress. The expedition departed Talkeetna on June 4. On June 14, nine members moved into the 14,200-foot camp. One member had returned to base camp a few days earlier. They acclimatized three days at 14,200 feet before ascending to a 16,200-foot camp on June 17. Ranger Roger Robinson at the 14,200-foot camp recommended that they take one or two rest days at 16,200 feet before making their summit attempt. At 0630 on June 18, the expedition departed for the summit.

Upon departing that morning, member Gary Keep experienced signs of AMS. The other eight members continued ascending while Keep soloed back to 14,200 feet. Keep was examined by Scott Damsey of the Ranger Patrol, where it was determined that Keep was experiencing AMS and dehydration. Keep was admitted into the medical tent at 1530, where he was administered oxygen, Diamox, and an IV. His condition improved, and he was released at 2400 on June 18.

The other members continued an extremely slow ascent up through the rocks of the upper Rib. By 1830, they had reached 19,000 feet, having only ascended 2,800 feet in twelve hours. Weather conditions remained good throughout the day. The first team of three, led by Martin Spooner (33), decided to deviate to the right of the normal route where a break in the Football Field cornice was found. This break could be attained by first ascending a 45-50-degree hard snow slope which led up to a 20-foot corniced section of ice that topped out onto the Football Field. Spooner was in the lead with Phil Whitfield (?) second and Steve Brown (25) third as they approached the cornice at 19,300 feet. At 1920, Brown lost his footing and fell, pulling Whitfield and Spooner with him. They tumbled approximately 300 feet before being stopped by large rocks. These rocks prevented an 8,000-foot fall off the south face. The other two rope teams were approximately 30 meters lateral of the team that fell and were resting when the accident occurred.

Spooner sustained injuries to both his ankles, while Whitfield received lacerations to his head. Brown received head injuries and was suffering from severe shock, delirium, and lapses of consciousness. The leader of the expedition, Justin Featherstone (29), attempted to call the Ranger Camp with their CB radio, but found that their batteries were too weak. Featherstone decided that he and Bougard would stay behind with Spooner and Brown and wait for a helicopter rescue. The other four, including Whitfield, would descend to 14,200

Expedition team unloading after flying into the Denali base camp (Steve Dundorf)

feet to report the accident. The four remaining each had bivi bags, down jacket, one shovel, one stove, one liter of fuel, and one day's food. They dug into the ice and built a small trench below several large rocks. This gave them some protection, but after the first night, they were unable to relight their stove due to the cold and windy conditions.

At 2030, the other four, including Whitfield, Johny Johnston, Ian Hayward and Nigel Coar, began their descent. They reached their 16,200-foot camp at 0030 on June 19. Hayward and Coar decided to stay at their camp while Whitfield and Johnston continued on down. At 0040, Whitfield and Johnston witnessed two Americans fall down the 45-degree snow slope on the West Rib known as the "Orient Express," coming to a stop at 15,800 feet. Whitfield and Johnston arrived at the accident scene at 0100 and provided assistance to the fallen Americans. Whitfield stayed behind with the Americans while Johnston decided to solo down for help. At 0220 Johnston reported both accidents to the 14,200-foot ranger camp. The Talkeetna Ranger Station was notified of the situation at 0240.

Weather conditions remained poor for a helicopter evacuation. A large ground team was assembled, first to evacuate the two Americans who were closer and more exposed. Once this was completed, a second plan would be developed to evacuate the British. During the American evacuation, Keep was escorted to his 16,200-foot camp with an NPS radio establishing a radio link from the ridge.

Brown's condition was deteriorating, so Featherstone decided the four would slowly begin their descent. All four attempted to descend, but Spooner experienced too much pain in his ankles to walk. Featherstone decided to take Brown down, leaving Bougard behind with Spooner. The foursome had only one rope, since their third rope had been damaged in the fall, so Spooner and Bougard opted to wait to be rescued. Unknown to Featherstone was the rescue operation of the Americans below his position. Weather at the 14,000-foot level and above began to improve so that at 1600, all four of the British climbers were observed through a spotting scope from the Ranger Camp. Shortly thereafter, Featherstone and Brown were observed descending. Brown was very unstable, requiring a constant belay and occasionally taking short falls. Weather improved enough to evacuate one of the Americans at 2230 by helicopter. At the same time, the two British were descending the "Orient Express," very close to where the Americans had fallen. At 2300 Featherstone and Brown fell 1,500 feet, ending where the Americans had landed at 15,800 feet. Featherstone had later commented that Brown had fallen into him, knocking him off his stance.

At 2300, Keep radioed from his 16,200-foot location to the Ranger Cap that they had witnessed the fall and would be providing assistance. Ranger staff also witnessed the fall. Plans were developed for their evacuation. Featherstone had broken his left leg in the fall, while Brown sustained additional head injuries. Brown was able to walk and decided to descend straight down the route to the 14,200-foot camp for help.

Rope team at 18,000 feet heading for the summit (Steve Dundorf)

Brown's descent was observed from the 14,200-foot camp. It was apparent his judgment was impaired. A well-marked trail led from their accident location to 14,200 feet, but Brown's decision to descend a direct course would lead him over an ice cliff and several crevasses. Brown was observed walking straight off a 20-foot ice cliff, then tumbling down a steep slope where he was buried in avalanche debris. He freed himself from the debris then continued his descent. Not 30 seconds later he was observed falling into a crevasse. Brown recalled that he attempted to climb 30 feet out, clawing with bare fingers and kicking with his crampons. Brown had lost his ice ax on one of his earlier falls, then lost his gloves on the fall down the Orient. After 30 minutes and several attempts, Brown clawed his way out to where he lay motionless by the lip of the crevasse. Eventually he stood up. Ascending rescuers shouted to Brown to stay put.

Because of Spooner's and Bougard's position at 19,000 feet, and the fact that all able bodied rescuers at 14,200 feet had exhausted themselves over two long nights of rescues, the shorthaul extraction became the more logical choice of getting them off if the weather cooperated. When the weather began to improve on the evening of June 21, Jay Hudson was able to spot the pair at 2035, where they were standing and waving, to everyone's relief. Establishing a good communication link with the pair was imperative in order to perform this short-haul mission. The initial cargo drop of radio and survival gear established this communication link. Once we could talk with the pair, we could judge their physical and mental capabilities, and get current weather conditions. The pair had previous training using a similar "Screamer Suit" (big diaper) in the military. This allowed the two to be extracted without a ranger involved in the short-haul. Weather was the primary factor in retrieving both at the same time, since conditions were constantly changing for the worse. This had been the first time in many days that the upper mountain was calm. At 0248 on June 22, the LAMA helicopter held a hover at maximum power until the two were hooked in. Then a slow descent was made directly to the Ranger Camp. At this point the LAMA had to refuel at 7,200 feet then return. When the LAMA, came back into 14,200 feet, the weather was already changing and it was

The Denali Lama pulling into base camp (Steve Dundorf '98)

Steve Dundorf on the summit of Denali, 20,320′—June 9, 1998. Gear includes Marmot shell, Bolle goggles and Grivel crampons

fortunate that the pair got flown off. This short-haul performed may be the highest short-haul rescue mission ever performed in the world. Pilot Jim Hood did an exceptional job in the decreased light and high altitude conditions.

Analysis:

Here is a good example of an inexperienced party getting in over their heads by attempting too difficult a route. The "Summit to the Sea" expedition was advised in Talkeetna to stay on the West Buttress since half the party had very little glacier or ice climbing experience. While at 14,200 feet, they were advised to take several days to acclimatize at their high camp, since their summit day would be a long one. Because of good weather, they chose to make their summit push the next morning. Their lack of acclimatization became apparent when one member became ill on summit day and the rest required twelve hours to ascend just 2,800 feet over technically easy ground. The experienced members in the party should have realized that their slow pace was a good indication that it was time to turn around. To make matters worse, after a long day they planned to attempt a difficult corniced finish to the West Rib—a route seldom tried even by very experienced parties. On top of all these concerns they left an ill member behind in camp, who was then forced to solo down because of his AMS. To further their predicament, the leader descended with a person who was unstable, down a route notorious for its falls. Belays and fixed protection should have been used. The leader was familiar with the route and conditions they had ascended the day before, which would have been the safer choice. Had they carried extra batteries for their radio, the pair could have been warned about the conditions in the descent route.

The two who survived four nights out in bivi bags were incredibly lucky. A strong will and determination kept them alive.

(Source—Roger Robinson,
Mountaineering Ranger)

PART III

ALPINE RESCUE

17

INTRODUCTION TO RESCUE OPERATIONS

The most important caveat that can be given concerning rescue work is, "be sure that you can be part of the solution and not become part of the problem." All too many would-be rescuers attempt to help people in distress only to become one more victim of the same situation. In fact, it is very hard for the average person not to leap to the rescue when a child is being swept down a flooded river or clinging to the side of a cliff. But in both instances, brave and well meaning, but untrained and ill-equipped individuals have lost their lives.

The same is true for professional rescue teams that may find themselves responding to a situation beyond their training and experience. Or the team that takes unnecessary risks just to recover a victim who is already dead. Granted, everyone wants to bring closure by recovering the body, but this should not be done with unnecessary risk to the living.

For these reasons, in all rescue or recovery situations, a **Risk/Benefit** evaluation must be made based on the urgency of the rescue and the capability of the rescuers. This is especially true in the unforgiving environment found in mountain and alpine rescue.

Mountaineering and climbing are by nature potentially dangerous activities, and with the rise in popularity of both, along with other extreme sports and outdoor activities, it is only logical to note that the number of rescues have increased proportionately.

Putting professional rescue teams aside for the moment, every climber who ventures into the mountains should have some form of formal training in rescue techniques and carry at least the minimal equipment to extricate him- or herself or a fellow climber from a life threatening predicament. Whether it is a thirty-foot "whipper" or a falling rock that leaves your partner unconscious, or a fall into an unseen crevasse, it may be solely up to you to make the difference between life and death for yourself or your partner. So although this section is directed primarily at professional teams seeking a rescue capability, it also contains information that will be of value to all wilderness adventure seekers.

SEARCH & RESCUE

There are many types of Search & Rescue (SAR) operations ranging from urban to maritime to desert to mountain—each with its own specialized techniques and related equipment and training. Even within the mountain rescue community, operations can range from basic low-land operations such as lost or over-due backpackers and climbers, to the more technical high angle rescues encountered when climbers become trapped on the mountain by injury or inclement weather. The former requiring a large number of competent volunteers with basic backcountry travel and camping skills, while the

LA Sheriff's Montrose SAR team is illustrative of the many response capabilities that a team must have. From left, team members are equipped for work on the fire line, mine rescue, air crew, ground search, snow & ice, rope rescue, mountain bike, law enforcement, swiftwater and support.

later may need all the skills and resources of a professional rope rescue team supported by skilled helicopter pilots—but both under the control or an organized Incident Command System.

Above the snowline operations become even more challenging when avalanches, crevasses, colder temperatures, rapidly changing weather, whiteouts and high winds are thrown into the equation. Even teams deployed by helicopter must be trained and equipped to spend a night or two on the mountain or be prepared to pack out a victim if the weather should change suddenly denying them air support.

Higher on the mountain, the altitude itself becomes a problem for both victims and rescuers. Climbers and trekkers can become disabled at elevations beyond the reach of many helicopters and even the responding rescue teams must be acclimated to operate efficiently above the 10,000´ mark.

With this diversity of rescue operations, the responding rescue team must be either very proficient at a wide variety of alpine rescue skills or specialize in one type of rescue and know when to hand the other operations off to more qualified teams.

A professional search and rescue organization may have a wide variety of capabilities ranging from swift water to mine rescue but not necessarily be qualified to operate in an alpine environment. To qualify for Mountain Rescue Association (MRA) certification for example a team is required to be proficient in a wide variety of rescue related skills and is formally evaluated in three areas of expertise:

- Basic Search & Tracking
- Technical Rope Rescue
- Winter Skills

Although many of the top teams in the country will have internal training standards for their members that meet and exceed the MRA requirements, not all teams have the time or resources to maintain such standards or even have a real need for all these skills.

TYPES OF MOUNTAIN RESCUE TEAMS

There are primarily five categories of SAR teams that operate in the mountains.

1. Ground Search Teams
2. High Angle Rope Rescue Teams
3. Helicopter Rescue Teams
4. High Altitude Rescue Teams
5. Military Combat Search & Rescue Teams (C-SAR)

Ground Search Teams—Ground search teams are limited to non-technical rescue scenarios where a climber or backpacker has simply become lost or injured in mountainous terrain and must be found, stabilized, packaged and evacuated. Even though these teams may not have a high angle rescue capability, they must still be proficient in all aspects of wilderness living and safe movement on snow and ice.

Two team members equipped for ground search and two for technical rope rescue

In most national parks and wilderness areas, these teams are usually lead by a core group of professionally trained personnel drawn from park rangers and local law enforcement, supported by a larger group of volunteer search personnel. Some specialists within the ground search team include trackers, K-9 handlers, horse mounted posses, mountain bike units, ski patrollers, snowmobilers and local mountain guides.

Rope Rescue Teams—When the missing person is a climber or the rescue requires vertical or high angle rope work to access and evacuate the victim, then a team skilled in this type of endeavor is called for. These are trained technical rope rescue specialists but not necessarily climbers since most rescues are initiated from the top of the rock, cliff or crevasse. However, there are several advantages to having skilled climbers on the team (such as in Yosemite), especially when air support is not available and the victim can only be accessed from the bottom of a cliff or mountain.

On many professional SAR teams, all team members are required to be proficient in rope rescue techniques and are tested on their skills on an annual or semi-annual basis.

Helicopter Rescue Teams—Helicopters are an invaluable resource to any rescue team and especially to those that must operate in remote areas that are difficult or slow to access on foot. Helicopters also have a distinct advantage in search operations in that they can cover a considerable amount of territory quickly and remain at a low enough altitude to see tracks in the snow or any other ground distress signals.

However, helicopter search & rescue operations are severely restricted in the mountains by several conditions including altitude, visibility, high winds, inclement weather, low clouds, fog and the safe operating range from refueling. Steep terrain, rocks, canyons, trees, deep snow and icy conditions also create problems for both the pilots and rescue teams.

LA Sheriff's Air 5 rescue helicopter manned by Emergency Services Detail (ESD) mountain rescue paramedics

Although helicopters can be used to simply ferry ground teams into and victims out of the mountains, the full extent of their value is only realized when manned by a dedicated helicopter rescue team such as LA Sheriff's Emergency Services Detail (ESD) or the elite French Peleton de Gendarmerie de Haute-Montagne. These teams are usually composed of mountain rescue qualified paramedics who have adapted their skills and techniques to take advantage of the helicopter as an aerial work platform.

High Altitude Rescue Teams—The elite of the mountain rescue teams are the high altitude rescuers who can operate above the normal 10,000′ operating range of most helicopters, and the 15,000′-17,000′ limit of even high altitude helicopters. These teams must be able to not only survive in the low temperatures and thin air that stop so many, but also be able to work and effect a rescue. They must be able to package and evacuate victims suffering from acute mountain sickness, frostbite and exhaustion, to a lower elevation where they can be extracted by helicopter (if one is available).

Several popular climbing areas around the world have teams pre-positioned at altitude to assist climbers during their summit bids. NPS mountaineering rangers patrol out of a station at 14,200′ in Denali National Park for those attempting Mt. McKinley (20,320′), and on Aconcagua (22,829′) the Argentinean National Police mountain rescue team can be found at Base Camp (14,337′) during the peak climbing seasons. Because of the non-technical nature of the terrain, this team can move up to high camp (19,520′) to retrieve a victim and return the same day.

On the Everest approach route, the Himalayan Rescue Association operates a clinic with Western physicians in Pheriche, also at around fourteen thousand feet. Although there is no actual rescue team, the staff does a tremendous job educating climbers and trekkers about the various forms of high altitude sickness before they move to higher elevations. (Many also turn back at that point and go no higher).

Members of the Argentinean national police mountain rescue team Grupo de Rescate who have made rescues in the Andes above 20,000 feet.

Navy Seal C-SAR team practicing downed pilot recovery from a CH-46

However, in reality, most high altitude rescues are carried out by climbers already on the mountain and acclimated to working at that altitude. A good example of this was the ill-fated 1996 Everest expedition that gained notoriety in Jon Krakauer's book "Into Thin Air". Beck Weathers and the leader of the Taiwanese team, both suffering from severe frostbite and other complications, had to be evacuated by members of three other climbing teams; first by the heroic efforts of Todd Burleson, Pete Athans and the Sherpas, and then assisted by Ed Viesturs, Robert Schauer, David Breashears, Jim Williams, Veikka Gustafsson and Araceli Segarra, with medical assistance at Camp II from climber and physician Ken Kamler.

Upon arriving at a pre-selected LZ (19,860´) above the dreaded Khumbu Icefall, each victim was heli-vaced out by Lt.Col. Madan KC of the Royal Nepalese Army (See Chapter 21 for more details).

Even though both Weathers and Gau lost limbs to frostbite, it should be noted that they were the lucky ones—15 less fortunate climbers lost their lives on Everest that year.

Military C-SAR Teams—Military combat search and rescue teams are trained to operate behind enemy lines and to rescue downed pilots. They are usually helicopter mobile units but can also be inserted by parachute, boat or on foot when necessary.

Every branch of the military maintains some form of C-SAR capability but it is the Navy, Air Force and Marines that have the most air assets to protect. The best known of these teams is the USAF PJs out of Kirtland, New Mexico, but Navy Seals also have a C-SAR unit out of Fallon, Nevada and the Marine Corps TRAP teams came into the spotlight with their rescue of F-16 pilot Capt. Scott O'Grady after 6 days of E&E in Serb held Bosnia.

Military para-rescue teams like the Air Force and Air National Guard PJs can also be deployed in a civilian support role and one advantage that some of their helicopters have over civilian models, apart from capacity, is a mid-air refueling capability.

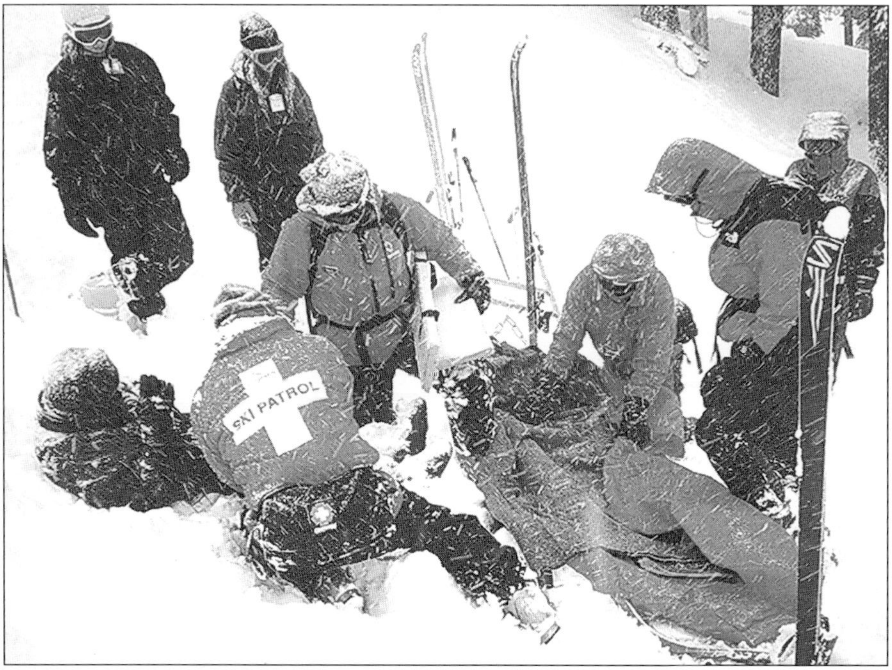

Ski patrollers stabilize an injured skier. Note the Conterra belt pack

Conterra Airway management pack being used by a moutain rescue team

18

SAR TEAM SELECTION

Once the mission has been identified, a feasibility study has been completed and the funds are on line, the selection and training of personnel can begin. To develop a long term, self-propagating program, especially if beginning from scratch, the first order of business is to select, evaluate and then train the in-house instructional staff. These should be selected carefully since the standard of this core group of instructors (initially some may have to be contract personnel) will set the tone for the team and all future training.

This period can also be used to evaluate mountaineering and rescue equipment and identify any deficiencies in the training program. A significant part of the instructors' final testing will be to have them develop and formalize the curriculum and training aids for the first batch of students to follow.

In the initial phase of program development, the instructional staff will have to be recruited from the general population with special attention given to those with prior experience and in excellent physical condition. However, once the program is up and running, future instructor candidates can be selected from the best of the students as they graduate training. But apart from getting competent instructors on line as soon as possible, it is also necessary to develop guidelines and standards for the selection and training of the team as a whole.

The foundation of any successful response team, tactical or rescue, is:

- Select personnel
- A selection process driven by the demands of the mission
- Comprehensive, reality based training
- High standards
- Competent leadership

If any one of these essential components is missing, the team is doomed to failure.

TEAM SELECTION

The selection of individuals for a mountain rescue team begins with the formation of a Selection Committee and identification of the desirable attributes of a candidate along with the standards to be used in the selection process.

The selection committee should be made up of senior personnel with the knowledge and experience to evaluate potential candidates. If an agency does not currently have experienced mountain rescue personnel, it may be necessary to bring in experts from other rescue teams or qualified mountain guides and instructors to consult to the command staff, develop the selection criteria and assist in the selection process.

ALPINE OPERATIONS

Initial selection should be based on:

- Prior Experience in mountaineering and rescue work
- Physical Fitness
- Psychological Suitability
- Self Discipline
- Commitment to team training and operations
- Compatibility with other members of the team

Obviously serious consideration should be given to the candidates with the most experience in mountaineering and mountain rescue work, because when it comes right down to it on the mountain, it is only experience and competence that will get the job done, not rank and seniority. In fact, one of the biggest mistakes an agency can make is to burden a team with senior personnel when they do not have the mission training, field experience or fitness to run an operation or lead a team.

Fortunately, the demanding environment found in the mountains is a formidable equalizer in itself and does an excellent job of identifying those who just talk a good story and those who can actually do the work. Those who have the skills and fitness to operate under extreme conditions with a lot of exposure will quickly become the natural leaders on any operation.

Whether it is mountain rescue or tactical operations, any individual aspiring to be an alpine operator must be carefully evaluated for fitness and endurance and then tested for aptitude and suitability for the exposure inherent in mountaineering. The larger the pool of candidates that the team has to draw from, the higher the probability that there will be individuals with prior mountain experience and aptitude, or the levels of fitness and commitment suited to mountain training.

SIZE & STRUCTURE

The general rule in special operations is that "a small team is a good team." The reasoning being that it is better to have a few highly trained, well disciplined, competent operators with proven experience under pressure, than a larger team that may include some "weak links" and less reliable individuals. Smaller teams are also easier to organize, equip, train, mobilize and command in the field. They are quicker to deploy, logistically easier to support, and operate with less confusion because there are fewer layers in the chain of command and communication.

The larger a team one tries to build, and this lesson was learned during the rapid expansion of US special forces during the Vietnam era, the more one must lower the standard to meet the required quotas. The Seal philosophy in BUD/S training is right on the money in that they believe that it is better to fail an entire class than allow one weak operator into the mix.

So the size of a team should be dictated by the minimum number of rescue specialists required to safely achieve the mission, but still allowing for adequate rest and leave time for all concerned. If one were to consider carrying an injured victim in a litter as a common rescue related task, then four would have to be the minimum for most operations, with six being more realistic.

But in the real world of compromise and bureaucracy, the size and structure of a mountain rescue team is governed by many factors not the least of which is available budget. Other factors are:

- Geographic location & Size of Area of Operation (AO)
- Type of terrain and accessibility
- Volume of public users and ease of access to recreation areas
- Type of activity: Climbing, alpine skiing, backcountry skiing, hiking, etc.
- Past history of rescue related activity
- Anticipated rescue activity based on increased land use
- Minimum number of personnel required to safely effect a rescue
- Existing standard of local volunteer resources
- Availability of assistance from neighboring teams or agencies
- Demand for rescue services by the local population
- Federal or State funding available for emergency services

Areas with heavy public usage, easy access and the potential for accidents require larger more active teams, but areas that have less rescue activity may be able to operate with no more than a four- or five-man rescue element.

In many mountain areas with significant rescue activity, local law enforcement may have a small core team of regular deputies trained in mountain rescue, with a large number of trained reserves and civilian volunteers on call. This adds a level of flexibility with an expandable capability without over burdening the budget.

In Los Angeles County for example, the LA Sheriff's Emergency Services Detail (ESD) supported by Aero Bureau's Air 5 rescue helicopter forms the core unit for all mountain rescues. But as soon as the first ESD team gets on scene and assesses the size and scope of an operation, they are able to mobilize any of eight highly trained reserve mountain rescue teams strategically located near the mountainous regions of the county. This gives them a total manpower resource of over 130 rescue personnel.

In other areas, mountain search and rescue is a joint operation involving Sheriff's Deputies, Park Rangers, Ski Patrollers, and various civilian volunteer teams that supply communications, four wheel drive, snow cat, helicopter and mountain guiding assets.

On an international level many National Police forces incorporate specially trained mountain rescue teams who also have collateral law enforcement duty securing their country's borders. In Argentina this would be the Gendarmeria National and in France the Peleton de Gendarmerie de Haute-Montagne. Most of these teams are relatively small and operate efficiently with fewer than 20 or 30 men.

Other considerations for group size may be the length of the climbing, skiing or recreational season. Also whether it is summer or winter rescue activity or both. In winter for example, much of the activity may be covered by seasonal ski patrollers, but come summer the rescue team may be responding to more climbing and mountaineering related rescues. Yosemite, the eastern Sierras and the Colorado Rockies are all good examples of areas that get heavy climbing activity in the summer months.

From an operational stand-point, it would need to be determined if the team is to be on call 24/7—24 hours a day, 7 days a week. This may require two or even three shifts to ensure coverage. And logistically, the type and capacity of the rescue helicopter may even dictate the size of the group. Smaller helicopters can only handle one or two rescuers along with a victim, while the larger birds may be able to handle a full complement of two pilots, two or three rescue specialists, a crew chief and a flight surgeon.

(Lonsdale)

19

BASIC MOUNTAIN RESCUE TRAINING

Learn it right and you'll do it right the rest of your life.
Learn it wrong and spend the rest of your life trying to get it right

Mountain rescue work can be a non-technical or technical endeavor depending on the terrain and type of accident history that an area may have. Since the next chapter will cover Technical Rope Rescue training, this chapter will concentrate on the less technical but equally important aspects of mountain rescue work.

This does not mean that basic mountain rescue does not require technical ability, agility and essential rope handling skills. Since most mountain travel carries with it a certain element of exposure, all mountain rescue personnel should be well versed in the basic principles of mountaineering, rope management, rappelling, belaying and moving as a rope team over steep terrain. In snow and ice country they should also be proficient in self-belay, self-arrest and crampon technique.

BASIC RESCUE TRAINING

There are fundamentally five levels of training in preparing an operator for alpine rescue work:

1. General Mountaineering & Rope Management
2. Basic Mountain Search & Rescue Techniques
3. Emergency Medical Skills
4. Technical Rope Rescue
5. Mission Specific Training

The first is outlined in some detail in Chapters 7 thru 13 and covers subjects that all climbers should be versed in such as wilderness living, mountaineering, glacier travel, avalanche awareness and basic survival skills.

Basic mountain search and rescue training is the next level and the foundation for all future training and operations. Concurrently, everyone should have training in basic emergency medicine, which can range from basic first aid and CPR up through EMT and full Paramedic certifications. However, even for groups that have doctors attached to their teams, the minimum for all professionals working in remote locations should be Wilderness First Responder (WFR) or similar certification. Tactical teams who have the luxury of attached fully certified EMTs, should ensure that those individuals up-grade their certification to W-EMT.

Technical rope rescue training gives the team the ability to safely access and rescue victims trapped on cliffs and ledges or in the bottom of gorges and canyons—covered in the next chapter.

Mission specific training covers such advanced specialties as cave and mine rescue techniques or helicopter and high altitude rescue operations.

Training for mountain rescue teams can vary in

both type and duration depending on the mission and structure of the team. In general, because of the diversity of subjects to be covered, training is usually structured in a series of programs or courses that may take up to a year to complete. However, there is also basic initial training that should be covered in one intense workshop lasting ten to fourteen days.

This initial training should encompass all the skills essential for basic mountain operations. Many of these have been covered in earlier chapters and should be reviewed as part of any program.

Basic Mountain Search & Rescue Skills:

- Mountain Travel, Living & Basic Survival
- Land Navigation, map reading and GPS
- Basic Rope Management & Rappelling
- The Incident Command System
- Risk Evaluation
- Radio Procedure
- Ground Search Procedure & Basic Tracking
- Emergency Medicine/Wilderness First Responder
- Packaging & Evacuation
- Basic Helicopter SAR Operations

After trainees have demonstrated common sense, initiative, a good grasp of individual skills such as rope management and basic rope rescue, there must begin a period of training where the emphasis is placed on teamwork. This should be designed to develop operational procedures and hone the group's ability to operate as a team. This training period is also used to further evaluate the individual's suitability for the team and to weed out the weak, the careless and the potentially dangerous.

The final part of any training program should be test and validation exercises based on reality and in the anticipated area of operation. At the end of all this, and it may take several weeks or several months depending on the standard and prior experience of the individuals coming into the program, will be the final certification of team members.

The training for Military C-SAR teams goes well beyond basic mountain rescue since they must also be trained to fight their way in and out of hostile ter-

Mountain rope rescue training (Charlie Shimanski)

ritory, even though anyone entering a C-SAR program should have already demonstrated proficiency in weapons, field-craft and small unit tactics.
To deal with down pilots trapped in military aircraft, they must also be able to:

- Identify armed weapons systems and live ordnance on the crashed aircraft
- Disarm ejector seats and canopy release systems
- Extricate the pilot and crew from the wreckage
- Destroy the aircraft and sensitive nav/coms gear with demolitions

They may also have to go into escape and evasion (E&E) mode if compromised by the enemy during the rescue, or worst case scenario, their own rescue helicopter is shot down or disabled.

WILDERNESS FIRST RESPONDER

The Wilderness First Responder (WFR) certification is the basic requirement for all professionals working in the mountains and highly recommended for recreational users as well. The program includes the universally accepted core curriculum of American Red Cross Emergency Response, CPR for the Professional Rescuer, and emergency wilderness skills in accordance with the guidelines of the nationally recognized Wilderness Medical Society.

The WFR program is usually a 10-day course that is best taught in an environment where realistic scenarios can be enacted and dealt with. The course should include but is not limited to:

- CPR & Airway Management
- Oxygen Administration
- Basic Human Anatomy & Physiology
- Patient Assessment
- Rescue Leadership & Incident Command
- Evaluation of life threatening injuries: Shock, Anaphylaxis, Wound Management
- Traumatic Injuries: Head, Spine, Chest, Abdomen, Pelvis, Legs and Arms
- Improvised Bandaging & Splinting
- Fractures & Dislocations
- Sprains & Strains
- Poisons, Bites & Stings
- Medical Emergencies: Asthma, Seizures, Diabetes, Heart, Lung & Abdominal
- High Altitude Illness
- Heat & Cold related Illnesses & Injuries
- Water Hazards & safety
- Improvised Extrication & Litter Evacuation
- Prolonged Patient Care

Other topics may be added or above topics expanded to make the program more applicable to local conditions and accident profiles.

Even if an individual has significant EMT type training, a wilderness medical course should still be a requirement. In an urban environment, everything that a medic needs is usually right at his fingertips, for example high tech ambulances with all the necessary supplies. So in this setting everything is measured in minutes.

However, in a wilderness setting you are less likely to have all those goodies with you since everything must be packed in and operations are measured in days not minutes. As such, wilderness medicine courses are based on the concept of improvisation and using whatever you can carry on your back. This is a totally different focus to urban EMT training.

Avalanche probe training (Charlie Shimanski)

Basic rappel training. Note the use of a safety prusik loop above the rappel 8

EQUIPMENT SELECTION

Although the mountain rescue specialist will require a complete inventory of both summer and winter mountaineering clothing and gear (See Chapters 3, 4 & 5), he must also be able to select only those items specifically needed for the mission at hand. As with conventional mountaineering, a lighter load means more comfort, faster movement and less fatigue. However, there are also additional rescue related items that must be added to a standard mountaineering load whether on a one day or multi day operation.

When procuring equipment for a team, a formal selection process should be developed to thoroughly test and evaluate everything that the team will be using. This is not a difficult task since most mountain professionals already know those items best suited to both mountaineering and rescue work. The selection process is also a valuable, educational part of rescue work that gives the team an opportunity to get first-hand experience with a variety of available technology. This is made easier by the fact that most manufacturers will supply items for evaluation knowing that if their product meets the approval of the selection committee, it could mean a large order or procurement contract.

In developing a formal selection process, the following should be considered:

- Research—Make a study of the equipment suited to mountain rescue work and particularly look at those items that are being used by other teams with the same mission.
- Individual Requirements—List those items essential for each individual on the team.
- Team Requirements—List the larger and more expensive pieces of rescue and support equipment required by the team. This could range from longer ropes to suitable vehicles.
- Test & Evaluation—Evaluate each piece of equipment under realistic conditions to assess its strength, weight, suitability and reliability for local applications.
- Standardization—Try to standardize equipment as much as possible
- Communications—Evaluate lightweight portable radios that have the range and rugged reliability suited to local conditions. They should also have frequencies compatible to other rescue assets in the region.
- Budget & Planning—Make a complete breakdown of the procurement costs of all essential equipment, its availability and delivery time.
- Quartermaster—Every equipment intensive activity requires a quartermaster to handle procurement, accept delivers and track the issuance of equipment to the team members.

Over all, the equipment used in rescue work is more ruggedly constructed than lightweight mountaineering equipment, but although considered more heavy-duty, it should still be light enough so as not to overly burden the team.

The key difference between mountaineering equipment and rescue equipment is the anticipated loading. For the mountaineer, equipment is designed to support the individual's weight (on average 200-250 pounds) with a considerable safety margin.

For the rescuer, equipment must be able to support not only his weight but also that of a victim and another rescuer. This brings the expected loading up to 600-700 pounds.

This increased strength must be built into the entire system from anchors and slings, through ropes, prusiks and carabiners to descenders and harnesses.

BASIC EQUIPMENT LIST

The required equipment for mountain rescue team members should be standardized so that on any given operation, the team leader knows exactly what each man can contribute to a rescue system.

Ground search personnel and non-technical rescuers will require little more than appropriate clothing and what would be routinely carried in the mountains. However, ropes, helmets, harnesses, 4 locking carabiners, a rappel device, 2 prusik loops and a first aid kit are always recommended.

In avalanche country, rescuers should also be

equipped with avalanche beacons, shovels, probes and snowshoes; and if deploying onto glaciers then crampons and ice axes are essential. If steep ice is to be negotiated then a second ice tool is also recommended.

For teams that have a high angle rope rescue capability, a more extensive list of equipment is required. In addition to clothing appropriate for the conditions, the following is the minimum personal equipment list for one of California's premier mountain rescue teams.

1 Approved Rescue Harness
1 Helmet, Rescue
1 Carabiner, Auto-lock
10 Carabiners, Standard Locking D
2 Carabiners, Large Locking D
2 Ascenders, Gibbs
1 Descender, Rescue 8 with Ears
1 Pulley, 2″ Prusik Minding
1 Pulley, 2″
1 Personal Rope, Min 3/8″ x 100′
2 Utility Cord, 8mm x 70″
2 Utility Cord, 8mm x 56″

1 Webbing, 1″ x 20′ Orange
1 Webbing, 1″ x 16′ Blue
1 Webbing, 1″ x 25′ Red
1 Webbing, 1″ x 12′ Yellow
1 Rescue Runner
1 Multi-loop Strap, Large

1 Gloves, Leather
1 Gloves, Winter with liners
1 Blanket, All-weather
1 Compass
1 Goggles, Winter
1 Goggles, Clear
2 Whistles, Safety
1 Maps, Set
1 Light, Handheld
1 Pen Light
1 Headlamp with Batteries for 12 hours
1 Light, Strobe

CMC Personal Rescuer Kit

Again, weight is always a consideration so the rescuer must select from the above list only those items that would be needed for a particular operation, knowing that other team members will have duplicate equipment in their packs.

To better organize their rescue equipment, many rescuers will keep the essential items that will be taken on every operation in one backpack, and all the additional equipment in a second bag or pack. Not knowing if an operation is going to be one day or several, it is advisable to have a rescue pack (3,000+ cu in.) loaded with rescue and medical essentials and a full sized multi-day pack (5,000+ cu in.) loaded with spare clothes and camping gear.

TEAM EQUIPMENT

Depending on the nature of the operation, and in addition to everything needed to equip an Incident Command Post, the following list of team equipment will be needed to supplement the individual equipment listed above.

- ❑ Radios & Batteries
- ❑ GPSs
- ❑ Stretchers with Stretcher Harnesses
- ❑ Victim Packaging materials
- ❑ Victim Warming materials
- ❑ Medical / Trauma Kits
- ❑ Backboard & C collars
- ❑ Oxygen Bottles
- ❑ 200′ Rescue Ropes

ALPINE OPERATIONS

- ❑ 300´ Rescue Ropes
- ❑ Anchor Straps
- ❑ Additional Webbing
- ❑ Additional Large Carabiners
- ❑ Additional 8mm Accessory Cord
- ❑ Tents & Tent Flies
- ❑ Spare Stoves, Pots & Fuel Bottles

The Denali 14,000-foot camp viewed from the Head Wall (Steve Dundorf 1998)

HIGH ALTITUDE MOUNTAIN RESCUE

On Denali, the mountaineering rescue ranger program that has been developed over the years is considered an excellent working model for other teams with a similar task. The National Park Service's (NPS) high altitude mountain rescue capability is considered on the cutting edge of mountain rescue work, but this capability did not come easily.

Initially, the responsibility, risk and expense of forming and maintaining a high altitude rescue capability were major management issues for the park. From the 1940s to the 1960s the NPS philosophy was to screen climbers and try to stay out of the rescue business. Not willing to risk the lives of their own personnel unnecessarily, the rangers interviewed climbing expeditions and inspected clothing and equipment with the hope of preventing teams going ahead that did not have the necessary levels of experience and preparedness to handle the extremes found in Alaska. They also required expeditions to arrange a back-up expedition team to come to their rescue—a task often handled by a volunteer group out of Anchorage called the Alaska Mountain Rescue Group.

However, by the late sixties the expenses incurred rescuing climbers had become quite significant, military assistance was not always timely, and realizing the humanitarian need to protect the park's visitors, the NPS was forced into the rescue business.

By the mid-1970s the Park Service seasonal staff included experienced climbing rangers hired to help execute rescue operations. To further improve their capabilities, in 1991 the NPS began leasing a high altitude helicopter which added greatly to the efficiency and safety of the rescue teams.

For additional assistance with rescues, NPS rangers can call upon the US Army Northern Warfare Unit, the Air National Guard, various Talkeetna air services, and other experienced guides and climbers on the mountain.

Today, there is a ranger camp with rescue and medical support capability at the Basin at 14,200´, the area where most teams acclimatize and regroup for two or three days before moving to the 16,200' Ridge Camp and 17,200´ High Camp.

From the Basin, the staff is able to monitor the radio traffic of expedition teams on the mountain as rescue teams composed of NPS mountaineering rangers and qualified volunteers patrol the park and respond to calls for assistance.

Even with this capability, the NPS and rangers do not guarantee rescue. Expeditions are encouraged to be self-reliant and to have sufficient supplies to weather out a storm at high camp, and sufficient personnel and resources to effect a self-rescue. The inclement weather, high winds and white-out conditions can make it impossible for mountaineering rangers to access a team in trouble and the high altitude rescue helicopter can only operate when conditions permit.

Conterra's ALS Extreme Pack is designed specifically for paramedics, EMTs and mountain rescue personnel operation in harsh environments. Its narrow cut and low volume hip belt also make it well suited to use on snowmobiles or skis.

The Conterra Crag Alpine pack is designed for serious ski professionals and alpine rescuers. With a 3,300 cubic inch volume and a weight of 2.4 pounds it is an excellent climbing and rescue pack.

Conterra Patrol III belt pack used by ski patrollers

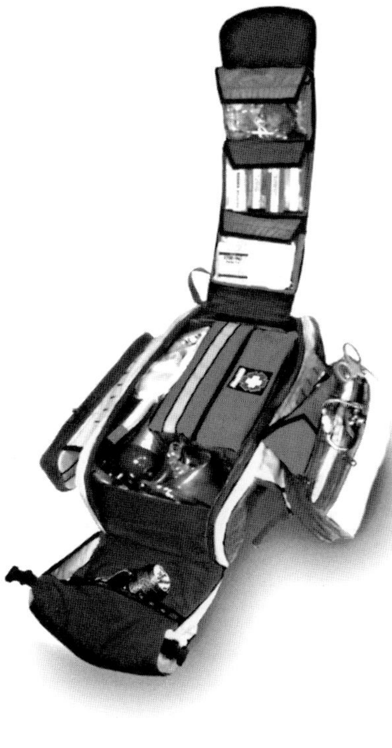

(Lonsdale)

20

TECHNICAL ROPE RESCUE

Search and rescue operations are a team effort that draws on the skills of many talented and dedicated individuals, all coming together with one purpose—to first locate and then rescue the victim. Although the rope rescue team may be only one small part of the overall rescue effort, they are the part that carries the greatest risk so requires the most training.

Once the victim(s) has been located and found to be in a position that is not easily accessed by the ground search team, the technical rope rescue team is brought in to perform the final stages of the rescue.

The operational skills required to execute a high-angle rope rescue are often dictated by the terrain, the season and the nature of the rescue. Although there are some differences between working on a vertical rock face versus a steep icy chute or couloir, most rescues come down to one of two options. The victim will be either raised to the top or lowered to the bottom in preparation for extraction. This makes the safe use of raising and lowering systems the foundation for many rope rescue programs.

VERTICAL ACCESS

Once a victim has been located and the situation assessed, the next step in the rescue process is to get a rescuer/paramedic down to the victim to ensure that the victim does not fall any further. Where possible, the rescuer may be deployed directly from a helicopter in which case he can execute a hoist pick-off. But in situations where a helicopter is not available or there is a chance of blowing the victim off the ledge, the rescuer will have to first rappel or be lowered down from the top of the bluff, crevasse or gully.

The advantage of rappelling is that it is quick, simple and one rescuer, with no more than a rope, a couple of carabiners and a rappel device, can set up an anchor and get down to the victim with little or no additional support. But once the rescuer has reached the victim, he is not well positioned to do anything except a pick-off and continued rappel to the bottom. However, in a crevasse or deep river gorge, continuing down may not be a viable option and ascending alone with an injured climber would be all but impossible.

By working as a team and being lowered to the victim, as opposed to rappelling, the rescuer is now better positioned to first stabilize the victim's situation and render immediate medical aid before being raised back up by the support team.

In situations where the victim has sustained serious injury, a second rescuer can walk a litter down to the accident site, being careful not to dislodge rocks, and then help package the victim for extraction.

SINGLE ROPE TECHNIQUES

Although it may seem to contradict the mainstream thinking and teaching about rope rescue, there are times when single rope techniques (SRT) are called for. A rappel rescue can be a quick solution requiring a minimum of equipment and manpower, particularly if the victim is in a precarious position and in danger of falling.

The decision to use single rope rescue techniques can also be influenced by weight considerations where an alpine rescuer can only carry one rope. When working from vehicles or in large teams there is always the convenience of unlimited access to additional ropes and equipment, but this is not the case when everything must be packed in.

Another situation that may justify SRT is when rappelling or ascending long drops and overhangs and there is a possibility that a belay rope would twist around the main line. There are also situations where it is more efficient to take advantage of a single long rope to rappel a victim off of the mountain, especially when the victim is in need of medical assistance that cannot be rendered on the side of a mountain or because inclement weather is moving in.

However, there is a trade-off between safety and efficiency. If a single rope system is to be used, it is important to ensure that every piece of the system is of the highest quality and strength. The anchor must be absolutely **bombproof**; the rope should be in good condition; anchor straps and carabiners should be double checked; and edge protection placed to better protect the rope. In addition, the rappeller should self-belay himself with a prusik and use the rule of two points of contact with the rope at all times—one being the descender and the other being a prusik. In this way, even if the rappeller is knocked unconscious by falling rock, he will not fall from the rope or accelerate to the ground at an uncontrolled rate.

If it is anticipated that single rope techniques will be used by a mountain rescue team, and there is a high probability of that in the real world, then instruction and training must be given in these techniques.

MECHANICAL ADVANTAGE

If sufficient manpower is on hand to lift the victim, a simple 1:1 hauling system can be used. This means that if the victim weighs 200 pounds, the hauling team must be able to pull over 200 pounds on the hauling rope. However, if the load is a victim and two rescuers, this load could jump to over 600 pounds in which case a mechanical advantage must be created.

The most common hauling system used in rescue work is the 3:1 mechanical advantage, also known as a Z rig, that can be assembled with two pulleys, a Gibbs and a couple of prusik loops. (See diagrams). A 3:1 mechanical advantage means that if the haul team can generate 200 pounds of pull, they should be able to raise a load weighing 600 pounds (less the drag and friction generated in the system).

Anyone who has had formal crevasse rescue training will know that this system can be modified to a 6:1 mechanical advantage by adding a 2:1 C rig to the 3:1 Z rig. (See diagrams).

In crevasse rescue, these raising systems are constructed with whatever carabiners and utility cord is available, but in professional rope rescue, every team member will carry sufficient hardware, pulleys, etc, to construct an efficient raising system more suited to the loads encountered in rescue work.

Rope rescue instructor demonstrates basic hauling system using a Gibbs ascender

RAISING & LOWERING SYSTEMS

Obviously lowering a heavy load is easier than raising one and requires considerably less manpower. In fact, one rescuer can easily set up and operate a lowering system using just a Brake Bar Rack or Rescue 8, but a raising system intended to raise two rescuers and a victim takes several rescuers to operate efficiently.

Once an operation becomes a team effort, it is critical that everyone on the team understand how to build and run the raising or lowering system. Any error in rigging or operation could result in an accident with fatal results.

The first rule in rigging a rescue system is **Check and Double Check** everything throughout the operation. This is especially important when responding under the stress of rescue mode or when cold and fatigued, all conditions that contribute to the possibility of mistakes. So whether working as a Buddy Team or having a Safety Officer, there should be at least two people checking every link in the system. This is best done with the **Touch System** where the rescuer physically touches every link in the system from the first anchor all the way down to the rescuer's harness. It is not enough to just look at a carabiner, Gibbs or belay device—they must all be physically touched, tugged and twisted to ensure that the rigging is correct.

The true test of whether a raising or lowering system is rigged correctly, is what happens if everyone lets go at anytime (also called the Whistle Test). If rigged and backed-up correctly, the whole system should cease to move. This is achieved by always having back-ups, prusik loops, or ratchets (Gibbs ascenders) that will stop the system running or falling if the belayer or hauling team should let go of the ropes or hauling lines. For additional safety, anytime a rope is used to support human life, the system should be supported by a second independent belay system.

However, all of this redundancy requires additional rope and hardware, making the system more complex so it becomes even more critical that everyone on the rope rescue team be proficient with both the set-up and operation of these systems.

High angle rescue operations can expose several members of the team to risk if sufficient safety is not built into the system

Anchor tending (CMC Rescue)

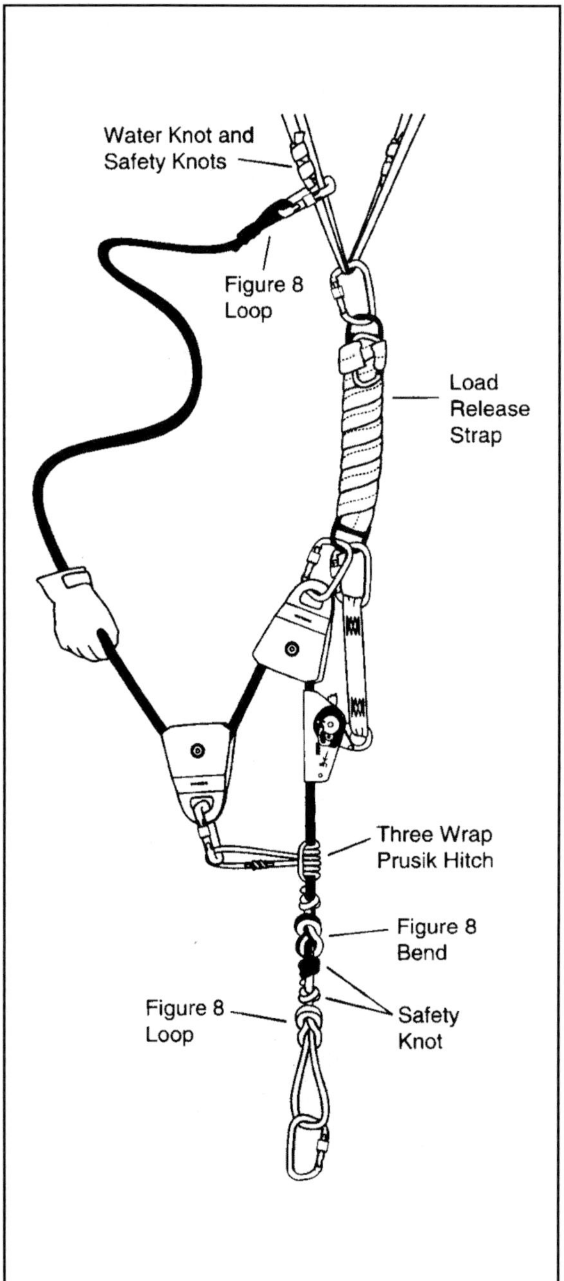

Where knots are used in the rescue system (CMC Rescue)

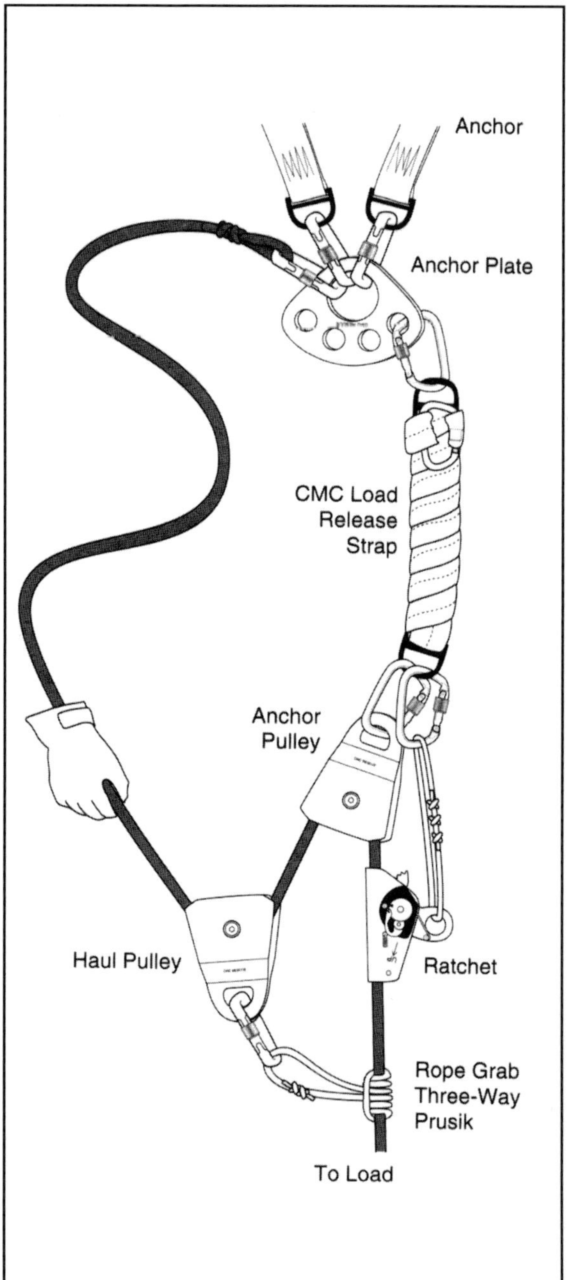

Parts of a 3:1 mechanical advantage (CMC Rescue)

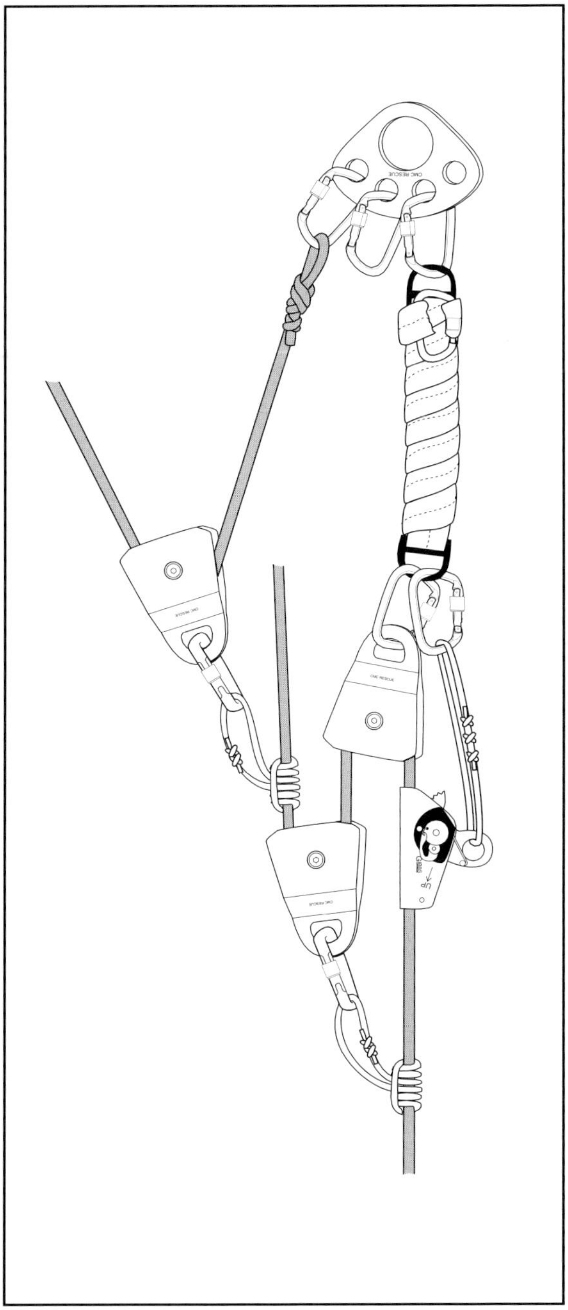

6:1 mechanical advantage made up of a Z-rig and a C-rig (CMC Rescue)

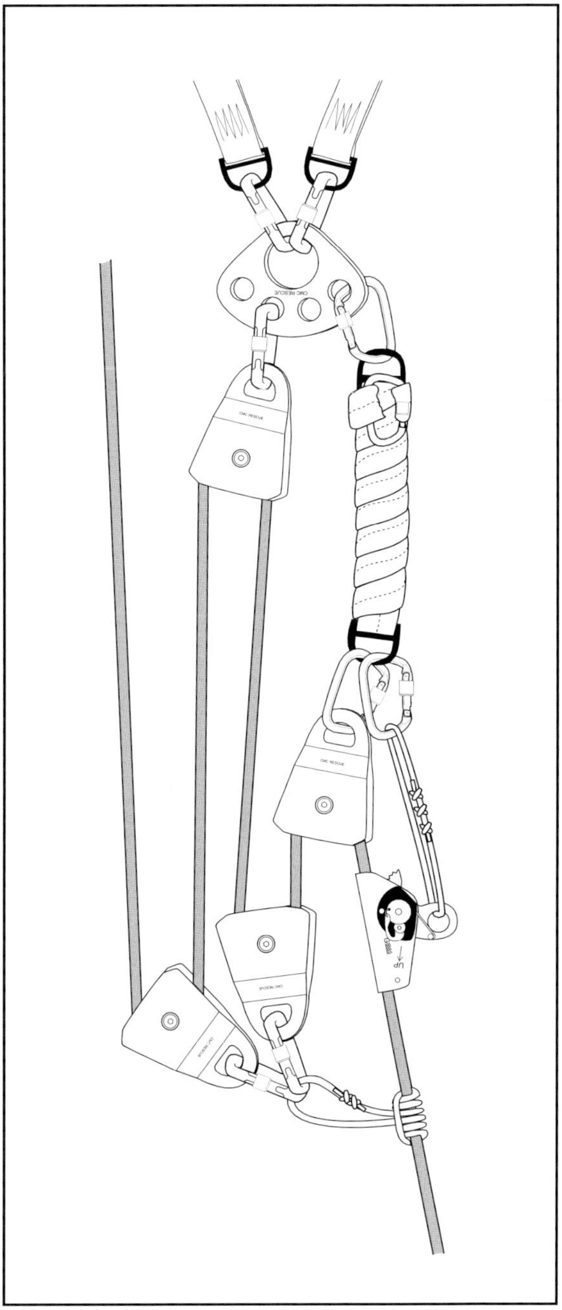

Simple 5:1 mechanical advantage (CMC Rescue)

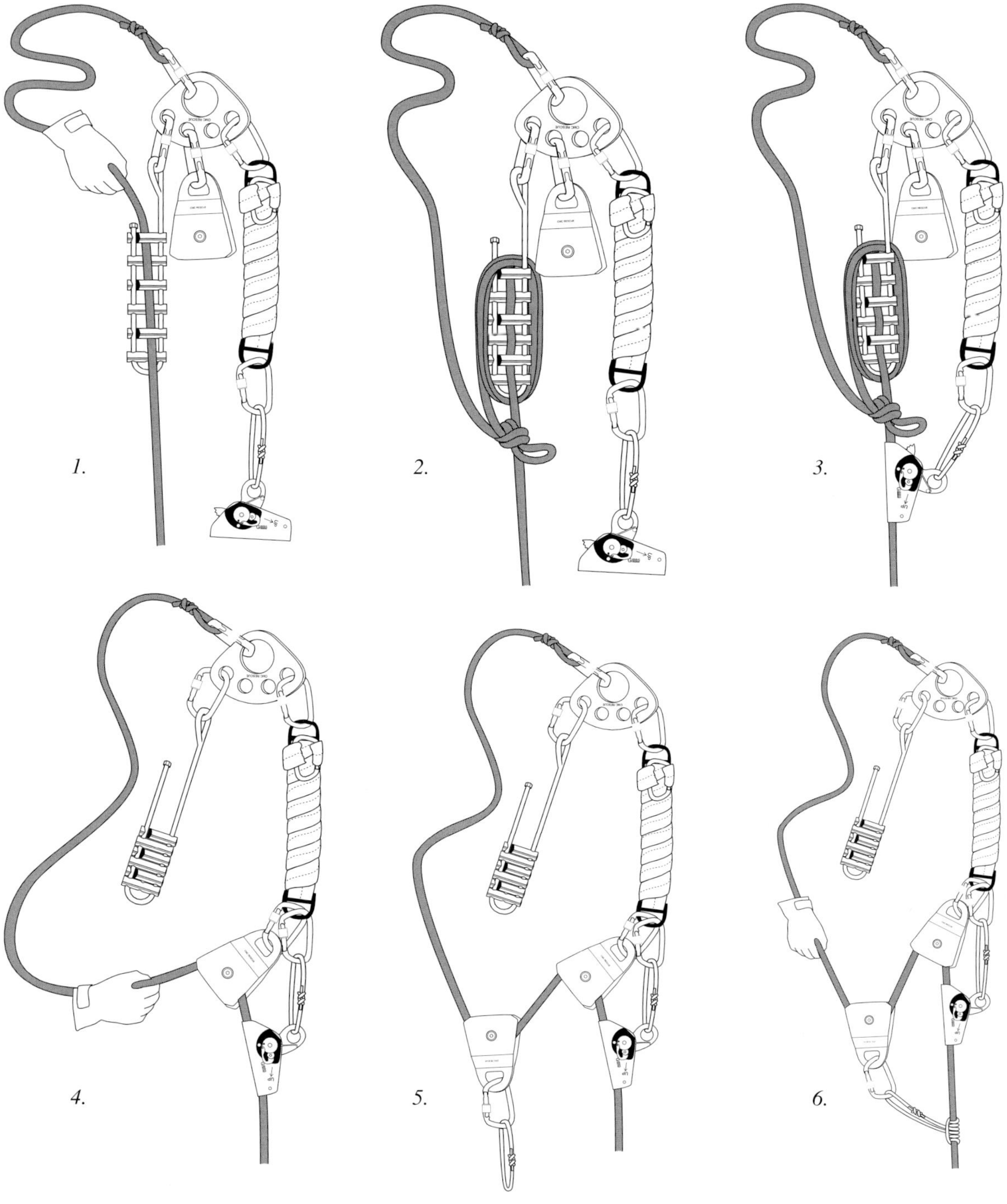

Changing a lowering system to a 3:1 raising system (CMC Rescue)

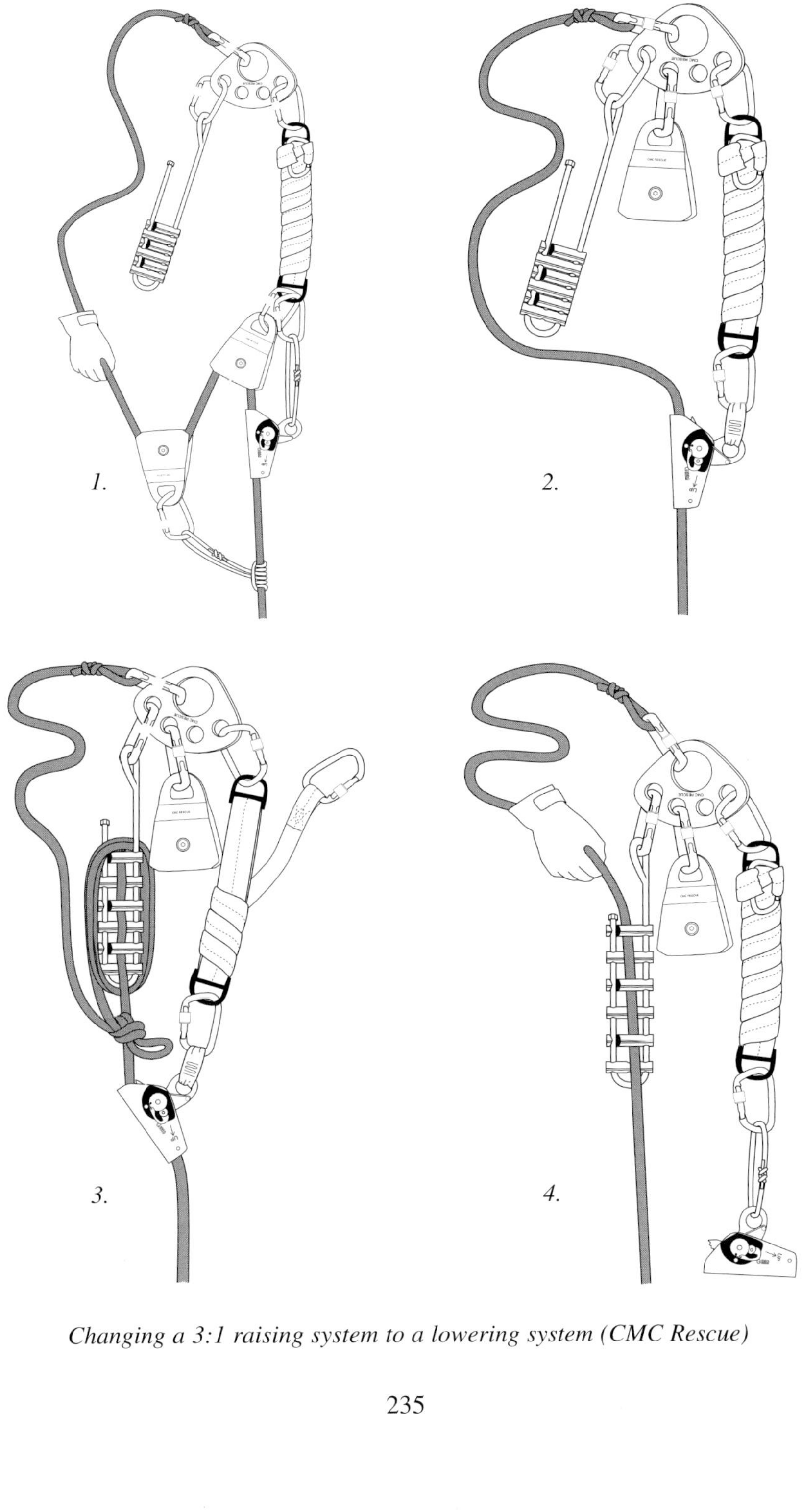

Changing a 3:1 raising system to a lowering system (CMC Rescue)

ROPE RESCUE PROGRAMS

The increased complexity of rope rescue systems requires that every team member be trained and drilled in their construction and operation. There is no room for the proverbial "weak link," so every rescuer on the team should be schooled in the same operational methodology. This is best achieved by selecting one school or instructor to standardize the training so that future trainees will receive exactly the same training as their predecessors.

Although the rigging, terminology and systems should be standardized, the format and duration of the programs may vary. Some groups prefer to train in a series of four or five day modules with multiple levels of certification, while others prefer the intensity of a single ten or twelve day package. For teams that must respond to both urban and mountain rescues in both winter and summer, it may be necessary to structure two or three programs to cover the full range of skills required.

However, whether urban or mountain, and immaterial of season, the fundamentals of technical rope rescue remain the same. The following represents the typical progression for a rope rescue program that would form the foundation for all other training. The first two days would concentrate on individual skills, with teamwork becoming more important as the program progresses.

Day 1—Classroom:

- Review of participant's mountaineering and/or rescue experience
- Thorough review of knots, ropes and rigging theory
- Introduction to rescue hardware and rescue equipment; its application, advantages, disadvantages, handling and care
- Rescue Rope selection, desirable properties, handling and care
- Practice and demonstration of proficiency in the essential knots used in rescue work—especially the Double Loop Figure 8, Water Knot and Prusik Hitch.
- Anchors & Anchor Systems suited to rope rescue

Day 1—Training Tower:

- Selecting, Rigging & Backingñup Anchors
- Basic Rappelling with Self-belay
- Use of the Brake Bar Rack
- Belaying & Lowering Techniques
- Pick-offs & Single Victim Rescues

Note: The first day on the tower is also intended to show the students their limitations: their comfort level with height; their physical strength limitations; and their ability to function under stress.

Day 2—Training Area

- Mechanical Advantage in Raising systems
- Construction & Use of 2:1 (C rig) and 3:1 (Z rig) simple raising systems
- Construction of 4:1 and 5:1 systems
- Compounding a 3:1 with a 2:1 to make a 6:1
- Changing from a lowering system to a raising system
- Changing from a raising system to a lowering system

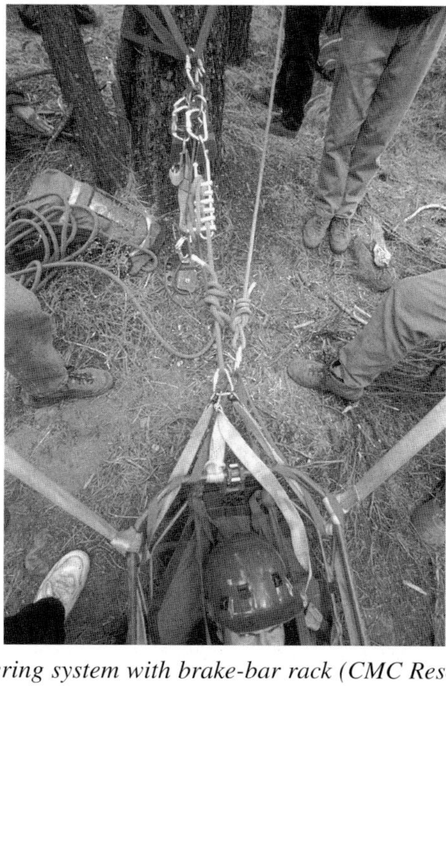

Lowering system with brake-bar rack (CMC Rescue)

Day 3—Training Area:
- Stretcher Rigging for Low Angle rescues
- Stretcher Rigging for High Angle rescues
- Victim packaging
- Raising and Lowering Systems with a Stretcher
- Knot Passing
- Stretcher Tending while Raising & Lowering

Day 4—Field Exercises:
- Natural & Artificial Anchors
- Use of Vehicles for Anchoring
- Field Communication—Radios & Whistles
- Over-the-Side Rescue Exercises with an emphasis on Teamwork
- Raising & Lowering
- Low Angle rescue
- High Angle rescue

Night 4—Field Exercises:
- Simulated Over-the-Side Accident
- Use of Lights & Communications
- Establishing Anchors & Lowering Systems
- Victim Access & Stabilization
- Victim Packaging & Raising

Low angle stretcher tending (CMC Rescue)

High angle stretcher tending with an A frame (CMC Rescue)

Day 5—Classroom:

This is the beginning of the transition from conventional rope rescue to mountain rescue and the need for lighter equipment and single rope techniques.
- Mountain Rescue Equipment
- Lightweight Ropes & Hardware
- Single Rope Techniques
- Advanced Anchoring Methods
- Load Sharing & Self Equalizing Systems
- Prusik Ascent & Rope Climbing

Day 5—Vertical Rock & Cliff Training:
- Rigging Anchors
- Rappelling & Lock-offs
- Victim Access & Pick-offs
- Lowering Systems
- Rope Climbing Systems

ALPINE OPERATIONS

Days 6 & 7—Mountain Training:
- Risk Assessment & Rescue Supervision
- Victim Access & Stabilization
- Individual Rescue Exercises simulating a climber or hiker trapped on a cliff
- Stretcher Rigging and Handling
- Raising & Lowering Systems with Victim and Two Rescuers.
- Full team, high angle, stretcher rescues on a vertical rock face.

By this time the class should feel confident in their rescue capabilities, rope handling and rigging. They should be able to select natural anchor points, set-up anchor systems, rappel a rescue medic down to secure the victim and assess injuries while the rest of the crew rig for the recovery. When ready, a stretcher tender walks the litter down the face and assisted by the medic secures the victim into the litter. The whole package is then raised and recovered.

Several similar exercises should be run to give everyone the opportunity to experience every role: victim, medic, stretcher tender, edge man and raising team.

As with all programs, Rope Rescue should finish with a final critique, written test and course evaluation. Field training exercises (FTXs) should also be scheduled to better prepare the team for more rugged terrain and under more realistic conditions, with special consideration given to moving a victim over steep snow and ice.

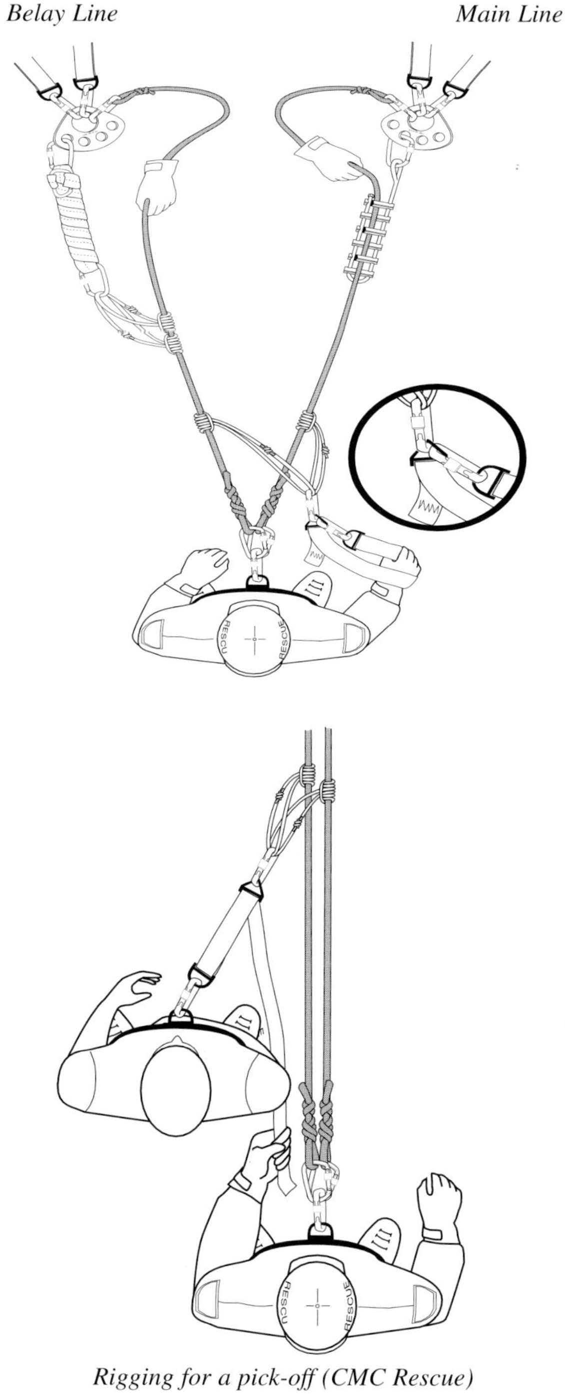

Rigging for a pick-off (CMC Rescue)

RESCUE EQUIPMENT SELECTION

Climbing ropes and hardware are designed to be lightweight and to handle the loads generated by a single climber, where as dedicated rescue equipment is usually built heavier and stronger to handle the daily use and abuse encountered with team loading. However, the heavy-duty equipment used by urban search and rescue teams (USAR), fire departments and for industrial applications is often too heavy to be carried in the field for mountain rescues —a good example of this being rope.

RESCUE ROPE—For most rescue applications, a ½″ (12.7mm) low stretch, kernmantle rescue rope is the first choice for teams that have vehicle or helicopter access to the accident site. But teams such as those governed by the Mountain Rescue Association (MRA) and National Park Service (NPS) that must carry their ropes in their packs, prefer the lighter 7⁄16″ (11.1mm) static rope. For working in rugged terrain, some mountain rescue teams even authorize ⅜″ (9.5mm) rescue rope which weighs only 4.8 pounds for a 100′ length as their individual rope system.

Similarly, many teams will utilize 300′ ropes when deploying from their vehicles, but will opt for a 200′ rope for most other rescue applications. The advantage of the 200′ rope being that it is fifty feet longer than the standard 150′ or 50-meter rope favored by climbers and mountaineers.

The following are ropes commonly used for rescue work, all with a recommended working load fifteen times less than the breaking strength.

CMC Rescue Lifeline ½″—NFPA 1983 Two-Person Rated

Maximum working load:	630 pounds
Minimum breaking strength:	9,500 pounds
Weight:	7.2 pounds per 100 feet

CMC Rescue Lifeline 7⁄16″—NFPA 1983 One-Person Rated

Maximum working load:	430 pounds
Minimum breaking strength:	6,500 pounds
Weight:	5.6 pounds per 100 feet

New England KM-III 7⁄16″ Static Kernmantle

Maximum working load:	540 pounds
Minimum breaking strength:	8,100 pounds
Weight:	5.4 pounds per 100 feet

CMC Rescue Lifeline ⅜″—NFPA 1983 One-Person Rated

Maximum working load:	330 pounds
Minimum breaking strength:	5000 pounds
Weight:	4.8 pounds per 100 feet

HARDWARE—Hardware is another area where the mountain rescue team must compromise between weight and strength. For most urban vehicle or helicopter supported operations, steel hardware and carabiners are favored for their strength and durability, but aluminum will always be the first choice of the mountain rescue teams.

Since rescue systems are always life support systems and must support significant loads, non-lock carabiners are not generally used. However, larger aluminum carabiners with at least a one-inch gate opening are favored for their ability to accommodate multiple ropes, straps or the bars on the sides of a Stokes litter. Each rescuer will normally carry 8-10 standard locking Ds and at least 2 large locking Ds.

Aluminum hardware, since it is more prone to damage from being dropped, should be inspected on a regular basis, particularly before and after each operation. If there are any signs of damage, the carabiner must be retired and destroyed. When in doubt, REPLACE IT!

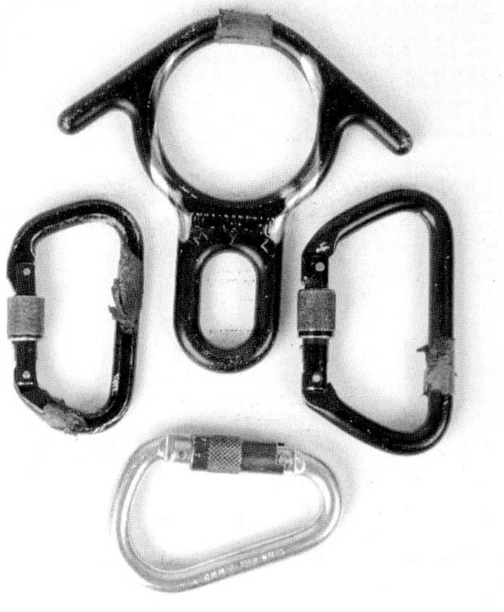

Aluminum Rescue 8, personal sized carabiner left and rescue size aluminum carabiner on right. Autolock rappel biner at bottom.

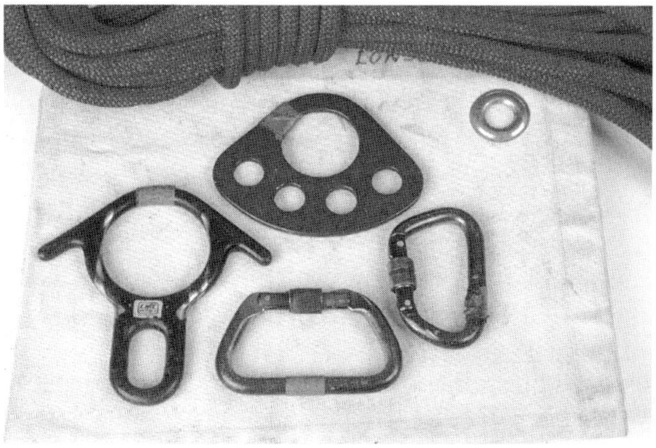

Steel Rescue 8 and steel carabiners for heavy-duty applications, CMC 7/16″ rescue rope, anchor plate, and canvas edge protection with tie-in grommet

DESCENDERS—The two most popular descenders or belay devices used in rescue work are the CMC Rescue 8 with Ears and the SMC Brake-Bar Rack. The former for individual rappelling, lowering and back-up operations and the latter for heavier loads and team rescue applications.

The Rescue 8 is a standard piece of equipment that all rescuers should be issued and can be used for all types of rappelling and rescue operations. The heavier brake-bar rack has the advantage of being able to add or subtract friction bars to match the load but is too heavy and bulky for many mountain rescue applications.

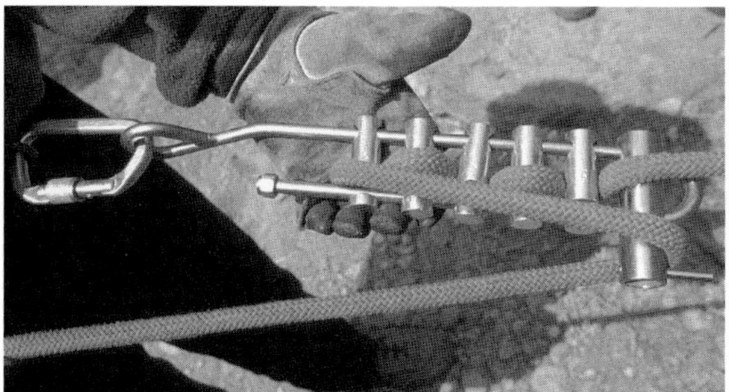

Brake-bar rack (CMC Rescue)

GIBBS ASCENDERS—The Gibbs Ascender has been in use with climbers and cavers for over 20 years and has built a reputation on simplicity and reliability. In rescue work they have become standard equipment in raising systems and are used as a mechanical prusik—however considerably more efficient.

The advantage of the Gibbs over Jumars or Petzl Ascensions is that they capture the rope in a closed system so cannot disconnect from the rope by accident. Available in aluminum or stainless shells, the Gibbs ascender features a high quality forged cam that works as a one-way ratchet to hold the rope in place as the hauling team re-positions for another pull (See diagrams). The rope is gripped in the shell by the cam which is held in place by a hardened push-button, quick release safety pin.

There are several models of the Gibbs ascender, not all of them suited to rescue loading. The light-weight #1 and #2 Gibbs are a single person climbing unit that can be either free running or spring-loaded and are factory tested to 1,000 pounds, but have a breaking strength of 2,550 pounds.

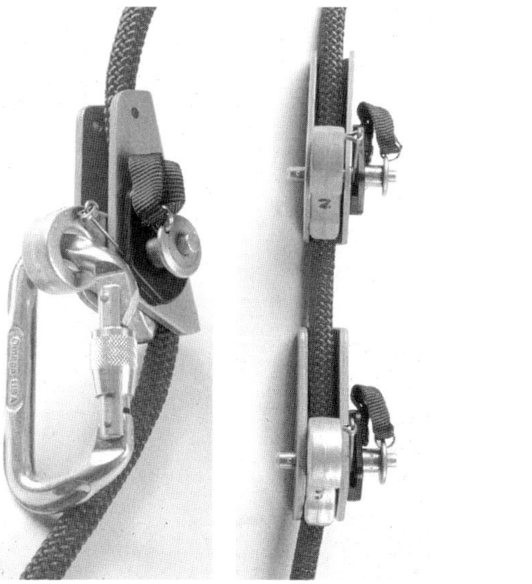

Gibbs Rescue Ascenders. It is recommended that a soft link be used between the carabiner and the Gibbs to avoid side loading the cam (see photo page 228). On right, notice the difference in wall thickness between the #2 sport model above and the heavier #3 rescue model below.*

For rescue applications and loadings, the more robust #3 ($\frac{7}{16}''$-$\frac{1}{2}''$ rope) and #4 ($\frac{5}{8}''$-$\frac{3}{4}''$ rope) are standard. Both have a thicker aluminum shell and are intended for litter raising and load hauling where extra strength is required. The #3 has a breaking strength of 5,000 pounds and the #4 has 5,650 pounds, but both will cause rope damage above 3,000 pounds.

Note: As with any mechanical device, there is always some knucklehead who can mis-use it or install it incorrectly. Proper training and practice are essential and caution must be observed at all times.

PICK-OFF STRAP—In addition to all the essential ropes and hardware, the CMC Pro-Series pick-off strap is one of the most useful items for doing a vertical evacuation of a victim. This adjustable strap allows the rescuer to rappel down to just above the victim, connect into the victims harness and then cinch up on the pick-off strap to transfer the victim's weight onto the rescuers harness. After connecting a second back-up, the victim and rescuer can then be lifted off or rappel down to safety.

Pick-off strap

PULLEYS—For crevasse rescue applications, a glacier rope team will usually carry two very small, light-weight pulleys, but for professional rescue

Prusik minding pulley and aluminum anchor plate

applications a more substantial pulley is needed. The recommended pulleys for most rope rescue applications are 2″ x ½″ CMC Rescue prusik-minding models with sturdy aluminum side plates or the Russ Anderson/SMC stainless steel variety. The prusik-minding feature is built into the squared side plates of the pulley that work to keep pushing the prusiks down the line as the load is raised.

With three pulleys a team can construct a 3:1 Z-rig with a directional change, or use the third pulley to make an additional C-rig for a 6:1 mechanical advantage.

ANCHOR PLATES—An anchor plate is a flat aluminum or stainless plate that allows a rescue team to bring four or more carabiners into the same anchor point without bunching them all together. This allows ease of access to the system and individual carabiners to be changed or adjusted while the system is under load.

EDGE PROTECTION—Edge protection is needed to protect any ropes running over a sharp or abrasive surface. These can be improvised with jackets, ground sheets and packs but the commercial manufactured units should be considered for professional applications. The heavier more expensive units like the SMC Edge Roller work extremely well in an urban environment, but compact CMC Pro-Series Edge Guards, Edge Protectors and Rescue Edge Pads are lighter and better suited to field operations.

STRETCHERS—The stainless steel Stokes Litter has been the standard for rescue work for years and is available in a Break-Away folding model for easier packing (with the CMC Litter Bac-Pac). Other useful accessories for the litter are the CMC Vertical Evac Stretcher Harness and Rescue Litter Shield to protect the victim's head.

Litter with patient shield (CMC Rescue)

Pick-off training

Patient packaging (CMC Rescue)

Litter movement over rough terrain (CMC Rescue)

Author walking a litter up a rock face (circa 1991)

SUGGESTED EQUIPMENT LIST

The following CMC Rescue equipment lists give a rope rescue team a broad range of high and low angle rescue capabilities, however many items may be unnecessarily heavy for alpine operations. Additional rope, web and carabiners may be required for more complex rescue scenarios.

Rope Rescue Team Equipment: It is recommended to have two rope kits, one for the main line and one for the belay line. The kits should be identical allowing either one to be set-up as either system. A standard length of rope is 200 feet, but longer ropes may be needed for some locations, or even shorter lengths to save on bulk and weight.

1 ½ inch (12.7) low-stretch, kernmantle rope
9 Carabiners, NFPA General rating
2 Carabiners, NFPA Personal rating
1 Anchor Plate
2 Pulleys
1 Gibbs Ascender
1 Brake Bar Rack or Figure 8 with Ears
2 Anchor Straps
3 One-Inch Tubular Web, 12-foot (3.7 m)
3 One-Inch Tubular Web, 20-foot (6.1 m)
3 8 mm Prusiks Loops, short, 56-inch (1.42 m)
2 8 mm Prusik Loops, long, 70-inch (1.78 m)
1 Multi-Loop Strap, Medium
1 Multi-Loop Strap, Large
1 Load Release Strap
1 Ultra-Pro Edge Protector or 2 Edge Rollers
1 Rope & Equipment Bag, XL

Rope Rescue Technician Personal Kit:
1 Helmet
1 Gloves
1 Eye Protection
1 Hearing Protection
1 Rescue Harness
1 Rescue Chest Harness (optional)
6 Carabiners, NFPA Personal rating
1 Headlamp
1 Chemical Lightstick
1 Multi-Loop Strap, Large
1 Figure 8 Descender or Brake Bar Rack
1 8 mm Prusiks Loops, short, 56-inch (1.42 m)
1 8 mm Prusik Loops, long, 70-inch (1.78 m)
2 Gibbs Ascenders
2 One-Inch Tubular Web, 12-foot (3.7 m)
1 Gear Bag with shoulder straps

Victim Kit:
1 ProSeries Victim Harness
2 Carabiners, NFPA Personal rating
1 Pick-Off Strap
1 Helmet
1 Gloves
1 Eye Protection
1 Hearing Protection
1 Headlamp
1 Gear Bag

Edge Protection Kit:
4 Edge Rollers
4 Ultra-Pro 2 Edge Protectors
1 Knot Pass Pulley
4 Edge Guards
4 Edge Pads, XL
6 20 ft. 6 mm cord (for anchoring edge protection)
1 Gear Bag

Stretcher Kit:
1 Stretcher
1 Litter Shield
1 Vertical Evac Stretcher Harness
1 Low Angle Stretcher Harness
1 Etrier with Tie-in
2 Gibbs Ascenders
6 Stretcher Straps
2 One-Inch Tubular Web, 12-foot (3.7 m)
4 One-Inch Tubular Web, 20-foot (6.1 m)
1 Stretcher Pack

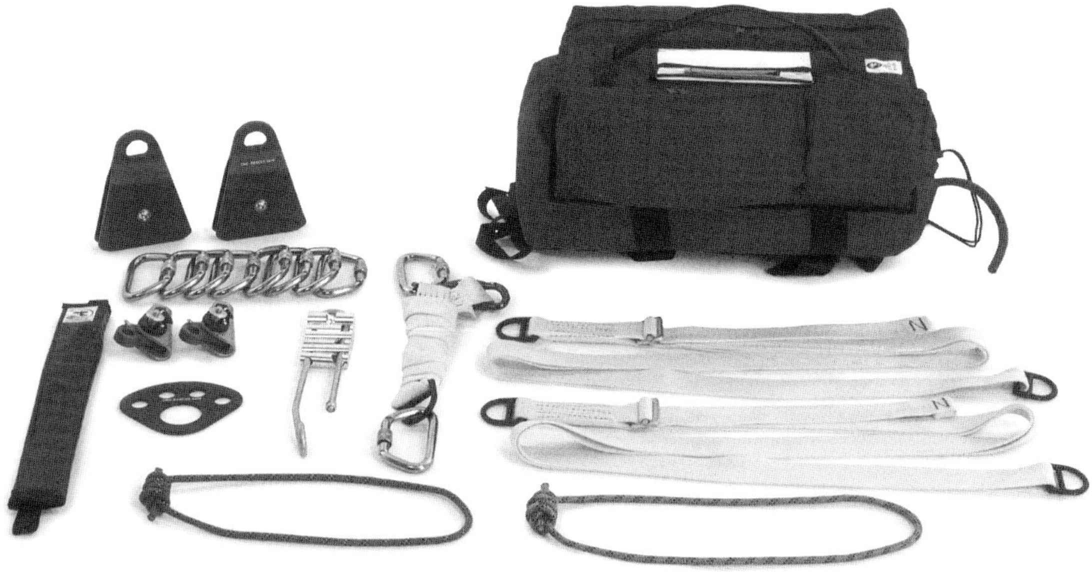

CMC Squad Rope Rescue Kit

CMC Rescue USAR litter package

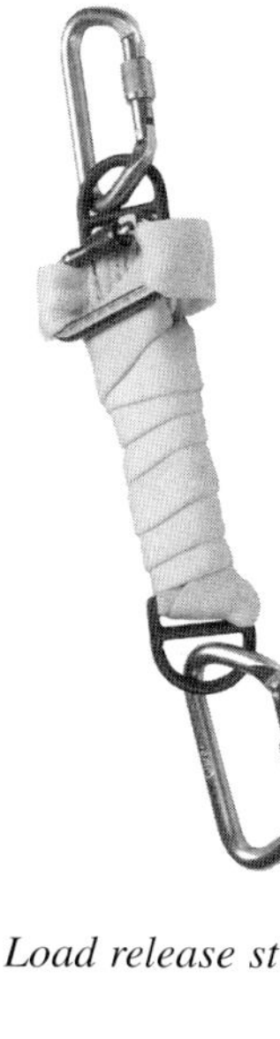

Load release strap

The load release system begins with a Munter hitch

The 8mm cord is then taken around the loop

Wrap the cord tightly around the loop

LOAD RELEASE HITCH

Secure with an overhand knot

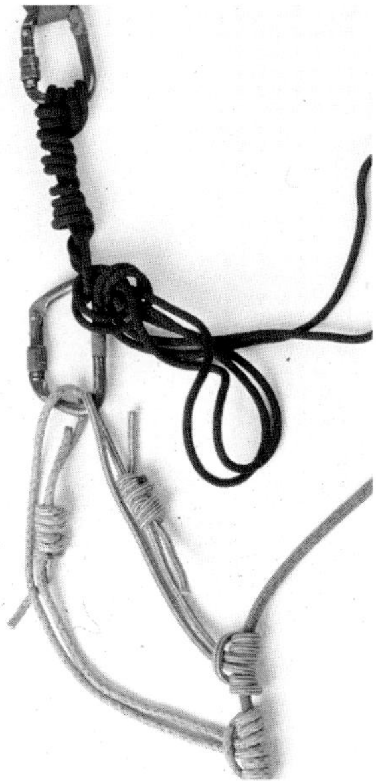

Load release hitch used in a belay system with two 8mm prusiks

COMMUNICATIONS

Effective communication between the Rescue Coordinator and the rescuers or Stretcher Team is an essential element of any rescue. This is often made more difficult by the fact that the rescue team is over the side and a hundred feet down from the raising or lowering system.

The most effective form or communication is to have the Stretcher Team Leader in direct radio communication with the Rescue Coordinator or supervisor of the lowering and raising system. TEA, the same company that makes the TASC and LASH communications systems for SWAT teams, also makes rugged low profile headsets, throat microphones and boom microphones suited to mountain rescue work.

In many situations direct voice communication is possible but voice commands should still be kept Short and Simple. An Edge Man will be necessary to keep visual contact with the rescuers and to relay necessary voice commands to the Rope Team Supervisor.

Standard Commands are:

"On Belay?"	Rescuer is asking if the belay system is set and ready?
"Belay On!"	Yes, the belay is ready
"On System?"	Is the main system ready?
"System On!"	Yes, the system is ready
"Down Rope"	Begin letting out rope(s)
"Up Rope"	Begin pulling up the rope(s)
"Faster"	Increase the rate of movement
"Slower"	Decrease the rate of movement
"Rope Free"	Rope is clear and no one is relying on it for support
"STOP!"	Stop all movement immediately
"Rope!"	I am throwing a rope down, watch out below
"Clear"	You are clear to throw the rope

Special Note: Anyone can command "STOP" at any time during a rescue and all movement must cease immediately until the problem has been remedied and/or the system checked and adjusted.

Some other useful commands used by climbers are:

"Rock!"	Beware of falling rocks or equipment / Cover your head
"Ice!"	Beware of falling ice / Take cover
"On Rappel"	I am clipped in and ready to rappel
"Take Tension"	I am tired and need you to hold me on the rope
"Ready to Lower"	I am ready to be lowered on the rope
"Lowering"	I am beginning to let out rope to lower you
"Belay Off"	I am secure and do not need a belay
"Off Belay"	I am no longer belaying you
"Off Rope"	I am off the rope and you are free to pull it up

When radios fail, batteries die or voice commands cannot be heard above the wind or ambient noise level, whistle commands can be used. The following are a set of standardized whistle commands for rope rescue that can be remembered by taking the first letter of each command to make the pneumonic SUDRH.

Stop	1 short whistle blast
Up	2 short whistle blasts
Down	3 short whistle blasts
Rope free	4 short whistle blasts
Help	Continuous blast

(Photo—Conterra)

21

HELICOPTER AND AIR OPERATIONS

The helicopter has become a workhorse of the mountain rescue community to the point that few present day operations are completed without some form of air support. The helicopter's ability to land in remote places or hoist personnel from confined areas, and then transport them directly to medical support units or hospitals, has resulted in the saving of innumerable lives that would have otherwise been lost.

In the early stages of a rescue operation, helicopters and fixed-wing aircraft afford an excellent search platform that can cover large expanses of wilderness in a short time. But the helicopter's real value is in its ability to insert rescue teams into remote areas of the mountains saving time, energy and injury, and then return to extract victims and expedite their movement to a medical facility.

For a ground team to carry a victim suffering from fractures and internal injuries several miles on a stretcher, over rough terrain may take several hours and further exacerbate the injuries. The same movement by helicopter could be done in a matter of minutes with considerably greater comfort for the victim, also removing the burden of a heavy stretcher from the ground team, and allowing them a much easier hump out.

Helicopters are such an effective tool, that Even if they are not part of a team's routine operational planning and movement, they should at least be available for on-call air support in the event of an accident or need for medical evacuation of a team member.

Unfortunately, many teams that could make good use of a helicopter do not have them since they are expensive to purchase, operate and maintain. In addition to these obvious recurring costs, time must be allocated to train the team in helicopter deployment and recovery methods, while the pilots will also need to gain experience in flying rescue operations in mountainous terrain. Each insertion and extraction technique whether it be heli-tac, basic rappelling, fast roping, skid riding, short-hauling or hoist recovery needs to be practiced many times by pilots and team members under both ideal and realistic conditions, before it is attempted on an operation. All of this essential flight time equates to increased operating costs.

TYPES & MODIFICATIONS

Not all helicopters are suited to SAR operations. Some are specifically designed for this role while others can be modified to support mountain rescue operations. Not all helicopters have the range, lift capacity and altitude capability to support alpine operations.

A mountain rescue helicopter should be powerful enough to handle high altitudes and out-of-ground effect (OGE) hovers, along with landings in tight, unimproved, improvised landing zones. It should also heave sufficient capacity and lift to handle both

ALPINE OPERATIONS

USMC CH-53 supporting mountain warfare training

LASD Sikorsky S-58T

Air Force Pararescue UH 60 Black Hawk

Sheriff's Bell / Huey

Bavarian police BK 117 in the Alps

Argentinean Lama at 14,000´

rescuer(s) and victim(s).

There are primarily three categories of helicopter used for mountain rescue, each with its own advantages and disadvantages. The first is the larger heavy-lift, military transport type helicopters like Sikorsky's CH-53 Sea Stallion, CH-53E Super Stallion, Boeing's Chinook CH-46E Sea Knight, the older HH-3E Jolly Green Giant and the HH-3F Pelican that was operated by the Coast Guard. In this category, the US Air Force Para-Rescue jumpers (PJs) make good use of the HH-53 (the extra H is for mid-air refueling capability) and the MH-53 (advanced navigational system) for their rescue needs. All are equipped with litters, recovery hoists and ample room for a full team with equipment.

The second category is medium lift helicopters such as the venerable Huey UH-1H (single engine) and UH-1N (twin engine); the newer UH-60A, UH-60L Black Hawk and the US Coast Guard Dolphin HH-65A. In the civilian market there are the Bell 412, 212, and older 205 (single engine Huey); Aerospatiale's AS 365 N3 Dauphin, and one of the leading European SAR helicopters, the twin-engine MBB BK 117.

The third category is smaller helicopters such at the MD 500E popular with US law enforcement; the Bell 206 Jet Ranger and 206L-4 Long Ranger; the MBB BO 105; the high altitude French SA315 B LAMA used by the NPS rescue rangers on Denali; the Alouette III; the Aerospatiale Squirrel AS 350 B3 that is well suited to extreme conditions, and it's twin-engine brother the AS 355N.

Large transport helicopters have the advantage of significant power and lift capacity; extended operating range and in-flight refueling; advanced navigational packages; a large stable work platform with ample space for a full rescue team; hoist recovery and the ability to administer medical care to a victim during the flight to the hospital.

The disadvantages of large helicopters are that they are more expensive to operate; burn more fuel; are more complex to fly; require larger LZs; have a very powerful downdraft and are not always available. The military also places restrictions on the maximum altitude that pilots can take these ships—usually 10,000′ for peace-time operations.

The advantages of small helicopters are that they are cheaper to purchase, maintain and operate; are easier to fly and maneuverable in tight areas; and are more readably available.

The disadvantages are that they have limited range and power; virtually no interior space for a victim; have no hoist; require external sling loading or skid mounted stretchers; and the victim cannot be monitored in flight.

So this brings us to medium sized helicopters that are best suited to all around rescue applications. Although not cheap, they offer an agency the ability to insert rescue teams and extract victims in relative comfort and safety. Many have two engines with sufficient power and range for most rescue applications and can be equipped with a hoist. They are also quite agile in the mountains and do not require a huge LZ.

DESIRABLE OPTIONS & ACCESSORIES

There are many factory options and after sales additions that make a helicopter a more valuable rescue tool:

- Advanced avionics & navigational equipment
- Sliding or removable doors
- Hard points for Rappelling or Fast-roping
- Recovery Hoist
- FLIR—Forward Looking Infra Red
- NVDs—Night Vision Devices
- Night-Sun remote external lighting system
- WSPS—Wire Strike Protection System (wire cutter)
- Internal & External Litter mounts
- Crew Chief's belt and retaining harness
- Additional external hard points for skid deployment
- Radio systems compatible with ground units

SAFETY & LIMITATIONS

Helicopters, by nature and reputation, are very sensitive to overloading, weight distribution, wind, weather conditions, temperature, and altitude changes. It is only through complex engineering, skilled piloting and the grace of God that they succeed in flying at all. Even though ships like the venerable Huey have proven themselves under less than ideal conditions, helicopters are in reality not rugged pieces of machinery that will sustain much abuse and operate only within a narrow set of limitations.

The ultimate responsibility for the loading, unloading, flight operations and safety in a helicopter rests with the pilot. Since only the pilot understands the flight characteristics, loading limits and performance limitations of his aircraft it is essential that there be constant communication between him and the alpine rescue team during training and operations.

All flights should begin with a briefing for all participants in the rescue. No matter what the team requests or needs, it is the pilot who will have the final say as to whether the requests are reasonable, possible or safe. With time and training a solid, professional relationship will develop between the pilot, crew chief and the rescue team.

The pilot may delegate many of the loading and unloading duties to a Crew Chief or Flight Engineer. This will include the boarding and deployment of rescue personnel and their equipment. Under these circumstances the Crew Chief's word is final since he is in constant radio contact with the pilot by means of a head-set and throat or boom microphone.

In situations where the helicopter is too small to accommodate a Crew Chief, then the crew chief responsibilities may be given to a member of the rescue team. This should be a member with considerable experience in helicopter operations and intimate familiarity with the dynamics involved in deployment and recovery procedures.

The following are some of the factors that affect and limit helicopter operations:

- Configuration and manufacturer's performance specifications
- Pilot's proficiency in mountain flying
- Density altitude—a combination of temperature and altitude which limit engine performance
- Bad weather, rain or high wind
- Poor visibility created by fog and low clouds
- White-out conditions which can cause vertigo for a pilot
- Air stability, turbulence, up and down drafts
- Night operations
- Available space and lift capacity which limits the number of rescuers and/or victims, and amount of rescue equipment
- Size of the LZ and height of obstacles that could damage the rotors
- Steep sloping terrain that can deny landing (approximately 8 degrees but this is also affected by cyclic range, wind and CG loading)
- Deep or unstable snow and ice
- Loose debris that can be sucked into engine intakes
- Effective operating range without refueling
- Hazards such as power lines and wires
- Altitude limitations related to air density and available oxygen
- Noise—From a tactical viewpoint, helicopters make a lot of noise and are a relatively easy target for ground fire, especially when hovering, landing or taking off.

VISIBILITY

Larger military helicopters can fly and navigate on instruments at night and in zero visibility, particularly those that support special operations missions. In addition to night vision and thermal imaging capabilities, these ships have advanced navigational packages which may include DME, VOR, Doppler, inertial navigation, GPS, terrain following radar, forward looking radar and sophisticated anti-collision systems.

In the civilian world, suitably equipped fixed-wing aircraft can also fly at night and in low visibility conditions under instrument flight rules (IFR). Both commercial and general aviation aircraft are tracked by air traffic control when flying IFR. But to do this, the IFR trained pilot must file an IFR flight plan and the aircraft must be equipped with sophisticated navigational instruments and a transponder so that air traffic control can identify and track their progress, so as to avoid mid-air collisions.

However, smaller civilian helicopters usually do not have a full instrument flight capability so operate under visual flight rules (VFR). Even though the helicopter will have the essentials like a compass, attitude indicator and radio, and possibly GPS, the pilot must be able to see the ground at all times to fly safely and navigate. VFR flight rules dictate that the pilot must have a minimum one-half mile visibility and 500-foot ceiling (the distance from the bottom of the clouds to the ground).

At night, VFR dictate that there must be a minimum of three miles visibility and the pilot must remain at least 500 feet below clouds, 1,000 feet above clouds and maintain a 2,000-foot horizontal distance.

Since weather changes come quickly in the mountains, and it is not unusual for clouds and fog to move in from below, there is always the possibility that a rescue helicopter pilot may find himself flying in low visibility conditions. So although most helicopter pilots will refuse to fly in poor visibility, a rescue helicopter should have an instrument flight capability with GPS and radar altimeter for no other reason than to be able to get back to base safely.

ALTITUDE

Apart from the density-altitude limitations on helicopter performance, operating at altitudes above 12,000´ pilots will require oxygen. Without it they are judgment impaired even though they may not realize it.

Prior to and during WWII, many military pilots who were overly confident in their ability to handle the effects of high altitude flying perished, not realizing that they were unable to detect the impairment and that their night vision was reduced to half that at sea level when flying at 12,000 feet.

This can also become a problem if the helicopter should become disabled at higher altitudes since the ground personnel will have had time to adapt, but the pilot will have probably flown directly from a lower elevation to altitude in a matter of minutes. Without oxygen or immediate evacuation to a lower altitude he will succumb to altitude sickness and possibly die.

From a mechanical standpoint, the helicopter itself is also affected by altitude. As the air density decreases or temperature increases, the rotor blades find it more difficult to generate lift so the engine must work harder to achieve the performance necessary for flight. But along with the thinner air at altitude, the oxygen so necessary for combustion engines is also decreasing in partial pressure making it even more difficult for the pilot to find the power necessary for safe flight and critical maneuvering.

One of the selection criteria for a mountain rescue helicopter is its ability to hover out-of-ground-effect (OGE) at altitude. The following table illustrates some of the operating limitations of a helicopter at altitude. Even though a helicopter may be able to maintain forward flight at its manufacturer's specified Service Ceiling, its ability to hover and make a rescue are severely limited. This is quite dramatic when trying to hover without the added advantage derived from being in-ground-effect (IGE).

This example is for a Bell 206L-4 with a Standard Configuration weight of 2,309 pounds.

Gross Take-off Weight (lbs) =>	3,600#	4,000#	4,450#
Service Ceiling	20,000´	20,000´	10,000´
Hovering OGE	16,600´	13,600´	6,500´
Hovering IGE	20,000´	17,900´	10,000´

The Huey UH-1N has a Service ceiling of 15,000´ and can hover at 12,000´ IGE but only 4,900´ OGE. The SA 315B LAMA is popular for high altitude rescue work because it can hover at 16,700´ IGE and 15,300´ OGE.

Note: These numbers can vary with loading, configuration, model, engine up-grades, performance, and air temperature.

The weight of the pilot, co-pilot, fuel, rescuers, rescue equipment and a victim need to be added to this for a gross take-off weight.

HIGH ALTITUDE RESCUE

One of the highest altitude helicopter rescues in history occurred in the Himalayas on May 13, 1996, with the medical evacuation of Beck Weathers and Makalu Gau after the tragic events that had unfolded on the summit of Everest that spring. Having barely survived their climb and suffering from severe frostbite, they were brought down from the South Col (26,000´) by the superman efforts of climbers and Sherpas from three different expeditions. After initial treatment by expedition doctors at Camp II (21,300´) who proclaimed that their frostbite was the worse they had ever seen, both were then moved down to an area just above the Khumbu Icefall with the hope of a helicopter evacuation.

To effect this difficult rescue, Lt Col. Madan Khatri Chhetri of the Royal Nepalese Army Rotary Wing took a French B2 Squirrel to over 19,800´, at the very limits of its operating capability, to land on a make-shift LZ between two crevasses. Stripped of all unnecessary fuel and equipment to further lighten his ship he was still only able to move one victim at a time down to Base Camp.

Having come directly from Kathmandu at 4,000´ the pilot had to stay on oxygen while performing this difficult extraction. If the helicopter had been disabled at this extreme altitude, the pilot would have had to stay on oxygen just to survive since he had not acclimated like the other climbers and rescuers.

This heroic effort on the part of the pilot saved the climbers and rescuers from having to negotiate the deadly Khumbu Icefall, with two victims who may not have survived the arduous and dangerous passage.

Another equally hazardous high altitude rescue occurred on Denali at 0248 on June 22, 1998. Pilot Jim Hood had to use maximum power to hold the NPS contract LAMA helicopter in a hover at 19,000´ as two victims clipped into the short-haul line. He than made a slow descent directly to the Ranger Camp at 14,200´ before having to fly to 7,200´ to refuel. The pilot did an exceptional job in the decreased light and high altitude conditions making this probably the highest short-haul rescue mission ever performed. (See Chapter 16, page 203, for the AAC account of this incident).

ESD Helicopter Rescue team loads medical and sensitive equipment in sturdy Pelican cases

HELICOPTER OPERATIONS

The helicopter is not the answer to all prayers for an alpine operations team. There are many instances where a helicopter is of no value, for instance, inclement weather, poor visibility, night operations (without special training and NVDs) or response to areas in tight canyons. Unless the rescue team is routinely stationed at the heliport there can be significant delays in getting a team airborne.

First the team must be assembled and transported to the helipad; the flight crew must be notified, assembled and briefed; pre-flight checks must be made; engines must be run up, weather checked and navigational calculations made; equipment and rescuers must be loaded; some equipment and personnel may have to be discarded because of weight limitations; and only then does the mission begin.

Now, providing the helicopter does not have mechanical, electrical, hydraulic or navigational trouble, and weather remains cooperative, the accident site must be found from the air; an evaluation of conditions made; and the rescuers deployed.

Whenever a rescue team is deployed from a helicopter there is always the possibility that the helicopter may have to leave because of low fuel, mechanical problems or deteriorating weather conditions. But the most common scenario is one where there is more than one victim to be evacuated and once they are loaded, the helicopter has insufficient space or lift capability to be able to take the rescue team as well.

This necessitates that the team be able to sustain itself on the ground until the helicopter can return. But if for some reason the helicopter cannot return, then each member of the team should have sufficient clothing, equipment and food to remain on the ground under inclement conditions for 24 to 48 hours or be prepared to walk out.

So even if rescuers routinely deploy from the helicopter with minimal medical and rescue gear, they should have fully loaded packs in the helicopter that can be dropped to them in just this eventuality.

HELICOPTER DEPLOYMENT

The easiest and safest method of inserting and extracting by helicopter is for the helicopter to actually land, but this requires a suitable landing zone (LZ).

In selecting an insertion point or suitable LZ for a helicopter there are five major considerations:

- Surroundings—Are they clear of obstacles
- Size—Sufficient to handle the size of helicopter and rotor span
- Shape—Sufficient clearance for landing and take-off with suitable wind direction
- Slope—Not more than the helicopter can handle or at least place one skid close to the ground
- Surface—Stable and firm enough for landing

For helicopters the size of a Huey UH-1 or smaller, a clear area about 100 feet by 100 feet is sufficient for landing. Some manuals recommend an area at least 33 paces long by 30 paces wide, with no obstacles higher than 1-2 feet and a slope not exceeding 8-10 degrees. But this is just for landing and does not allow for the take-off.

The tail rotor is almost invisible when spinning and a significant danger to the unwary

Contrary to popular belief, helicopters do not take-off vertically, especially at altitude when already running at maximum power. So to leave the LZ, the helicopter will need a take-off path that is long enough to clear any obstacles by at least ten feet when climbing out at about a 15-degree angle. For this reason, LZs on exposed areas like mountain tops and ridgelines are preferable.

Since a head wind will help the pilot on both landing and take-off, and cross winds are difficult to deal with, the LZ should also be selected with wind direction in mind.

If operating in steep snow country, the LZ and flight path should be selected so that the helicopter does not trigger any avalanches that would endanger ground search teams, rescuers or victims who may be in the vicinity.

Any obstacles should be removed from the LZ but only the area directly under the skids where the helicopter will settle needs to be actually leveled. Loose debris and equipment should also be removed or kept well clear of the LZ.

Rescuers who are not loading or unloading from the helicopter should position themselves forward of the cockpit and about 100 feet away so that the pilot can see that they are clear. This also gives the rescue team eye contact with the pilot so they can exchange hand signals when it is safe to approach.

For locations where a suitable LZ is not available and landing is not an option, there are several other insertion methods.

- Skid step-off on sloping ground or a ledge
- Hover jump
- Rappelling
- Hoist lowering
- Short-haul insertion

HELICOPTER RAPPELLING

Most rescue teams that routinely deploy from helicopters prefer to simply jump from the helicopter in a low hover or use the hoist to be lowered into tighter areas. However, if several rescuers must be put into an area where the helicopter cannot land, then rappelling is faster and more efficient than using the hoist which is often quite slow due to its low gearing.

Rappelling is suited to SAR operations in mountainous, uneven terrain, heavily wooded forests and areas covered in deep unstable snow—all areas where a helicopter cannot land. Rappelling is also a skill that all mountaineers and rescuers already possess so will need only slight modification when deploying from a helicopter.

If a rescue team is to be deployed by rappelling there are a few points worth considering. Firstly, all team members should be trained in rappelling and rigging on land before attempting helicopter descents. The noise level, increased costs and increased stress in a helicopter hovering at 100 feet make for a very unsatisfactory learning environment.

ESD crew chief rigs a rappel rope for Sheriff's mountain rescue deputies

A local fire tower, cliff or bridge can be utilized for valuable low stress training with an emphasis on making a smooth, safe descent with no violent or unnecessary braking that could rock the ship or cause the pilot to move off station.

After a lecture on safety, helicopter hard points, rope rigging, hand signals and helicopter flight dynamics by the pilot, crew chief or rappel master, the team should be ready for their first helo-rappel. Because of the high ambient noise level in helicopters and the fact that most rescuers will be wearing ear-plugs, there should be hand signals for, "Stand-by," "Snap-in," "Get in the door," "Get on the skid", "Hold" and "Go!".

The descent should be smooth with no aggressive jumping from the skids that will rock the ship unnecessarily and make it harder for the pilot to hold position. The same is true for violent braking or premature drops before the team member(s) on the other side of the ship is ready to go.

Upon landing, the rappeller should immediately pull sufficient rope through his rappel device to cre-

A helicopter rappel should be a smooth descent with no unnecessary jumping or hard braking

ate enough slack to disconnect both his carabiner and rappel device. He should then signal to the rappel master or crew chief that he is clear and the next rappeller can be dispatched or the rope can be dropped or pulled up.

The helicopter should never attempt forward flight with the rappel rope hanging since this can cause the trailing rope to fly up into the tail rotor with predictably disastrous results.

HOIST OPERATIONS

Frequently in the mountains, the ability of a rescue helicopter to land is restricted by confined areas, canyons, trees, rocks or steeply sloping terrain, with even deep, soft snow posing a problem. So the alternative to landing is a hoist insert and recovery if the ship is equipped with a hoist, or short-haul line to a more suitable landing area if the ship does not have a hoist.

Hoisting seems logical and sounds simple and this has long been the standard procedure world wide, but up until February 1998, the FAA regulations limited helicopter hoist recovery operations to military aircraft. After considerable battling, the long awaited decision came down that suitably equipped civilian SAR helicopters are now permitted to perform hoist recoveries.

Hoist operations require, you guessed it, a power driven hoist, sufficient cable (100´-270´ of 3,200-pound cable) and some form of recovery sling, screamer suit, jungle penetrator, litter, rescue diaper or basket. Ideally the hoist should have a 600-pound rating but that may be dictated more by the strength of the helicopter's airframe and the location and method that the hoist is mounted than by the power of the hoist itself. Hoists can have a single speed controller that brings in cable at about sixty feet per minute or a variable speed controller with a maximum speed of up to 300 feet per minute.

Military hoist and jungle penetrator on a PJ CH 53

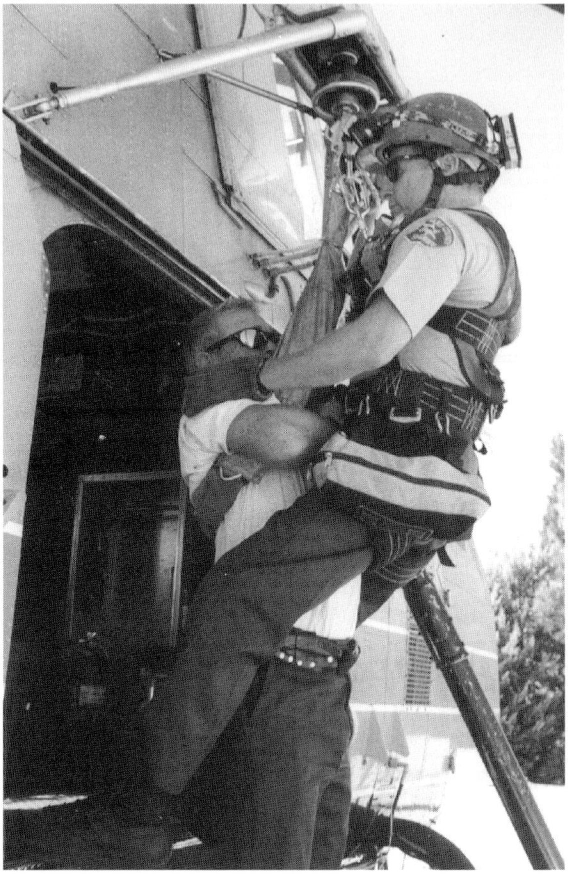

Sling technique for getting the victim into the door so that he can be assisted by the crew chief

Rescue slings are commonly used in mountain rescue but litters and baskets are preferred for victims who have sustained injures to the back or chest or are unconscious. Particular care should also be taken when using a sling to hoist victims who are in a weakened state, tired, incoherent or lapsing in and out of consciousness, since it is possible for them to slip from a sling. If the sling must be used in these cases, then a rescuer must accompany them on the hoist and ensure that they are secure throughout the ride.

Whichever method is used, and since the load can pendulum under the helicopter while being hoisted, it is important to have sufficient area (50 feet) so that the load can clear any obstacles such as trees. It is also recommended to have a tag line attached, particularly to litters.

There is a potential for injury when using a hoist so everyone should receive training in its application and operation, particularly the hoist operator, crew chief and rescue team leader. Hand signals should be established between the hoist operator and the rescuer on the hook for use when radios are unavailable or unnecessary. Usually signals given with the hand open and the palm flat are for helicopter movements and signals given with the thumb or index finger are for the hoist.

Everyone working around hoists and steel cables should also be educated about static electricity. If being picked up, or picking someone up, it is necessary to first ground the hoist cable to discharge any

ESD deputies perform a hoist recovery in the Angeles Mountains

static electricity that may have built up during flight. Failure to do this could result in a severe electrical shock.

If hoisting near a downed aircraft it is also important to ensure that the point where the hook or hoist cable is to be grounded is well clear of fuel spills. Failure to do this could generate sparks and result in a fire or explosion.

The actual technique to be used will often be dictated by whether the victim needs to be hoisted in a litter or just in a harness. To pick a stranded or uninjured climber off of the side of a mountain, the standard procedure is for one rescuer to be lowered down to the victim on the hoist cable. Then, without disconnecting himself from the cable, the rescue slips a sling or harness onto the victim, gives the hoist operator the thumbs up, and they are both hoisted back into the helicopter.

One problem with this technique is that the victim is subjected to the down wash from the helicopter's rotors and there is a risk of blowing the victim off of his perch. The down wash may also dislodge rocks, ice or snow from above, and the high noise levels make communication with the victim all but impossible.

In situations where the pilot cannot maneuver close enough to the face because of rotor diameter, the rescue team can deploy onto the top of the mountain or ridgeline where they can set-up anchors and rappel down to the victim. However, when working on very exposed rock faces or ridgelines, the rescuers need to securely anchor themselves and the victim before signaling the helicopter to approach. If not secured they could easily be blown from the perch by the 100mph downdraft

When a litter extraction is called for, the standard practice is to first lower one rescuer with the trauma kit to stabilize the victim and then the second rescuer with the litter. Once on the ground, the rescue team disconnects the litter allowing the helicopter to back-off so that the rescuers are then free to stabilize and package the victim without having to work under the down wash and high noise levels of the rotors.

When signaled, the helicopter moves back in to make the recovery. The litter is sent up stabilized by a tag line from the ground, and/or one rescuer rides up with the litter.

Smoke or Mk 13 Day / Night flares are also very useful in helicopter rescue work. When ready for a pick-up the team can pop smoke to let the pilot know that they are ready, but more importantly, it will also give the pilot the wind direction and approximate speed.

The helicopter should not hover over the rescue site after the drop or during the early stages of the operation. This will only stress the victims, raise wind-chill factors, whip up a snow storm, dislodge rocks, eliminate the possibility of voice communications and subject the rescuers to a fierce down draft from the rotors. The pilot should maneuver down wind or orbit so that he can observe the rescue site and be prepared to move in on the rescuers' signal.

The rescue team must always be aware of the load rating for the hoist, since some civilian models are considerably less than their 600-pound military counterparts. A 600-pound hoist can lift two rescuers with their equipment or one rescuer along with a victim and a litter. Ideally that leaves the second rescuer to be hoisted with the packs, ropes and climbing gear. If there is any doubt, err on the side of caution and do not overload the hoist.

Rescuers should also be aware of a situation that

can occur with helicopters called **settling with power.** This can occur when a helicopter is preparing to hoist a victim or a rescuer out of the mountains, is hovering out of ground effect and already at maximum power. In this configuration, and with the slightest change in conditions, the pilot can find himself in a situation where the helicopter begins to settle and he has insufficient power to prevent the unexpected descent. Obviously this is a dangerous situation for both the helicopter and the rescuer who may have just hooked into the hoist cable below.

The pilot has only one option to save his ship and that is to begin flying and gain some translational lift from the airspeed generated by forward flight. This means that the rescuer must be able to unhook very quickly or he is going for the ride of his life—possibly being first dragged through the rocks and trees. The crew chief also has the option of cutting the cable with bolt cutters but that is extreme and he simply may not have time. If the rescuer and victim do get jerked away by the sudden transition to forward flight, the crew chief should immediately hoist them up and get them inside to evaluate their injuries.

BASIC HELICOPTER SAFETY RULES

1. Stay **well clear** of operational helicopters unless directed to do otherwise
2. Do not approach the helicopter unless **cleared** to do so by the pilot or crew chief
3. Always approach from and leave to the **front** of the helicopter so that the pilot can see you at all times
4. Stay well clear of the **tail rotor** and never walk under the tail boom
5. On sloping ground always approach from and leave to the **lower down hill side** to clear the rotors
6. Keep your **head down** when moving under the rotors. As they slow the tips will dip lower.
7. **Secure all hats** and loose items of clothing and equipment around helicopters
8. Ropes should be **coiled** and secured and lighter items like sleeping bags, tent flies and parkas should be **secured** in packs or bags
9. Keep skis, ski poles, ice axes, radio antennae and any other long items **parallel to the ground** when approaching or leaving a helicopter. Also cover or protect sharp points.
10. Positively **no smoking** within 100 feet of a helicopter
11. **Do not touch** slings, hooks, external loads or any part of the helicopter until it has made contact with the ground and discharged static electricity
12. Never stand **directly beneath** a helicopter during take-off or landing unless authorized to hook sling loads
13. **Helmets, eye protection and ear protection** are recommended when working in or around helicopters
14. Be careful **entering and exiting** a helicopter so as not to interfere with any controls, control cables or the pilot
15. All medical and rescue equipment carried in the helicopter must be **secured**
16. **Do not load** any equipment into the helicopter without the pilot or crew chief's knowledge and authorization. Only they can judge the best weight distribution and tie-down points
17. Dedicated ground crew should be wearing **hard-hats with chin straps,** goggles, ear protection and bright colored vests or jackets. The ground crew coordinator should be in radio contact with the pilot
18. Use only dedicated **hard points** for rappelling and back-up the system to at least two other points
19. Weapons should be kept **on safe and pointed downward** when traveling in helicopters
20. The pilot must be notified if carrying any **explosives, grenades or pyrotechnics** and all must be handled with extreme care

22

ALPINE RESCUE OPERATIONS

Alpine rescue operations can range from the very simple to extremely complex depending on the nature of the rescue, the elevation, the terrain, accessibility of the location, the time of day, the season and the current weather conditions. There are in fact so many variables that a single chapter cannot do justice to all that is entailed, but it will at least give a team some idea of the scope of planning that can go into these types of operations. So, at the risk of over simplifying this subject, the principal operational stages of a mountain rescue are:

- Locate the accident site
- Access the victim(s)
- Stabilize any injuries
- Transport the victim(s)

However, there are many other levels of preparation, planning and support that must be in place to make the above happen in a safe and efficient manner. There must be an element of pre-planning, driven by the known local variables, that can be modified to suit most operations; a workable command structure with good communications; all supported by suitable personnel with the right training, equipment and experience.

A team that has a high operational tempo and is doing rescues on a regular basis will have set procedures based on experience, often gained through both past successes and mistakes. A full-time rescue team like this can usually mobilize, respond to the location, make an on-site assessment, effect the rescue and transport the victim to whatever medical support is called for.

Volunteer teams however, or teams that get very little activity in their area, need to be a more methodical in their approach to a rescue since they will not have the confidence in their SOPs that only comes from actual experience. There may also be a considerable disparity in the standard of technical mountaineering skills possessed by each team member.

It is important to differentiate at this point between a wilderness rescue and an alpine rescue. On a wilderness rescue in non-technical terrain, an area can be saturated with searchers, rescuers and rescue assets with little or no danger to the team. However, on an alpine rescue operation the victim may not be as easily accessible so only skilled mountaineers can be deployed. It is important to keep in mind that the same conditions that contributed to the accident, steep ice or rock, pose an equal danger to the rescuers.

This is best exemplified by the typical alpine rescue (if there is such a thing as typical) that would be encountered by a volunteer mountain rescue team in Colorado, for example. I pick Colorado because of the fact that there are some 54 peaks in the Rocky Mountains over 14,000 feet that each year draw thousands of hikers and climbers of varying experiences trying to bag as many as possible. In addition, most local mountain rescue teams are made up of volunteers who although very professional, are also of varying skill levels and rescue experience.

TYPICAL RESCUE SCENARIO

The average recreational hiking or climbing group will leave the road or trailhead before dawn to take advantage of the firm snow, with a plan to summit by early afternoon and be off the mountain by dark. Now, knowing that most accidents will occur after summiting and on the way down, or at least not until one member of the group is high enough on the mountain to be exhausted and make a mistake, it is safe to make the observation that most accidents occur in the afternoon. This is most often a fall on either rock or steep ice at an elevation of 12,000-14,000 feet, but could also be because of a sudden change in weather with either whiteout conditions or dangerously strong winds.

At this point, unless someone in the group has a radio or a cell phone that can reach the nearest town, it will probably take two to three hours for one of the party to get off the mountain and notify authorities of the accident. Similarly, if the accident occurs on the backside of a mountain or in a ravine, one climber may have to climb back to the top to get a signal out.

So late in the afternoon (4-5 pm) the mountain rescue team is notified and begins a call-out, but since the volunteer members all have regular jobs or may be off doing things with the family, it takes time to assemble a full team. Fortunately, as with many teams, there are always a couple of strong climbers who are close at hand and better prepared to respond than most.

Knowing that it is matter of life and death on the mountain; knowing that the victim may be in critical condition and may not make it through the night; and knowing that helicopters are slow to mobilize and do not fly after dark; the first pair of rescuers takes a radio and heads up the mountain to assess the situation immediately. However, already being late in the day and rapidly approaching darkness, the incident commander holds back other arriving personnel and assets until he gets a report from the first rescue team.

Even though they are in good shape, the rescuers move slowly because of heavy packs loaded with rescue equipment, medical supplies and emergency bivouac gear. By six or seven they find themselves climbing after dark and do not reach the victim until nine or ten o'clock that night.

By this time, apart from the injuries sustained in the fall, the victim and possibly fellow climbers are extremely cold and thirsty. They have been on the mountain since dawn (16-17 hours), had been traveling light and had not anticipated this possibility by carrying even minimal gear like bivi sacks or a stove. The rescuers immediately stabilize the victim's injures and get him or her into one of their sleeping bags, and then melt snow to make hot drinks for all present. The injured climber has neck, back and head injuries, all common in a fall, but the rescuers do not have a backboard or oxygen—both items that would help this victim but were too heavy for the initial climbing team to carry.

The rescuers report their exact location to the incident command post and notify the Incident Commander that it will take a full team effort, complete with stretcher, backboard, lowering systems and anchors, to move the victim down to an area suitable for a helicopter extraction.

The Incident Commander, not wanting to risk more injuries on the mountain at night, decides to launch the main body of the team at first light. This means that our first two rescue climbers are in for a cold night on the mountain along with the victim and any other climbers that may have stayed to assist. The team members at the command post now have a couple of hours to plan the next day's operation, prepare their ropes and rescue gear, assign team responsibilities, divide up loads and get a couple of hours sleep.

Leaving just before dawn, the main rescue team climbs hard through the morning and reaches the accident site between ten and eleven. After checking on their team mates and giving everyone a chance to hydrate, they package the victim and begin the long slow process of belaying and lowering the stretcher down to a site suitable for a helicopter to at least hover against the steep slope.

Provided a helicopter is available and the pilot is skilled in mountain rescue, the victim may be off the mountain by two or three, twenty-four hours after the accident.

Keep in mind that the entire rescue effort is predicated by the rescue team receiving the initial notification, and that is only possible if the victim's friends were able to get a call out or one was able to descend for help. If the weather had closed in, or the friend didn't make it out, or was caught on the mountain after dark, then the team would not get a notification until a friend or family member reported that the climbers were overdue—probably later that night.

In these situations the victim(s) will now be required to spend the night on the mountain alone without medical assistance, warm clothing, sleeping bags, bivi sacs and warm drinks. The lesson for climbers is to plan for the worst-case scenario.

For the mountain rescue team, there are several learning points:

- Expect to get calls in the afternoon, especially on weekends, so have your gear packed and ready to go.
- If at all possible, it is important to access the victims as soon as possible since they will seldom be prepared to survive a night on the mountain.
- Only skilled rescue mountaineers, familiar with the terrain, should attempt an unsupported climb, especially late in the afternoon.
- A two man rescue team cannot carry everything they would possibly need, so should concentrate on essential medical supplies and emergency rewarming and bivi gear.
- An Incident Commander is wise to avoid working at night with large numbers of rescuers of varying skill levels. At least in daylight, the stronger members can support and encourage the less experienced rescue climbers.
- You cannot always count on helicopter support so should not delay the initial response.
- It takes significant manpower to bring a seriously injured climber down a steep ice chute, glacier or rock face.
- Have a plan before the team is committed to the rescue. Only by having a plan will a team be able to identify the risks involved and the equipment, manpower and skill levels required for a safe resolution.

PLANNING

The primary goal in any planning is to eliminate confusion and this is best achieved by having standard operating procedures (SOPs) for a broad range of anticipated rescue scenarios. Different teams may have different procedures but within one team everything should be standardized. Teams that may be required to work together should also attempt to standardize their equipment and procedures, and if possible, this should extend to all regional and state rescue assets.

PRE-PLANNING

The very nature of most rescue situations requires a rapid response since lives may be at stake, and a rapid response can only be initiated safely if there is are pre-established rescue procedures in place.

This plan does not need to be very detailed as there are too many rescue scenarios and variables involved. In addition, the accident site may carry the rescue team many miles from the command post or any additional support assets, so the actual planning of the recovery will have to be done on site with the resources available.

Pre-planning is based on the standard operating procedures of the team, which should in turn be based on nationally accepted practices for mountain rescue. This pre-planning is not a mission specific plan but more an overall procedure designed to increase efficiency and eliminate confusion for the team as they respond to most anticipated call-outs.

Pre-planning is only effective if both command and rescue personnel make an effort to stay as close to the pre-established plan as possible with only slight modifications to suit the situation.

A basic rescue plan should include the team call-out procedures; a list of team assets and available resources; and the standard operational procedures for the more common rescue scenarios encountered on a local level. Risk assessment and management should also be built into the plan at several levels.

A part of the pre-planning should also take into consideration which members of the team are most qualified for certain types of operations, particularly those requiring a high level of technical expertise. The criteria for rescuers to be called may include:

- Training and experience in technical mountain rescue
- Physical fitness and emotional stability
- Ability to make accurate assessments of the risks involved
- Experience in helicopter deployment
- Ability to get down from any site if they are left by the helicopter
- Competency in emergency medical skills
- Ability to choose safe routes in avalanche prone terrain
- Mountaineering skills
- An appreciation of the risks involved

Another factor that should be considered in preplanning that will directly affect a team's ability to respond quickly to an incident is the team's state of operational readiness. This can include:

- Availability of qualified personnel
- Overall physical fitness of personnel
- Individual equipment readiness
- Level and frequency of training
- Condition & Location of support equipment
- Response time to the Incident Command Post
- Readiness, Training and Experience of Command Staff
- Efficiency of Dispatch and Call-out procedure
- Availability of support assets

RISK MANAGEMENT

The single most important consideration on all rescues is the safety of the rescuers. So acknowledging the fact that mountain rescue work is hazardous duty under any circumstances, the overall risk of an operation can be reduced with:

- Competent Incident Command Staff
- Experienced Leadership in the field
- Comprehensive Standard Operating Procedures
- Physically fit rescue personnel
- Mature judgment and common sense
- Professional initial training with regular re-qualifications
- Quality, well maintained rescue equipment
- An overall professional approach to rescue work
- The ability to resist external pressures

No matter the pressure, a rescue team must maintain a professional and methodical approach to all rescue operations. An operation should not be mounted, accelerated or influenced by external forces such as:

- Friends and family of the victim(s)
- Press and media coverage
- Administrative pressure
- Political pressure from local politicians
- Peer pressure by potential "cowboys" on the team
- Competition from other teams that may create a "Race mentality"

PHASES OF AN OPERATION

As with any complex operation, the simplest approach to rescue work is to break an operation down into a more manageable list of tasks that can be delegated to qualified individuals and handled in a logical progression.

The following is a list of tasks or phases in a rescue, some of which must be handled in sequential order while others can be handled simultaneously to expedite the operation. For example, when the initial call comes in and essential information is collected, different individuals can be assigned to notifying the rescue team, organizing the command post, contacting support groups, preparing equipment, and placing air assets and trauma centers on stand-by.

Phases of a Rescue Operation:

- Initial Notification of Incident
- Collection of essential information
- Team Call-Out
- Command Post Set-up
- Team Response
- Review of SOPs
- Preliminary Risk Assessment
- Initial Briefing
- Logistical Planning
- Equipment Preparation
- Coordination of Air Support
- Contingency for Loss of Air Support
- Ground Search Considerations
- Location of the Victim
- On-site Risk Assessment
- On-site Planning
- Victim Access & Stabilization
- Victim Packaging & Extraction
- Victim Transportation
- Team Evacuation
- Post Operational De-Briefing
- Equipment Maintenance & Demobilization

INITIAL NOTIFICATION

The first responsibility of an Incident Commander is to the safety of the rescue personnel and this begins with accurately assessing the type of rescue operation, the location and the conditions.

The following is a list of questions that need to be answered before the call-out is initiated since the answers to many of these will dictate which team members will be alerted and what equipment they will need to bring.

- What is the nature of the rescue?
  - Overdue climbing party
  - Lost hiker or climber
  - Trapped or injured climber
  - Crevasse rescue
  - Avalanche
  - Aircraft down
- Is the situation life threatening?
- Where did it happen?
- How long since it happened?
- What is the elevation of the accident site?
- What are the snow conditions?
- What is the current weather?
- What is the weather for the next 24 hours?
- What rescue assets have already been mobilized?
- What rescue assets are available?
- Is helicopter support available?
- What are the potential dangers for the rescue team?

ADMINISTRATIVE CONSIDERATIONS

Large search and rescue operations can be quite complex as indicated by thc following list of Incident Command System SAR Forms generated by the Bay Area Search & Rescue Council. (The current edition can be down loaded from www.basarc.com)

However, any efficient incident management system should be scaled to fit the size of the incident, so not all of these forms are required on smaller operations. But having a list similar to this will give the Incident Command staff a foundation for the planning and flow of an operation.

SAR Forms:

- General Briefing [Notes & Instructions]
- Missing Person [Notes & Instructions]
- Team Assignment [Notes & Instructions]
- Team Debriefing [Notes & Instructions]
- Team Debriefing—Dog supplement [Notes & Instructions]
- Team Debriefing—Area Search supplement
- Team Debriefing—Equestrian supplement
- Team Debriefing—Tracking supplement
- Team Debriefing—Hasty Search supplement
- Individual Resource Availability
- Interview Log
- Radio Log
- Clue Log
- Clue Report
- Investigative Plan
- Lost Person Questionnaire
- Area Search Planning Worksheet
- Incident Briefing

- Incident Objectives
- Organization Assignment List
- Incident Radio Communications Plan
- Medical Plan
- Incident Organization Chart
- Incident Status Summary
- Agency Check-In List with SAR Skills
- Check-In List
- Unit Log
- Operational Planning Worksheet
- Radio Requirements Worksheet
- Radio Frequency Assignment Worksheet
- Support Vehicle Inventory
- Air Operations Summary
- Demobilization Checkout

OPERATIONAL CHECKLISTS

The following checklists cover many of the essential considerations in operational planning.

THE CALL-OUT

❏ Phone & Pager procedure
❏ Essential Equipment Required
❏ Anticipated Response Time
❏ Assembly Point
❏ Command & Control

COMMAND POST

❏ Select a suitable Location for the CP
❏ Man the CP with an Incident Commander & Communications Officer
❏ Designate a Press Liaison Officer
❏ Test & Activate Communications Net
❏ Coordinate Arriving Personnel & Assets
❏ Establish minimum number of rescuers required to launch operation
❏ Initial Brief of the Rescue Element on the situation

ADMIN & LOGISTICS

❏ Assign Search Areas
❏ Designate Rescue Elements
❏ Collect essential Maps & Aerial Photographs
❏ Issue Radios, Call-signs and Frequencies
❏ Perform itemized Equipment Checks
❏ Distribute Emergency Rations & Water
❏ Pack Personal & Team equipment
❏ Organize suitable Transportation
❏ Arrange EMS support

COORDINATION OF AIR ASSETS

❏ Initiate Air Search if necessary
❏ Coordinate Helicopter to insert rescue team and extract victim
❏ Brief pilots and team members on heli-tac procedures
❏ Establish call-signs and common frequencies
❏ Arrange for refueling and ground crew support
❏ Make contingency plan for emergency evacuation of injured team members

ON-GOING SUPPORT CONSIDERATIONS

❏ Notify additional support personnel as needed
❏ Expand and improve the Incident Command Post
❏ Arrange hot food for prolonged operations
❏ Organize additional ground transportation
❏ Activate mutual aid plans with other agencies if necessary

APPROACH CONSIDERATIONS

❏ Estimate Time & Distance to the accident site
❏ Consider Type of Terrain & Snow conditions
❏ Temperature, Weather & Visibility
❏ Team Fitness & Equipment suitability for conditions
❏ Feasibility of a Helicopter Insert
❏ Closest Accessible Roads
❏ Land Navigation, Maps, Compasses, GPS coordinates

AT ACCIDENT LOCATION

❏ On-site Risk / Benefit Evaluation
❏ On-site Planning & Communications
❏ Victim Access & Assessment
❏ Victim Stabilization
❏ Victim & Rescue Team Recovery

CASUALTY EVACUATION

❏ Medical Considerations
❏ Victim Packaging

❑ Spinal & Head Injuries
❑ Patient Monitoring
❑ Ground Evacuation Procedure
❑ Helicopter Evacuation Procedure

POST RESCUE
❑ Rescue Team Recovery
❑ Operational Debrief
❑ Demobilization
❑ Lessons Learned
❑ Changes to Equipment or SOPs if necessary

Note: For more detailed information on this subject, get copies of the operations manuals for professional teams working in similar conditions and any training publications put out by organizations such as the Mountain Rescue Association and Emergency Response Institute.

The author mapping a gorge high in the mountains of central Taiwan as part of a US search and rescue mission in search of a missing American student

Even though this operation was primarily a river search, the team had to rappel down into the gorge each morning and climb out in the evenings

(Photo—Charlie Shimanski)

"ACCIDENTS IN NORTH AMERCIAN MOUNTAINEERING"

Edited by Jed Williams, and published by the American Alpine Club and Alpine Club of Canada.

FALL ON ICE/SNOW, UNABLE TO SELF-ARREST, WEATHER

Alaska—Mount McKinley, West Buttress

(What follows is a synopsis of two incidents, one that happened as a result of the second party attempting to rescue the first.)

At 1130 on May 24, 1998, Jason Sinnes and Daniel Rowarth (29)—Burrito Brothers climbing team—decided to descend from the 17,000-foot campsite on the West Buttress route of Mount McKinley, where they had spent the previous night. They felt that weather conditions (which had been generally poor for the duration of their trip) were once again deteriorating. Nuke Vanderbeek (33) and Tim Hurtado, National Park Service volunteer patrol members, had been at 17K for three days and had spent the two previous nights there. The NPS team was feeling well with "only slight headaches the night before." They had originally planned to attempt the North Summit of the peak. Since the weather was getting worse again, they had delayed their departure for the NPS camp at 14,000 feet in hopes of being able to assist with the delivery of a helicopter sling load to 17K. After learning that the helicopter mission had been canceled, Vanderbeek and Hurtado began their descent at 1300. The climbers both had "full summit gear" on and carried 30-40 pound packs. Each wore climbing harnesses, crampons, and over-boots. They were each carrying one snow picket and one ice screw. They were not carrying bivouac gear, since they had used the patrol equipment available in the NPS cache at 17K. Winds at that time were steady at 30 mph with gusts to 60-70 mph. They discussed the need to be very careful with foot placement since "a misstep could be fatal."

Moving steadily, Vanderbeek and Hurtado overtook the Canadian pair some time during the next hour around 16,900 feet. Sinnes and Rowarth considered themselves fortunate to be in "close proximity to the rangers" and were glad that "every few hundred feet the rangers would wait for us." At a point just above Washburn's Thumb, the large, well-known granite monolith along this section of the route, Vanderbeek's crampon came off his boot. (Over his climbing boots he had been wearing Forty Below neoprene over-boots and Lowe Footfang crampons.) Sinnes and Rowarth caught up to them again at this point.

After getting his crampon back on, Vanderbeek, accompanied by Hurtado, climbed down below the Thumb, followed shortly thereafter by Sinnes. After down-climbing the steep, fixed section near the Thumb, Rowarth rejoined Sinnes and they regrouped at a flat area of the ridge. According to Sinnes, Rowarth said that, "he felt fine, not tired or anything." They continued down the ridge, staying slightly to the lee side. Sinnes described it as "slightly off the beaten track, steep, and slightly technical." Sinnes stated that this section "caught him [Rowarth] by surprise and he lost purchase with his feet." The approximate time of Rowarth's fall from the ridge was 1400.

Vanderbeek and Hurtado were about 100 feet away from Rowarth when he fell. Hurtado had turned around to "see how Jason and Daniel were doing" and saw Rowarth 50 feet below the ridge crest, catapulting downward. They could not see very far down from their vantage point so Vanderbeek traversed over to the fall line and found Rowarth's ice ax. Hurtado retrieved the radio from Vanderbeek's pack, but could not get out from that location. Vanderbeek therefore hiked over the ridge to the south so that he could transmit. Ten to twenty minutes later Vanderbeek returned to Hurtado and Sinnes' position. Hurtado's understanding of the situation and the plan at that point was:

- ❏ Vanderbeek and Hurtado had been given permission from Mountaineering Park Ranger Daryl Miller at 14K to establish contact with Rowarth.
- ❏ Adrian Nature was going to be dispatched from 14K to 16K to provide a necessary communications link at 16K since they would presumably be out of radio contact once they started down off the south side of the ridge crest.
- ❏ Upon further contact from Vanderbeek and Hurtado, additional resources would be available from 14K.

Hurtado stated that visibility was 200-feet, winds were 30-35 mph gusting to 40-45 mph and higher. It was not very cold and it did not appear to be unreasonable to be traveling in these weather conditions. At this point they met two climbers, Rowan Laver and Gordon Cox, ascending the ridge to put in a cache at 17K. Cox agreed to assist Sinnes back down to 16K Hurtado and Vanderbeek consolidated gear for the descent down toward the Peters Glacier. According to Laver, Vanderbeek could "see something down on the glacier." Vanderbeek and Hurtado took a shovel, tent fly and poles from Sinnes and added this to their other light bivouac equipment. Hurtado and Vanderbeck stopped briefly and had a discussion of how best to proceed with the descent. They talked about using a rope, but felt that rope work would be slower and that they thought they could perform a self-arrest under the present conditions. Hurtado recalled that Vanderbeek seemed more concerned with what he [Hurtado] thought, as he was the less experienced climber. They began to descend the ridge at 1430, winding their way down from Rowarth's ice axe, which they left in place.

Hurtado reported that the terrain that he and Vanderbeek encountered during their descent consisted of "600-800 feet of hard snow" with occasional patches of ice "not more than 10-30 feet long." They were in voice contact with one another descending 10-15 feet apart, and left and right of the fall line, which was marked by occasional patches of blood. Hurtado stated that he and Vanderbeek "maintained continual voice contact, determining each other's fatigue and comfort level" and that Mike had relayed pointers regarding different descending techniques. Hurtado's goggles became fogged and his visibility became limited to about three feet up or down. In Hurtado's words, "I was mostly feeling my way down the terrain." They now encountered an extensive ice patch and Hurtado, facing in, became much more deliberate with his tool placements. (Each carried a single, 70 cm ice axe.) He described the ice as being in the 40-50 degree range and as being "very brittle and the first swing often dinner-plated." While descending, Hurtado was not aware that the ice patch was extensive, but rather assumed that it was another occasional patch of ice which would be followed by snow, which had been the pattern to this point. They descended this particular ice patch for 10-15 minutes when Hurtado requested that they head for some rocks that they could see in order to stop and reevaluate. Vanderbeek replied, "Okay." Hurtado stated that, at that time, he felt "it might be wise to rope and use a running belay." They continued down toward the rocks and once again made voice contact. Two or three minutes later, at 1345, Hurtado heard "the sound of nylon on ice." He heard nothing else and shouted for Vanderbeek and received no answer. He removed his goggles, could see much better and realized that his position was precarious. He was on a "400-500 foot sheet of hard, blue water ice that was 45-55 degrees." As he continued to descend toward the rocks for 10-15 feet, the full gravity of the situation hit him. Hurtado decided that the best course of action was to place his one ice screw and anchor himself to that. This accomplished, he ate, drank and began to chop a ledge on the ice sheet for himself. Hurtado realized that Nature was on his way and that he could wait for a rescue.

At 1426 Adrian Nature left the 14K NPS camp

carrying little more than a radio and personal gear (excluding sleeping bag and bivouac equipment). He reached 16K one hour and fifty-six minutes. There he met up with Rowan Laver. They roped up and proceeded up the ridge to where Rowarth had disappeared. Nature spoke of being in "rescue mode" as opposed to "climber mode." He differentiated between the two by stating that when in "rescue mode," he was roped up with a partner, and they safeguarded their travel through the use of running belays or fixed anchored belays, depending upon terrain and conditions. Nature's plan was to check in with 14K at 15-minute intervals. Gordon Cox was enlisted as Nature's replacement to serve as the vital communication link between 14K and the second rescue team (Nature and Laver).

At 1646 a support party consisting of four additional volunteer patrol members left the 14K NPS camp bound for 16K. The group consisted of Dr. Mark Elstad, Linda Davis, Dean Giampietro (volunteers on Daryl Miller's patrol), and Dr. Colin Grissom (volunteer on Billy Shott's patrol with extensive Denali experience, as well as a great deal of other high altitude climbing experience from around the world.)

After they reached the accident site, Nature and Laver attempted to contact Vanderbeek and Hurtado by radio but were unsuccessful. They descended one pitch through the rocks, and screamed for Vanderbeek and Hurtado between wind gusts (60-70 mph). During one relatively calm period they heard someone scream, "Help!" They contacted Miller at 14K at 1730, relayed that they had voice contact, and requested permission to begin down-climbing. Permission was granted to try and contact Vanderbeek and Hurtado. They down-climbed one pitch and requested permission to continue, which was granted.

They continued their belayed descent into gradually steepening and progressively more technical terrain. According to Laver it was five full rope lengths down to Hurtado's position (750-800 feet). They encountered hard, blue ice for the final 200-300 feet.

At 1751 Jay Hudson left Talkeetna Airport for the McKinley area to provide an additional radio relay link during the operation and at 1829, an Air Force C-130 aircraft was requested by the Incident Command Post (ICP), located at the Talkeetna Ranger Station. At 1914 the Nature/Laver rescue team reached Hurtado's location and they continued down roped together as a party of three. As they descended through the rock band, they found Vanderbeek's sunglasses and ice axe. As they neared the bottom, they noticed a blood trail that was one foot wide and that led directly to the body of Daniel Rowarth, located one rope length from the bergschrund on the flat floor of the glacier. Vanderbeek's pack was found approximately 50 feet up-glacier of Rowarth's body, while Rowarth's pack was found about 150 feet down-glacier. Both packs were strap-side down, and the waist, shoulder, and chest buckles and straps were intact and not broken.

Communications with the ground team were pretty good at this point as both Jay Hudson and the Air Force C-130 were circling in the area. An attempt to utilize the NPS exclusive-use contract helicopter (SA315B LAMA) to assist with an aerial search proved fruitless when mountaineering ranger Kevin Moore and pilot Jim Hood decided to turn back from the SE Fork of the Kahiltna in whiteout conditions. Adrian Nature described the weather conditions in terms of some of the worst Alaskan/Denali storms he had experienced. On a 1-10 scale with 10 being the worst, he described the overall condition as an "8".

After Nature's team's official search efforts were fruitless, their main focus shifted to that of self-preservation. They managed to retrieve a tent and a sleeping bag from Rowarth's pack. Their instructions from Colin Grissom, leader of the support team at 16K, had been to "come out of the hole." They attempted to do so, but in the swirling winds and whiteout, they found that they had traversed too far to the west down the glacier.

At the 14K camp Daryl Miller had enlisted the help of Dave Langrish and John Elwell (Firestarters Expedition) and at 1836 had led the trio carrying a 600 foot rope and overnight gear to the base of the fixed lines and back to the 14K camp. The group returned at midnight, traveling in "hideous" weather after having hooked up with Gordon Cox and

Jason Sinnes on their descent to 14K. Miller had also contacted Ryan Hokanson and Kirby Spangler, a pair of strong climbers who had originally been at the 14K camp to attempt the Cassin Ridge. Hokanson and Spangler had returned exhausted at 1600 from an aborted attempt on the upper West Rib and declined Miller's original request for assistance, saying that they needed rest and rehydration. They did offer to assist the following day.

(AAC Editor's Note: Over the next four days, search efforts continued. Twenty NPS personnel, eleven volunteers, and seven emergency hires participated. Extreme weather marked the first two of these days, then it cleared. Rowarth's body and the two packs were recovered, but, though many small items belonging to Vanderbeek were found, including his facemask, there were no signs of his body. The search was called off on May 29.

An investigation team was assembled, and the part of their analysis pertaining to the cause of Vanderbeek's fall and some of the subsequent actions follows below)

ANALYSIS

The accident was an unwitnessed slip on a 45-degree ice slope. Whether a crampon came off, the wind blew Vanderbeek off the ice, or he slipped from his ice axe or crampons is unknown. The decision to descend the slope unroped in high winds with limited visibility was the major contributing factor. Descending unroped and unprotected by a rope and anchors produced a situation where there was no way to arrest the fall, given the terrain encountered. Mike Vanderbeek was never located and is presumed to have come to rest in either the bergschrund or a crevasse on the Peters Glacier.

The following factors were analyzed for their contribution to the accident:

Communications and Decisions. The decision to descend and look for Rowarth was the first critical decision in the incident. Immediately after Rowarth's fall, Vanderbeek radioed Daryl Miller at the 14,000 foot camp to report the accident. After a short discussion, Vanderbeek told Miller he could not see the bottom, but if he could "take a look" a short distance lower he could probably spot him. Miller discussed with Vanderbeek the importance of maintaining communications and that Adrian Nature would head up with a radio to act as a human repeater on the ridge. Miller felt that Vanderbeek and Hurtado would take a short look and try to determine what had happened to Rowarth, but would wait for Nature to maintain good communications.

After speaking with Miller, Vanderbeek descended off the ridge to where Hurtado and Sinnes were waiting. It was Hurtado's understanding that they had permission to descend and begin a search for Rowarth.

The physical communications network between the Talkeetna Ranger Station, the 7,000-foot camp and the 14,000 foot camp consists of hand-held FM radios and cell phones. Generally, communication between the three areas is good. Communications are more difficult in remote spots on the mountain. The park currently has sufficient radios to outfit each team of rangers or volunteers. However, there are not enough to supply each one on the mountain. In addition, CB radios are used by the Base Camp Manager and climbers. The ranger patrols monitor the CB broadcasts.

Communications between the Vanderbeek/ Hurtado team and the 14,000 foot camp disappeared as the team descended the side of the ridge toward the Peters Glacier. Only Vanderbeek had a park radio, which was strapped to his chest. After he fell, Hurtado was unable to communicate with Nature, who was making his way up the ridge. Only by screaming was Hurtado able to make his position known to Nature when he reached the accident site.

Witnessed Fall. Many of those interviewed surmised that the impact of witnessing Daniel Rowarth's fall created a great sense of urgency to aid the victim. This "rescue fever" is not uncommon in life and death situations, whether among firefighters in burning buildings, or on mountainsides. The heightened sense of urgency produces a desire for speed and a sacrifice of personal safety. The objective facts do not indicate any sense of panic or any initial poor judgment or rash decisions. On the contrary, the opposite was indicated, and Vanderbeek and Hurtado discussed their decisions and proceeded accordingly. Nevertheless, the impact of a witnessed fall cannot be ignored. Several of those interviewed used the term "rescue mode" when speaking of both the procedures and mind-set they were operating under during the search. This meant, among other things, that they would use all necessary precautions, rope up, set adequate anchors, maintain constant communications, and follow a line of authority in decision making.

The Rope. The decision not to use the climbing rope was the second critical decision in the incident. Vanderbeek and Hurtado had a rope available, stowed in Vanderbeek's pack. They made a mutual decision not to use the rope. The ground where they started the search was fairly easy, and it looked from their vantage that they could descend safely without it. As the ground steepened, they continued to discuss roping up, asking each other if they still felt comfortable on the steeper terrain. When the snow changed to ice, Hurtado finally asked Vanderbeek to move toward the rocks so that they could rope up.

Weather. The wind was blowing steadily at an estimated 30-40 mph; gusts well above 40 mph were described by witnesses. The temperature was well below freezing. Visibility was variable due to the blowing snow and clouds. At times the Peters Glacier was visible 2,000 feet below, and at other times the climbers could see only a few feet ahead. Hurtado describes how, even wearing goggles, he had extremely limited visibility, and after the goggles began to ice up, he could see only three feet and often felt his way down.

Terrain. The route from 16,000 feet to 17,000 feet on Mount McKinley for the most part follows the ridgeline, winding on snow between granite rocks. The route stays to the north side of the ridge. The slope toward the Peters Glacier is 45-50 degrees and about 2,000 feet high. Depending on the wind and snow cover, ice and rock appear discontinuously along the slope. Most private parties do not use a rope along this section. Guided parties rope their clients up. A section next to "Washburn's Thumb" has fixed ropes into which most parties clip a carabiner or ascending device for protection. The altitude at Washburn's Thumb where the initial accident occurred is approximately 16,500 feet.

Equipment. Both Vanderbeek and Hurtado carried one ice screw and one snow picket each, a standard number for the West Buttress route. This would not have been enough protection for a safe descent using the rope 2,000 feet down a steep slope. The leader would have to have placed several pickets or ice screws on a 150 foot lead down the snow slope to have safely anchored the belayer and protected the second person climbing down-just the opposite technique of leading up a similar slope. The lack of protection may have been a factor in the decision not to use the rope; a fall of one of the two would have been fatal to both without a sufficient number or pickets or ice screws.

Each carried one standard 70 cm ice axe. It is not possible to self-belay with one ice axe as it would be with two ice tools, where one tool would always be attached to the ice while placing the other. In using one tool, whenever it is extracted from the ice, the climber is without protection. During the interview, Ryan Hokanson stated that he was, "very surprised that Mike had down-climbed the ice with one axe—shocked."

Each wore crampons. One of Hurtado's crampons slipped off his boot at the start of the descent, just out of camp. He explained that their over-boots made it difficult to have the bail of the crampons grab the boot welt securely. Vanderbeek's crampon slipped off just before Washburn's Thumb, and he re-attached it. This may have been an immediate cause of his slip on the ice. One crampon was found among the rocks on the fall line. The crampons have two front points, one of which was broken. It was analyzed by a metallurgist at Black Diamond Equipment, who determined that the break occurred as a result of a high impact, most likely during the fall, thus ruling it out as a cause of the fall. Beyond the first short distance on the moderate angle snow slope, the team did not have adequate equipment to protect their descent.

Experience. Mike Vanderbeek was a very experienced mountaineer with a long resume of difficult climbs on various high altitude peaks, including Mount McKinley. Vanderbeek had significantly more experience than Hurtado, particularly in the Alaska Range; however, Hurtado had climbed on McKinley previously. Hurtado, a third-year medical student, had participated in SAR incidents with the El Paso search and rescue team. Vanderbeek did not list any SAR experience or training in his resume. Their level of experience and expertise would not lead one to expect to find them in a compromising situation.
(Source: Daryl Miller, Mountaineering Ranger, and Investigating Team, consisting of Ralph Tingey, Reynold Jackson, Jay Liggett, and Jay Cabler.)

TEA headset and Bollard helmet suited to alpine operations

LA Sheriff's Montrose Mountain Rescue practicing snow and ice skills

(Photo—Royal Marines)

PART IV

MILITARY & LAW ENFORCEMENT OPERATIONS

INTRODUCTION

This section of the book will give law enforcement agencies a clearer picture of the training and logistics involved in developing an alpine or cold weather operations capability. To do this I have profiled the schools and training of the US mountain warfare community since they would be considered the subject-matter experts for agencies seeking out advanced training in this area. The experience that these training centers and their instructors have amassed on both mountaineering and cold weather war-fighting is a significant resource for every branch of the military and civilian law enforcement in need of that type of information.

These same mountain warfare schools and their staffs will find this section an easy to read guide and an overview of mountain warfare as it is approached by other branches of the military. It is only by looking beyond the green walls of their own schoolhouses and gaining knowledge of each other's training methods and philosophies that the military as a whole will be able to move toward standardization of both training and operational doctrine—both growing necessities in a time of joint task forces and multi-agency, international peace-keeping actions.

"The only thing harder than getting a new idea into the military mind, is getting an old one out." —Captain Sir Basil Liddell Hart, 1944.

23

MOUNTAIN WARFARE

A Historical Perspective

It is a common axiom that "those who do not learn from history are doomed to repeat it." It is also a known fact that a commander does not have enough time to make all the mistakes himself so must try to learn from the mistakes of others.

Both of these carry particular truth in mountain warfare, since alpine operations place troops in an unforgiving environment where the first mistake may well be their last. Winter campaigns throughout history are a grim reminder that the unprepared and ill-equipped fighting man will be in a world of hurt if forced to operate in cold or mountainous environment. So hopefully this short study of mountain warfare history will prove to be illuminating and set the tone for the planning, preparation and training needed to succeed in future conflicts.

As far back as 218 BC history tells us that Hannibal lost thousands of men and scores of his elephants while crossing the Alps to do battle with the Romans. Two thousand years later, military leaders were still losing troops to the environment as seen in the 1812 Russian Campaign when Napoleon lost over 450,000 men during the French army's winter retreat from Moscow.

During WWI (1914-1918) avalanches killed tens of thousands of soldiers, with some historians putting the number at 40-60,000 for the 3 winters of the war. According to one account, in December of 1916, 6,000 Austrian and Italian soldiers were killed in a 48-hour period.

Again during WWII (1939-1945), many deaths and casualties could be attributed to the high mountains and cold European winters. Thousands of Italian soldiers, including Alpini mountain troops, froze to death in the mountains of Greece and Albania, while Germans on the Russian front in 1942 lost 14,000 men to amputation alone.

It was during these same cold winters, particularly the Battle of the Bulge in December, 1944, that the US military came to realize the real threat of cold related injuries and quickly developed thick felt-lined boots for frontline troops.

The US Army 10th Mountain Division trained for three years before finally being committed to the war and in the closing months earned distinction in Italy by spearheading the allied drive through the Northern Apennines and into the Po River Valley. (The 10th Mountain was deactivated after the war, then re-activated in 1948 as the 10th Infantry Division. It was deactivated again in 1958 but came back to life in 1985 again as the 10th Infantry Division, but was re-designated the 10th Mountain Division (Light) in 1987 and served in the Gulf War in 1991 and Somalia in 1993.)

During the Korean Conflict (1950-53) the ill-prepared American and UN forces sustained thousands of casualties especially during the retreat from the Chosin Resevoir. Again, freezing conditions and inadequate clothing and footwear were the primary problems.

However, cold, exposure and avalanches were not the only killers in mountain warfare—altitude has also accounted for more than its share of deaths and casualties.

Doctor Charles S. Houston's book "Going Higher" recounts several instances of soldiers being incapacitated by altitude. In October, 1962, during a border conflict with China, the Director General of the Indian Armed Forces Medical Services issued a booklet containing the following orders: "Previously acclimatized individuals of high altitude areas, returning to altitude after leave, and individuals who have had a previous attack should be aware of the possibility of acute high altitude pulmonary edema whenever they are involved in a rapid ascent to an altitude in excess of 9,000 feet. Gradual acclimatization should be observed. Undue physical exertion at high altitude should be avoided during the first 48 hours after arrival. Persons suffering from viral infection should not make trips to high altitude".

In 1962 the Chinese crossed their troubled border with India high in the northwestern Himalayas. Maneuvers and skirmishes followed at altitudes as high as 12,000′ to 16,000′ in very difficult conditions. (Anyone who has trekked or climbed in the Himalayas will know how hard it is just to hike and climb let alone fight a war).

The Chinese soldiers worked and fought well at altitude since they had been living on the high Tibetan plateau for many months and were acclimatized to 15,000 feet. The Indian troops on the other hand, were hastily moved to altitude by truck and aircraft from the lower plains. The result was predictable—it is said that the Indians suffered more casualties from altitude sickness than from enemy action.

On a more positive note, General Inder Singh and other Indian doctors soon published the most extensive collection of cases of altitude illness yet reported. A few years later, Singh wrote a paper on acclimatization that was to become a milestone in our progress to understanding altitude sickness. Sujoy Roy, an Indian cardiologist, also published a book of respiratory and circulatory measurements made on highland natives and new arrivals.

This was no doubt a wake-up call to military commanders around the world who hopefully learned never to attempt a military action utilizing unacclimatized troops deployed too rapidly to altitude. This had not been a problem for mountain troops in either World War or Korea, but after the Indian experience the US Army became actively involved in high altitude research.

In 1979, Soviet troops lead by Spetznaz special forces invaded Afghanistan only to find themselves embroiled in a guerilla war with Mujahideen rebels. In extremely rugged terrain, with mountain peaks up to 24,000′ and cold winters, the Soviet troops sustained significant casualties with over 6,000 in 1980 alone. Even their armored Hind helicopters were vulnerable to rocket and small arms fire from guerillas positioned on peaks and ridgelines above their planned flight altitudes. US military advisors, working out of Peshawar, also had to deal with the cold and altitude when they made the long overland trek into Afghanistan through the mountains near the famous Khyber Pass.

FALKLANDS WAR—April, 1982

The first operation in the Falkland campaign was a plan to retake South Georgia—1400 kilometers to the SE of the Falklands in the South Atlantic—approximately 55 degrees South Latitude. D Squadron Mountain Troop (SAS) would be the first to insert onto the massive Fortuna Glacier to make a covert approach across the back of South Georgia to establish an OP on the five settlements and Argentinean positions.

After two failed attempts because of poor visibility, high winds and white-out conditions, sixteen men including Captain John Hamilton were flown from the HMS Antrim in three Wessex helicopters and deposited on the glacier.

In 50-knot winds gusting to 80, and each man carrying 80 pounds of gear, they roped up into four-man teams and began pulling their pulks (sleds) loaded with 200 pounds of additional equipment

toward the rim of the inhospitable glacier. The terrain was rough and dangerous with deep snow filled crevasses that collapsed under foot. The GPMGs soon froze up and after taking 5 hours to cover less than a half-mile in miserable conditions, the team leader, a sergeant with considerable mountaineering experience, decided it was too dangerous to go on. They sheltered in shallow trenches hacked out of the icy glacier with their ice axes since the small two-man tents were just blown away by the wind. 5 men tried to sleep in one tent and the others had to shelter under the sleds.

After a night of howling katabatic winds and flying snow, and realizing that progress was impossible, at 1000 Hamilton had the radio operator contact Antrim and request help. Finally at 1330 three helicopters reached them in a 15-minute weather window homing in on their emergency beacons and green smoke signal.

However, soon after take-off one of the three crashed into the surface of the glacier in white-out conditions but luckily only one man was injured. The other two helicopters returned to pick up the remaining men but on take-off one of those tipped over when it hit a ridge of ice concealed by the white-out conditions.

The remaining helicopter dropped its passengers on the HMS Antrim and took five attempts to finally recover the men stranded on the ice. The Wessex, which was only designed to carry 5 men, was severely overloaded with 17 aboard so all their gear had to be abandoned. Even then, it could not hover over the deck so did an intentional crash landing, fortunately with no injuries.

After re-thinking the plan, 10 days later four SAS Troops, an SBS team, ten Royal Marines and an Artillery Officer, 75 men in all, went ashore at various points and headed directly for Grytviken. The Argentinean garrison surrendered and Mountain Troop, D Squadron was the first to raise a flag over the settlement.

In continuing sub-arctic conditions on the Falklands, both the SAS and SBS ran long term

intelligence operations with little or no re-supply and were once forced to make two weeks of rations last a month. But for all the British success in the Falklands, there were also some bitter lessons to be learned—especially concerning the inadequacy of existing glove and clothing systems. One particular company of Royal Marines had a near 98% casualty rate with some form of cold weather injury, particularly to hands and feet, as did most ground troops serving in that campaign but to a lesser extent.

Cold and wet conditions are the two primary contributing factors—both very difficult to avoid when moving rapidly over exposed terrain in fog and sleet and living in muddy, rain-soaked fighting positions. Conditions made worse by freezing South Atlantic winds and a lack of Gore-tex, fleece, a "layering system" approach to clothing or waterproof gloves.

The Royal Marines never-the-less proved their ability to operate effectively under these conditions and cemented their place as Britain's premier cold weather warfare unit.

The biggest tragedy of the Falklands war came on May 19. While moving SAS personnel from HMS Hermes to Intrepid (an assault ship) a Sea King helicopter ditched from 400 feet into the sea. It capsized and filled with water killing 20 of the 30 men in the freezing water, including both squadron Sergeant-Majors.

MOUNTAIN WARFARE TODAY

Many wars have been fought in the mountains teaching us that the very nature of the mountains and their sometimes brutal, often unpredictable, weather patterns have caused more casualties on both sides than any direct enemy action. Time and again, we have seen how a poorly trained, poorly equipped unit can be completely annihilated before ever coming into contact with the enemy.

As with rivers, mountains form natural barriers that have come to represent the borders between many countries. They therefore also become the location for many border conflicts and the source of refuge for guerrillas and resistance groups. So with the United States' growing commitment as a global peacekeeper, it is inevitable that we will continue to have troops deployed in mountainous terrain well into the next millennium.

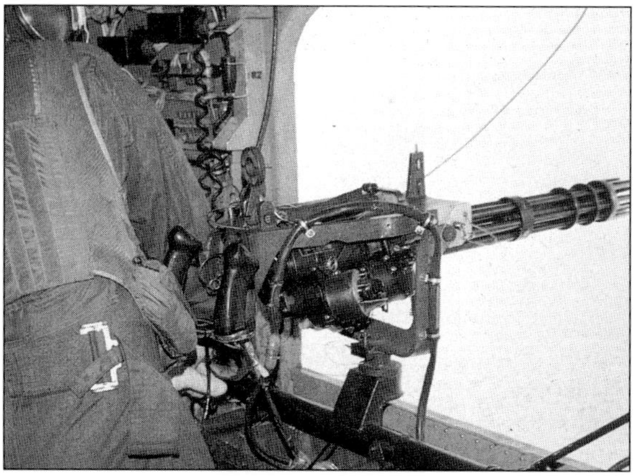

PJ CH53 mounted with a Minigun for protection during downed pilot rescues

Mountains are also a hazard for low flying aircraft in any operational theater, requiring the military to have a suitably trained combat search and rescue (C-SAR) capability. As with any air insert, there is always the possibility that after deployment of the C-SAR team, the rescue aircraft may not be able to stay on station because of inclement weather, low fuel or enemy fire, requiring that the team survive and evade for several days until a recovery operation can be launched.

To meet these demands, present and future, the military must maintain several different capabilities when it comes to war fighting in the mountains.

- Larger military units, battalions and companies, must be able to move and operate efficiently in the mountains without losing troops to the elements.
- Squads and platoons of mountain-trained troops need to be equipped to prepare the way and set rope installations for larger units and follow-on forces.

- Special operations units need to be able to operate in small teams for weeks or even months in an alpine cold weather environment on reconnaissance or direct action missions.
- Snipers must be trained to live for extended periods of time in the mountains to observe and report on enemy movement, and to engage pre-selected targets or targets of opportunity.
- Mobile training teams (MTT) need to have instructor level expertise in military mountaineer to be able to prep troops for deployment or train and guide local militia and resistance movements.
- Pilots need to be trained and helicopters equipped to support operations in a high alpine environment.
- C-SAR personnel, PJs and mountain-trained medics need to be able to locate, access, stabilize and evacuate all of the above.

To fill all these needs, the DOD operates several military cold weather and mountain warfare programs in the United States, designed to meet the military's global defense commitments. Apart from the defense concerns in Alaska and its proximity to the former Soviet Union, most US military programs are geared to a specific geographic region. The army programs are currently oriented toward the troubles in Eastern Europe, Bosnia and Yugoslavia while the Marines continue to train for their NATO commitments and the possibility of another North Korean or Chinese conflict, along with one of the Navy Seal Teams that also has that geographic responsibility.

Special Forces must maintain a broader spectrum of specialized skills including mountain warfare for operations in potential hot spots like Afghanistan, current base of terrorist Osama bin Laden, along with Bosnia, Pakistan, India, Africa, and to assist with drug interdiction programs in the mountainous regions of Peru, Bolivia, Colombia and Mexico.

"For those who want to win there is never enough time to train".

Commemorative plaque at the army mountain warfare school - "Dedicated to the courageous men of the 10th Mountain Division who Climbed to Conquer—25 July, 1981."

MARINE CORPS
UNITED STATES MARINE CORPS
MOUNTAIN WARFARE TRAINING CENTER
MOUNTAIN WARFARE
TRAINING CENTER
PICKEL MEADOW

24

MOUNTAIN WARFARE TRAINING

The origin of formalized mountain warfare training within the US military can be traced back to 1941 when officers from the National Ski Association and National Ski Patrol System petitioned FDR and the military volunteering their services to recruit and train army personnel in military skiing and mountaineering.

Although several military units had participated in winter warfare exercises in Alaska, New York, Washington and Wisconsin, they did not have standardized training or equipment. This was to change by the beginning of 1942 with the recruitment of Olympic class ski instructors and accomplished mountaineers; the formation of the 10th Mountain Division; and the establishment of Camp Hale, Colorado (9,480′) as the center for mountain warfare.

Note: Camp Hale was officially decommissioned in 1965 but still carries significant historical and sentimental value to those who trained and served there. For more information on the 10th Mountain, read "Soldiers on Skis" by Whitlock and Bishop.

There are currently almost a dozen schools within the US military system that cater to the needs of the mountain warfare community and several other foreign schools that are utilized for advanced training. Army, Navy, Air Force and Marine Corps all maintain schools or detachments to meet the needs of their own personnel and for cross training with other branches of the military.

One of the biggest and best known of these is the US Marine Corps Mountain Warfare Training Center (MWTC) located near Bridgeport, California, in the Toiyabe National Forest. It is also commonly referred to as just Bridgeport or Pickle Meadow.

This facility is a model of modern mountain warfare that allows for a "near realistic" combat training environment under the watchful guidance of MWTC instructors who are considered amongst the best in the military mountaineering business.

Most other branches of the military have relatively small schools or training detachments that train limited numbers in the technical aspects of military mountaineering and cold weather operations. In contrast, MWTC is a large facility dedicated to not only mountaineering but also mountain warfare training at the battalion level. Their training philosophy is to not only give the individual Marine the ability to live and move efficiently in the mountains, but also to give the company or battalion some exposure to war-fighting in mountainous terrain.

The two basic programs at MWTC are 28-day summer and winter mountain warfare packages. The summer package covers all the skills required to live and fight in the mountains including land navigation, terrain appreciation, wilderness living, climbing, survival, knots, rappelling, river crossing and rope management.

Marines who show an aptitude for climbing in the summer package are moved into an 8-day Assault Climber program where they learn to scale difficult obstacles and set ropes for the rest of the company or battalion. Assault Climbers are trained in more advanced rope management skills, belaying, rock climbing, placing protection and leading. They also become skilled in a number of rope installations designed to help move a large force with all its equipment over vertical terrain.

The winter package at MWTC places more emphasis on cold-weather survival, mountain living in a snow environment, patrolling and mobility on snowshoes and skis, skijoring, and cold-weather weapons training. Marines that show an aptitude for skiing are given additional training in the Scout-Skier program.

MWTC's state-of-the-art indoor training wall

One of the toughest programs at MWTC is the Mountain Leaders Course (MLC)—an intense six-week summer or winter program. On average, this course is made up of 40% officers and 60% NCOs, and is designed to give the successful Mountain Leader the skills and experience required to command and lead troops in an alpine environment.

MWTC also has an excellent survival package that is ideal for aviators, along with specialty programs tailored to suit the needs of medical personnel, engineers and civilian law enforcement. In addition, one of the most valuable specialty courses run at MWTC is the sniper satellite package that gives Marine scout-snipers the opportunity to gain first hand experience in slope-angle shooting and to log essential data in a realistic mountain environment. Apart from all the standard mountain warfare subjects the sniper package includes high-angle shooting, snow camouflage, hide preparation, stalking, tracking, mobility, advanced navigation and survival skills.

The US Army's primary training center is the Army National Guard's Mountain Warfare School (ARNG-MWS) near Jericho, Vermont. Established in 1983 and located at the Ethan Allen Firing Range this is also home for 3rd Battalion / 172nd Infantry, the only designated mountain warfare battalion within the army system. (The 10th Mountain Division keeps its name out of tradition but is no longer a dedicated mountain warfare unit).

The strength of the MWS training program lies in the instructors who are all accomplished mountaineers with both expedition and international mountaineering experience. The army also permits them to remain for several years in an instructional billet to take advantage of their experience. This is opposite to the Marine Corps' practice of rotating MWTC instructors back to the fleet after 3 or 4 years so as to get that expertise out into the field with combat units, and to give others the opportunity to gain instructor level training and experience.

In addition to the climbing and rope installation

ANG-MWS rappel tower

ANG-MWS artificial ice climbing wall

training done on the MWS reservation, advanced skills are practiced at Smugglers Notch on the side of Mount Mansfield (4,393´), Vermont's highest peak, and after completing both the 15-day winter and summer programs, soldiers are designated "Military Mountaineers."

If one were to identify any weakness with the Army's training, it would be the lack of tactical application brought about by the fact each class is made up of soldiers from several different units—Rangers, 10th Mountain, Special Forces, Reservists, and National Guardsmen. This mixed-bag approach is fine for technical mountaineering training but it negates any possibility of doing tactical drills or combat scenarios since each group has different small unit SOPs. Trainees are also not required to arrive at MWS with combat gear or weapons so only get to climb and train with rubber rifles.

At Ft. Greely in Alaska, the Army has their Cold Regions Test Center (CRTC), once run by a large contingent of both military and civilian personnel. CRTC has several ranges for weapons testing and the Bolio Lake facility used for testing individual and general use equipment; everything from sleeping bags and tents to field kitchens and anything used by a soldier to fight and survive in the cold.

Also in Alaska is the Northern Warfare Training Center (NWTC) whose motto is, "We Battle Cold and Conquer Mountains". NWTC provides training in arctic survival, navigation of land and waterways, military skiing, mountaineering and river crossing operations. Emphasis in their summer training is placed on mountain warfare, river navigation, mountaineering and glacier crossing techniques. During winter, training emphasis is placed on teaching soldiers how to live in the field, use and maintenance of clothing and equipment, snowshoeing and military skiing.

Ft Richardson in Anchorage is home for the Light Fighter Academy for 6th Infantry Div (Light); a multi-faceted academy that performs Arctic Light Unit Training and Arctic Light Individual Training, Combat Leader Course, Air Assault School, Sling Load Certification and Rappel Master Course.

Conducted at Ft Wainwright & Ft Richardson, the Arctic Light Individual Training is a five-day course given to all new soldiers in the division upon arrival. Snowshoeing, skiing, rappelling, glacier traversing, drown proofing, knot tying, and setting up the 10-man tent are a few of the skills introduced at these installations.

US Army Special Forces have considerable flexibility in their mountain warfare training and will send personnel to all of the above programs for training. 10th Group SF operates their own school at Ft Carson, Colorado and also send qualified individuals to overseas schools for advanced training. In addition, the Rangers have their own school at the Dahlonega Training Facility in Georgia where they run a basic 5-day mountaineering package.

The Naval Special Warfare Group (NSWG) maintains a detachment in Kodiak, Alaska where they run the Northern Maritime Training Course designed to give SEAL platoons a cold weather capability. This extremely intense 24-day program gives Seals the skills to infiltrate, operate and exfiltrate in sub-arctic conditions.

The SEALs also operate their own in-house rock

US Navy Seals training with their detachment in Alaska (NSWG)

climbing programs and contract with civilian mountain guides and instructors for more advanced training. Like any special operations unit, Seals have multiple uses for climbing skills including ship boarding and assault, scaling high-rise buildings, and aircraft takedowns.

The Marine Corps Special Operations Training Group (SOTG) also has a mountain warfare capability on both coasts that takes a Marine to Lead Climber level in 5 weeks. 1st SOTG trains at Camp

Pendleton, Bridgeport and Joshua Tree, California; while 2nd SOTG does their first 2 weeks at Camp Lejuene, North Carolina then moves to West Virginia for 3 weeks, culminating in a company night assault.

To qualify as an SOTG or MWTC mountain warfare instructor, a Marine must have completed both the 6-week summer and winter Mountain Leaders Courses (MLC) before applying.

Apart from conventional climbing and mountaineering, SOTG is also tasked with developing climbing methods for scaling coastal cliffs, ship boarding (VBSS) and buildings, along with training

Marines training with the Barrett 50 cal at ranges out to 2,000+ meters during the mountain warfare sniper package

Soldiers and Marines attending the Austrian International Rock and Ice Course (Sgt Clearman)

for helicopter deployment. They are also available as mobile training teams (MTT) for the winter training of units on float just before deployment.

The US Air Force Para-rescue jumpers (PJs) train out of Kirtland, New Mexico and maintain a full inventory of alpine rescue skills and related equipment. Their primary mission is the rescue of downed pilots, as is the NSWG/Seal C-SAR detachment out of Fallon, Nevada.

On an international level, the US military has sent individuals and small teams to train with the Austrian Mountain Troops, Canadian Mountain School, Norwegian Jaegers, Italian Alpini, French Mountain School, Argentineans, Chileans, Peruvians, Belgians and British Royal Marines. Select SF personnel who speak German are even fortunate enough to go through the German's internationally renowned Bergfuhrer Mountain Leader Course.

Stepping outside the system, the military has taken advantage of the training from several civilian programs including those run by American Mountain Guides Association (AMGA), Alpine Skills International (ASI), Outward Bound, Alpine Ascents (AAI), Jackson Mountain Guide Schools and National Outdoor Leadership School (NOLS).

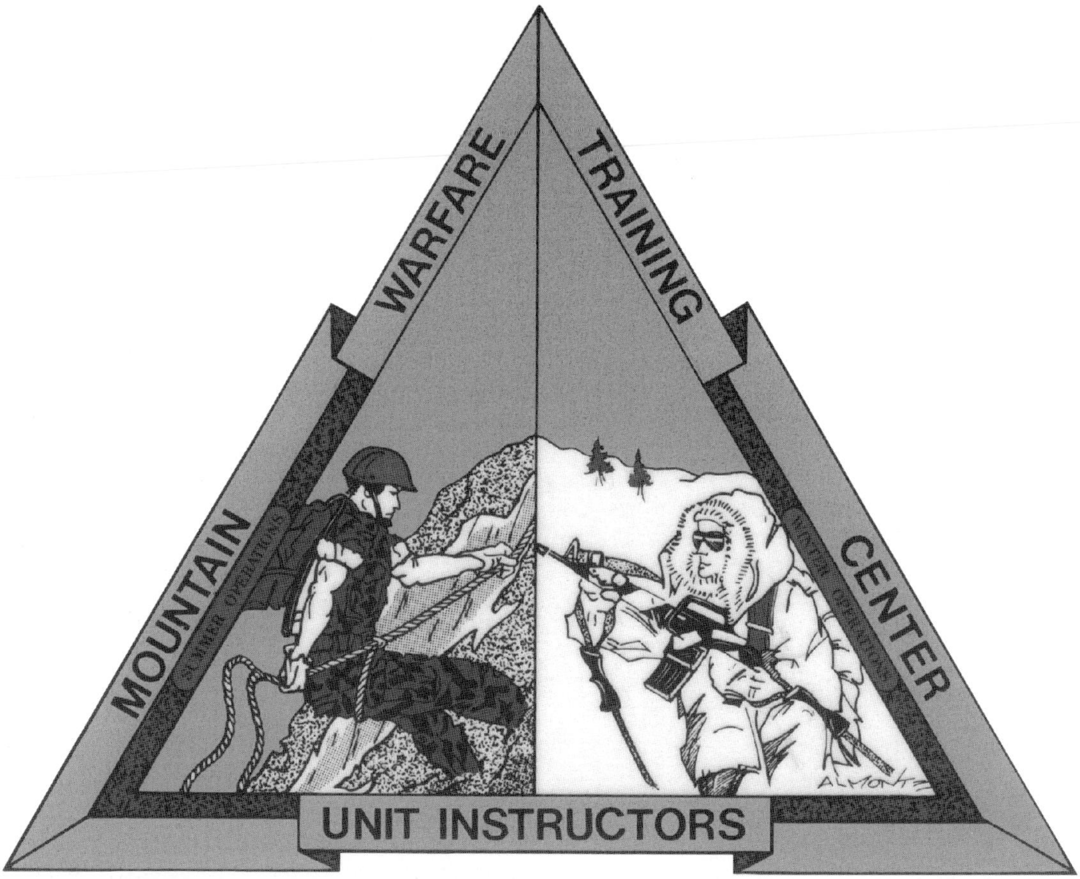
MOUNTAIN
WARFARE
TRAINING
CENTER
UNIT INSTRUCTORS

25

MILITARY MOUNTAINEER TRAINING

Mountain Warfare training is a tremendous confidence builder and probably one of the single most practical courses in the military inventory, supplying the individual fighting man with useful skills that will have applications throughout his military career.

Each branch of the military approaches mountain warfare training from a different direction reflected in a wide variety of training standards, certification and related manuals. However, as of 1998 there is an effort being made to standardize the training and certifications amongst all US mountain warfare assets. The US Army 90-6 Mountain Operations manual is under revision and representatives of each branch of the military are trying to develop a Universal Military Mountaineering Publication. They also intend to create Universal Individual Training Standards (UITS) to enable military commanders to quickly and easily estimate not only their own war fighting capabilities but also those of other units in a joint operations environment.

Until this happens, USMC, Army, Special Forces and Rangers will continue to run their own programs, but all with the same fundamental goal—to give the individual fighting man the best training and equipment available to operate efficiently in the mountains. To achieve this goal they must take the soldier or Marine's existing combat infantry skills and apply them to the mountain environment.

TRAINING LEVELS

The USMC currently requires an individual to attend both a summer and winter mountain warfare package (27-28 days each) to be qualified as a Military Mountaineer. This gives a Marine exposure to both seasonal extremes and experience in most aspects of mountain travel from balance climbing and rope installations to snow shoeing and skiing.

However, this can be a big burden to the operational commander who must now lose his battalion twice to get a single certification, with the reality being that snow skiing is used so infrequently that it is hard for the commander to justify the time and expense of sending his troops to MWTC twice. In addition, snow shoeing and basic crampon techniques can easily be taught in the field by a qualified military mountaineering instructor if this capability is required for a particular movement or area of operation.

The next level of training for the Marine is Assault Climber, a tough 8-day program within the 28-day summer package for Marines who have shown an aptitude for climbing. Within a Marine rifle company, basic manning calls for a minimum of 8 assault climbers, which is also reflected in the company MACK kits that have eight sets of assault climber gear.

Other advanced levels of training at MWTC include Scout-skier, Mountain Leader, Instructor, and within the ski program, military skier (MS), military ski instructor (MSI) and military ski exam-

Marines helo insert into the mountains under the supervision of an MWTC instructor

Rappelling with full packs should be first negotiated with safety belayers below

iner (MSE). On the 6-week Mountain Leader Course (MLC), the class will also be taken up to Mt Shasta for glacier travel and crevasse rescue training. MWTC instructors get their advanced training by working with local mountain guides and ski schools such as the Alpine Skills Institute (ASI).

The ARNG-MWS runs summer and winter packages (15 days each) to qualify the soldier as Basic Military Mountaineer, or to earn what the army calls an Echo designator (SQI-"E"). Various army specialties carry different designators, for example, Infantry is 11B (Bravo), Paratroopers P, Rangers V, and Instructors H.

The next level of training in the army system is Lead Climber, where soldiers with aptitude are trained 10 at a time with a 2:1 student/instructor training ratio.

The third level is military Mountain Guide, which is not used much since it is more a product of experience than training. To achieve this level, a soldier would need to be first a military mountaineering instructor, and then gain two to three years additional field experience training and operating under a wide variety of alpine conditions. This should include glacier, avalanche and big mountain experience.

Special Forces use a basic Level 1, 2, and 3 to grade their military mountaineers with their premier program being the Mountain Master Trainer Course.

For the Rangers, the progression begins with Basic training at Dahlonega (5 days) followed by sustainment training on the climbing wall and Mt Yonah climbing lanes. To progress to an Intermediate (Assault Climber) level they go outside for training at MWTC or Vermont and possibly an expedition to Denali for more realistic experience.

The ideal manning for the Rangers is to have a minimum of 2 Assault Climbers and 5 Basic Mountaineers per platoon and 1 Mountain Leader per company.

To become an instructor, a Ranger must be an assistant instructor for 3 months before going in

One rope river crossing at night

front of a class, plus complete the Infantry Trainer's Course and Combat Life Saver; then gain at least 9-12 months experience before becoming an instructor.

The one possible downside to the MWTC programs is the Marine Corps attitude toward all forms of specialized training. Even though it takes two to three years for an individual to become a competent mountain warfare instructor, the Marine Corps rotates instructors out of MWTC every 3-4 years with a significant loss in expertise and experience. However, the Marine Corps logic is that by rotating that Marine back to the fleet or a combat unit, he will take that expertise into the field with him. This also gives another Marine the opportunity to gain instructor level training. Both are very valid arguments that reflect the Marine Corps philosophy that every Marine is first and foremost a rifleman with an essential combat mission.

ESSENTIAL MOUNTAIN WARFARE SKILLS

Mountain warfare training is designed to build on the existing field skills of the soldier or Marine, but cannot make the assumption that he is proficient in all aspects of outdoor living and travel. Initial mountain warfare training also needs to include conditioning runs and hikes with a pack to prepare the trainees for the steep terrain and altitudes encountered in mountain warfare.

In addition to physical conditioning, all mountain warfare training programs should begin with a thorough review of the following field skills:

- ECWCS (extended cold weather clothing system) use and care
- Mountain Living—tents, stoves, cooking, sleeping systems
- Mountain Navigation—map, compass, GPS
- Water Procurement—sources, boiling, filtering, purifying
- Health & Nutrition
- Wilderness Medicine—heat & cold related injuries; mountain sickness
- Cold Weather Survival—shelters, signaling, stress management
- Knots & Rope Handling—coiling, deployment, recovery
- Basic Rappelling

Soldiers prepare an A frame to support a suspension traverse.

BASIC MILITARY MOUNTAINEER

Even though each branch of the military has their own method of classifying an individual's training level in mountain warfare, the general consensus is that the first standardized level should be that of Military Mountaineer and should take a minimum of 2 weeks of training.

It is also the recommendation of military mountaineering subject-matter experts that the Military Mountaineer designation should not require ski training since the US military has never skied in combat. As stated earlier, it is also a burden to the battalion commander to have to send his people to two schools for one specialty designation. Even after 14 days on skis at the MWTC winter ski package, the average Marine is not very proficient on skis, in fact most are quite terrible and the skills are perishable. The preference is to simply give them snowshoes, which require little or no training time.

In addition to the subjects listed above, the Basic Military Mountaineer training should include:

- Identification & Care of Mountaineering Equipment
- Essential Knots & Hitches
- Rope Selection, Care, and Handling
- Rappelling & Belaying
- Balance Climbing
- Basic Anchors
- Use & help build Rope Installations
- Top-rope Climbing
- Climbing as #2

The Basic Military Mountaineer should be taught to construct and operate the following rope installations under the supervision of an assault climber, mountain leader or guide.

- Basic Rappel station
- Fixed Lines
- Prusik Ascent System
- Vertical Hauling System
- Suspension Traverse
- Casualty Evacuation System
- 1, 2 & 3 Rope Bridges

War fighting skills such as live-fire training should be incorporated into all facets of training

Skijoring is taught as part of the winter package

ASSAULT CLIMBER

It is not feasible to train an entire battalion in all aspects of mountain warfare—so those with an aptitude for climbing and rope work are selected for specialized training as assault climbers. Their job is to move ahead and construct various rope installations to facilitate the movement of follow on forces over difficult terrain, ranging from coastal cliffs to mountain ridges and glaciers. When leading a cliff assault, assault climbers should be able to first climb the cliffs as a two man climbing team, and then establish climbing lanes with the use of fixed ropes, wire ladders and hauling systems.

The essential skills for an Assault Climber are:

- Supervise & Establish Complex Anchors
- Supervise & Establish Belays
- Perform a Rappel Pick-off
- Execute a Multi-pitch Rappel
- Lead Climb at 5.5 level of difficulty in combat boots
- Supervise & Construct Basic Mountaineering Installations
- Vertical Hauling System
- Suspension Traverse
- 1, 2 & 3 Rope Bridges
- Supervise, Establish & Lead Fixed Ropes
- Handle Medical Evacuations in Steep Terrain
- Perform a High Angle Rescue
- Analyze the Risks involved in any operation

Marines prepare a ladder installation for follow-on forces

Note the belayer is anchored to a large rock to prevent being pulled into the cliff face in the event of a fall by the climber

Note that all climbing is done in issue combat boots

Lead climber working a crack

Prusik ascent training using Texas prusik system

MOUNTAIN LEADER

The position of mountain leader (ML) is a critical part of mountain warfare since the unit commander will be looking to the ML as a subject-matter expert capable of dispensing sage advice on all aspects of alpine operations. This will include troop preparedness, equipment suitability, mobility techniques, route selection and operational tactics suited to the mountain environment.

A basic mountain leader course is a 6-week program, and if fortunate enough to attend both summer and winter packages, the officer of NCO will have amassed a total of 12 weeks training in military mountaineering and mountain warfare. Although these are excellent programs and longer than many other specialty courses, they do not give an individual the time to amass the experience necessary to function as a mountain guide.

In addition, mountain warfare doctrine generally calls for a minimum of one ML per company, but that is like having one mountain guide for a group of up to two hundred. This is far more than one man can safely supervise in the mountains. A more realistic ratio would be for each rifle platoon to have its own ML assisted by at least two assault climbers, and to then have the senior ML attached to the headquarters platoon to advise the Commanding Officer.

The mountain leader should be capable of the following:

- Risk assessment in hazardous mountain terrain
- Lead glacier travel and crevasse rescue operations
- Instruct and supervise high angle snow and ice climbing
- Supervise and execute various mountain rescue techniques
- Organize an avalanche rescue
- Organize and instruct ski training and over-snow mobility
- Perform duties as a basic mountaineering instructor

All of the above is a rather tall order for someone who has only had 12 weeks of training that is not supported by considerable field experience. The ML will be further hard pressed if the entire company has not had formal mountain warfare training in the preceding twelve months.

To avoid unnecessary accidents and casualties, military units who are not current and are anticipating operations in the mountains should request a mobile training team (MTT) of mountain warfare instructors be sent into the field with them. These specialists can issue and fit the cold weather and mountaineering equipment, run refresher programs and assist the ML in bring the troops up to an acceptable level of operational readiness.

INSTRUCTORS

Since the mountain environment can be remote, inhospitable and as cruel as it is beautiful, mountain troops must be in excellent physical condition, highly motivated and professionally trained. The training must be comprehensive and to be of value must reflect reality—"For as we fight, so must we train". The responsibility for all this falls squarely on the shoulders of the instructors within the mountain warfare community.

Mountain warfare instructors are the heart and soul of mountain warfare training so must be selected carefully. At a minimum they should have com-

A right of passage for all graduates of the Royal Marine's Mountain Leader course (Royal Marines)

Mountain warfare instructors are the heart and soul of any program and carry a tremendous responsibility for both the safety and training of the troops under their supervision. They are referred to as "Red Hats," which distinguishes them in the field from the white helmets worn by students.

pleted both summer and winter mountain leaders courses and shown considerable aptitude and personal motivation for mountaineering. They should also be encouraged to strive to improve their skills and experience. This is best done through exchange programs with other branches of the military and overseas mountain warfare units, participation in big mountain expeditions such as Denali, and by attending advanced training outside the military system with nationally recognized instructors.

However, maintaining expensive specialized training assets such as mountain warfare is of little value unless they are a force multiplier for the overall combat capability of larger conventional ground forces. Mountain warfare must be more than just a military version of Outward Bound. Using the Marine Expeditionary Units (MEU) as an example, sending a battalion out for mountain warfare training must give the MEU commander the capability to operate more effectively well beyond the coastal region and up into the mountains. It must be one of those specialties where a relatively small amount of training can have a significant impact on a unit's mobility and combat readiness. Only by giving Marines practical war fighting skills in rugged terrain, will the mountain warfare instructors prove their value to the system.

MOUNTAIN GUIDES

The title of Mountain Guide is one of the hardest to bestow on anyone, military or civilian, since this position carries with it the responsibility for the safety of everyone under his or her leadership. Individuals will blindly follow guides into the mountains assuming that they know their business, will keep them safe, and are expected to not only bring them home alive, but to make their adventure a pleasurable experience.

In the presence of a mountain guide, other members of a group will pay little attention to navigation, route finding, weather conditions or the difficulty of terrain, further assuming that the guide is doing all that for them. They will accept the guide's advice on all matters to do with climbing and safety and will rope-up and follow a pitch without question.

This makes the position of Mountain Guide the single most important job on the mountain, earned more through experience, demonstrated expertise and a proven safety record than any formal certification. It is also most definitely not a positioned derived from rank.

At a minimum, military mountain guides should have been mountain warfare instructors with at least 2 years experience in advanced mountaineering, along with formal training in avalanche rescue, glacier travel and big mountain expedition climbing. Even then, the aspiring guide should limit his guiding operations to the types of terrain where he has experience, remaining very aware of his own strengths and weakness. His ego must be set aside for the safety of the troops he is leading.

By civilian standards, 2 years would be considered marginal with many guiding services requiring experienced climbers to serve long internships and apprenticeships with senior guides on the routes they intend to guide. A strong climber must demonstrate that he has the **caretaker personality** necessary to be a professional guide since a single accident can destroy the reputation of both the guide and the guide service. Again, the certifications and individual climbing skill mean less than actual experience successfully guiding a specific peak or region.

However, in an organization that is anal about certifications, it is recommended that the military look to the standards and training of the American Mountain Guides Association (AMGA) to validate future mountain guide certifications. Organizations like the AMGA and its affiliate schools and instructors have the combined knowledge, teaching skills and experience to train and certify guides beyond the Mountain Leader level. They also require proof of a significant climbing and guiding resume to go along with professional membership and certification.

Professional Mountain Guiding is one of the most physically and mentally demanding professions on this planet. When a guide is doing everything right, nobody really notices, but when a guide makes a mistake, people can get seriously hurt or killed. The guide must be able to shoulder the extra burden of a heavy load, but maintain sufficient reserves to execute a rescue. He must be able to assess the skill level of each individual and select the appropriate routes. Only by constantly monitoring the condition of his group will he know when and how hard to push and when to slow down and rest.

The joy of leading a difficult pitch must be sacrificed for the satisfaction of leading the other climbers through safely. He must inspire confidence by making things look easy, while constantly assessing the risks and being alert for any possible dangers. And most of all, the mountain guide must have the good judgment and courage to turn a climb when he believes the route is unsafe, it is beyond the capability of his group or conditions are deteriorating. It is always preferable to argue with a bruised ego than to have to drag a casualty off the mountain.

One does not become a mountain guide in the civilian world for the financial remuneration since it is not significant. Guiding is more a vocation or a calling for hardy individuals who derive their rewards from working in the mountains, seeing others learn and then helping them to achieve their goals. Similarly, the military mountain guide must find personal satisfaction in leading troops safely through difficult terrain that would have otherwise stopped an advance or caused unacceptable casualties.

MWTC rope locker with North Face 4-man tent on display.

Below: The full content of the assault climbers' MACK box

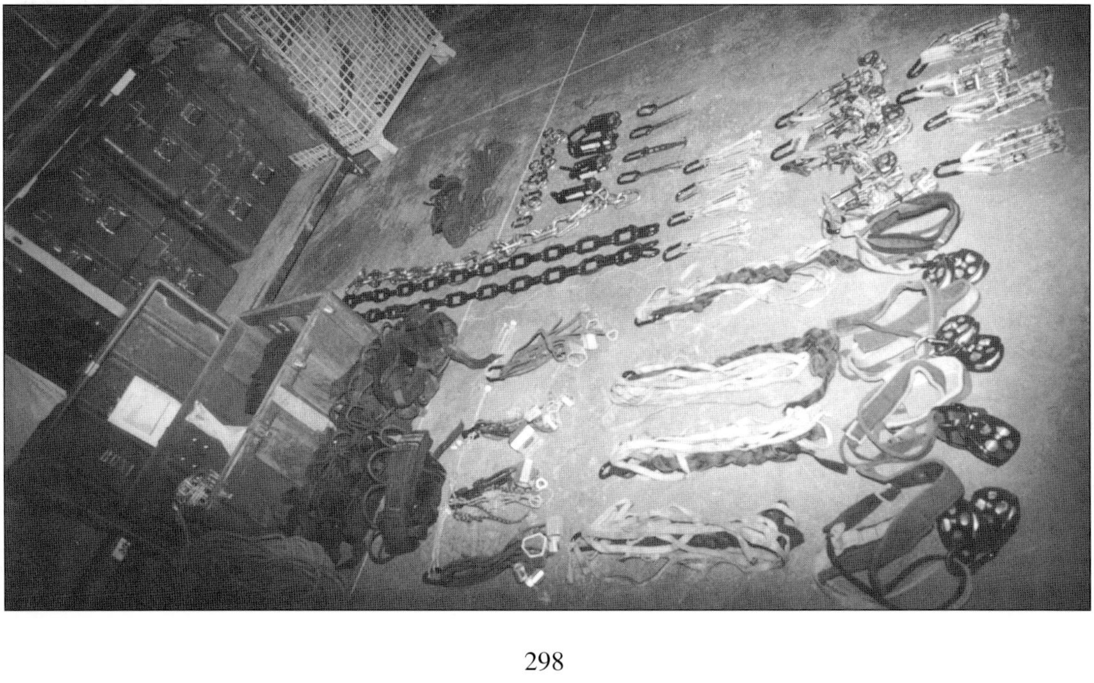

26

MOUNTAIN WARFARE EQUIPMENT

"Weapons are an important factor in war, but not the decisive one; it is man and not materials that counts"

—Mao Tse-tung

We are fortunate in the United States in that we have access to some of the best equipment in the world, even in the bureaucratic world of military procurement where the "low bidder" is often the first choice of the Pentagon pencil pushers. The same is true for many of the European mountain warfare units found in France, Switzerland, Austria, Germany and Norway where cold weather alpine activities are a way of life. However this overall quality of equipment in the US and Europe is not indicative of the standards in all countries.

When operating with foreign military units, particularly in semi-third world regions, it is not unusual to find that their soldiers are poorly equipped and that their issue equipment is of a significantly lower quality. Immaterial of their level of training, this equipment deficiency can severely compromise a military operation or movement in the high alpine regions encountered on many borders.

While doing research for this book I accompanied a very well equipped US military group on an expedition to South America, however the disparity in equipment with their local counter-parts became immediately evident when we moved higher in the mountains. The host mountain unit was tasked with guiding the US group throughout the expedition, but as the teams moved up through 17,000′ and spent their first night at that elevation, the South Americans began having problems that could be directly attributed to poor equipment. Even though they were in their own area of operation and had a good appreciation for the terrain and the weather, their tents and sleeping bags were simply not designed to handle the harsh conditions on the mountain. So within 24 hours one of their members had to be evacuated with pneumonia and the others quickly withdrew to the relative comfort of the forward operating base at 14,200′.

I noticed the same problem with a group in Nepal that was very strong and moved fast at lower elevations but fell apart and became walking casualties above 15,000′. The problem there was also substandard footwear, clothing and sleeping systems, compounded by the fact that they were lowlanders and not as acclimatized to altitude as the local Sherpas.

Within any mountain warfare school, or combat unit that has an area of operation that includes alpine regions, there must be some form of on-going test and evaluation of gear suited to military mountaineering and their perceived mission.

Much of the army's equipment development comes through the labs at Natick Soldier Center (NSC) in Natick, Massachusetts whose mission statement is to "Maximize the individual war-fighter's Survivability, Sustainability, Mobility, Combat Effectiveness, Quality of Life, and to treat the war-fighter as a System."

For the Marine Corps, test and evaluation is done at Marine Corps Combat Development Command (MCCDC, pronounce "McCidic") at Quantico, Virginia.

Both Natick and MCCDC routinely consult with subject-matter experts at the schoolhouse level and send equipment into the field to be evaluated under more realistic conditions. However, both these groups like any large bureaucracies move slowly and needed equipment can often take months or years to get to those who need it most. There is nothing more frustrating for the end user, the man freezing his butt off in the field, than to know that much needed items are in the system but delayed awaiting approval before they can become general issue.

Special operations units, who are fortunate in that they have their own discretionary budgets, will often by-pass this system by purchasing small quantities of whatever they need direct from manufacturers and distributors. This "brown bag" purchasing is done to meet "mission specific" demands or under a letter of exigency in time of war.

However, for the average grunt, state-of-the-art equipment is not always available in the system so they are forced to purchase personal equipment to augment their issue gear—especially warmer head gear, glove liners, socks, wide-neck water bottles, compasses, small waterproof flashlights, multi-tools, energy bars, etc. Fortunately, in the last two years there has been a quantum improvement in military equipment development and procurement in this area so many of these items are now becoming issue. For example, there is now a wide-neck canteen bottle in the system and the Marine Corps is issuing OR Pro mitts and gaiters for mountain operations.

ESSENTIAL COMBAT EQUIPMENT

Whichever avenue a group takes to procure better equipment, there must be a formal test and evaluation process that ensures the group is getting the best equipment that their budget can afford (See Equipment related chapters in this book). Detailed lists should also be complied of exactly what is needed for clothing, tactical load-out, sustained field operations, over-snow mobility and basic mountaineering.

Basic **Combat Clothing System** that would be standard for all mountain and cold weather operations:

- ❑ ECWCS Camo Parka & Trousers
- ❑ Cold Weather Pile Trousers & Jacket
- ❑ Cold Weather Hat or Balaclava
- ❑ Polypro or Capilene Long Underwear
- ❑ Insulated Boots with appropriate Socks and Gaiters
- ❑ Gloves with liners or Trigger-finger Mitts
- ❑ Over-whites—Parka, Trousers and Mitts
- ❑ Helmet w/ Camo & White cover

Additional pocket items that should be carried at all times:

- ❑ Pocket Knife or Multi-tool
- ❑ Whistle
- ❑ Pressure Bandage
- ❑ Sunscreen & Lip-block
- ❑ Sunglasses or Glacier Glasses
- ❑ Survival Kit—fire starter, signal device, water purification, energy snacks, etc
- ❑ Notebook & Pencil
- ❑ Lighters
- ❑ Flashlight w/ red lens
- ❑ 550 parachute cord (20´)

Tactical or Assault Load carried in the load bearing vest (LBV), butt pack and backpack. This is the Line 1 equipment that must be carried for short duration missions, patrolling or the final assault phase of an operation.

- ❑ Extra insulating layer, socks and gloves
- ❑ ECWCS / Gortex outer layer (if not being worn)
- ❑ LBV with 2 quarts of water and first aid kit
- ❑ Rations for time away from bivouac site (1—3 MREs)
- ❑ Isoppor mat (Ridge-rest) for ground insulation
- ❑ Skis, poles and/or snowshoes

- ❏ Mission essential gear as required:
  - ❏ Primary weapon with accessories and ammunition
  - ❏ Extra ammunition, demolitions and pyro as mission dictates
  - ❏ Optical gear, binoculars, scopes, NVDs, rangefinders, etc
  - ❏ Communications equipment w/spare batteries
  - ❏ Navigational equipment, maps, compasses, GPS

Full Combat Load (Level 2) that would be carried on longer duration missions and equipment essential for an unplanned bivouac.

- ❏ Large Pack—Vector, SF Elite, Gregory
- ❏ Sleeping bag w/waterproof (WP) stuff sack
- ❏ Bivi sack
- ❏ Snow shovel to dig shelters, fighting positions and rescue avalanche victims
- ❏ Stove—MSR Whisperlite w/pump and spares
- ❏ Fuel bottle w/fuel (White gas)
- ❏ Thermos
- ❏ Poncho for expedient shelter

Note: The last five items can be carried one per two-man team to reduce overall weight.

The Existence Load (Level 3) is any equipment that is not immediately needed and can be brought up to the forward operating area when the situation allows. In civilian terms this would be base camp gear.

- ❏ Extra insulating clothing & underwear
- ❏ Extra socks
- ❏ Extra gloves, liners and mitts
- ❏ Toiletries, sewing kit, etc
- ❏ Reading material
- ❏ Flashlight w/batteries

Group Equipment for a 4-man fire team, usually carried in the sled:

- ❏ 1 ECW tent complete w/fly, poles, stakes and stuff sack
- ❏ Extra fuel for stoves
- ❏ 1 case of extra MREs
- ❏ 2 Shovels and pioneer gear as required—hatchet, rope, etc
- ❏ 2 pr climbing skins for skis
- ❏ 1 whisk broom
- ❏ 1 team cook set
- ❏ Candles
- ❏ Trash bags
- ❏ Avalanche Probes & Beacons (if necessary)

All of the above gives the fighting man the ability to move, shoot and communicate but still does not give him the capability to climb or negotiate cliffs or steep mountains. To fill this need, the military pre-packages everything that the military mountaineer or assault climber will need. For the Army this is the SOFME kit (Special Operations Force Mountaineering kit), and for the Marines it is in the form of four large containers called a MACK (Marine Assault Climbers Kit).

According to the manual, "the MACK is a comprehensive collection of climbing equipment that enables the Marine rifle company (reinforced), approximately 200 men with organic equipment, to negotiate an average 300-foot vertical danger area. The kit contains sufficient climbing equipment to equip four 2-man climbing teams plus the additional items necessary to supply the remainder of the rifle company. The climbing teams use their equipment to conduct two-party climbs over vertical obstacles and establish fixed rope installations and vertical hauling lines to facilitate movement of the remainder of the company."

MACK container #1 contains all the Climbing Team equipment and #2, #3 and #4 contain the company climbing equipment. The itemized inventory includes:

- ❏ Carabiners—Aluminum Non-locking
- ❏ Carabiners—Steel Locking D
- ❏ Nuts—Wedge & Hex—Size #1–#10
- ❏ Camming Devices—Size #00–#4
- ❏ Pulleys—Rescue
- ❏ Belay plates
- ❏ Ascenders
- ❏ Tool—Nut removal
- ❏ Climbing Rack bag

Gore-Tex Gen 2 ECWCS parka

Wide mouthed canteens

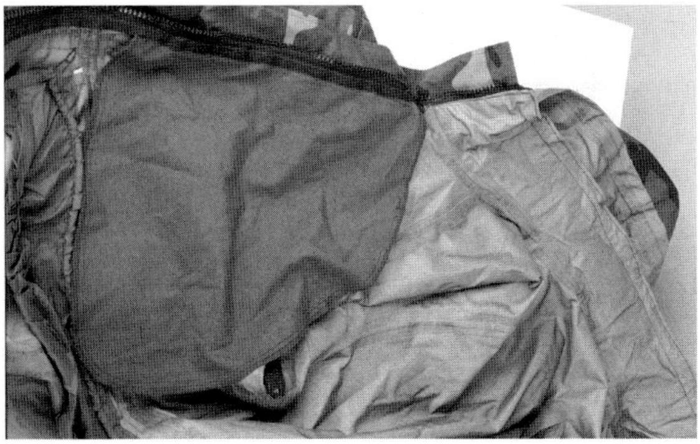

ECWCS features fully taped and sealed seams, large map pocket and integrated snow skirt

Eagle's enhanced load bearing vest with many well designed features such as mesh map pockets

Plastic mountaineering boots from Kolflach and Civetta

Leatherman and Gerber multi-tools, Benchmade's AFO automatic folder, and Cold Steel's tactical fixed-blade knife

- ❑ Climbing Harnesses—Black
- ❑ Ropes—Dynamic 11mm x 165´ (OD)
- ❑ Ropes—Static 7/16″x 165´ and 300´ (Black)
- ❑ Rope Bags
- ❑ Accessory cord—Type I (5.5mm Kevlar)
- ❑ Accessory cord - Type II (7—8mm)
- ❑ Webbing—Nylon (1" tubular) Black or OD
- ❑ Ladder—Cable

The Assault Climber 2-man Climbing Team Rack contents:

8 Carabiners, Locking
40 Carabiners, Oval non-locking
2 Gear Slings
2 Nut Picks
2 Belay Plates
2 Ascenders
2 pair Gloves
2 Sling Ropes
1 Dynamic Rope, 165´ OD
1 Static Rope, 165´ Black
2 Rope Bags
12 Stoppers—Size #1- #12
10 Hexentrics—Size #1–#10
9 Cams—Size #00–#4
2 3″ Sewn Runners
2 6″ Sewn Runners
2 12″ Sewn Runners
2 24″ Sewn Runners
2 48″ Sewn Runners

COMMERCIAL EQUIPMENT

Military development and procurement, as it relates to mountain warfare, has come to realize that there are numerous commercially available items that meet all the requirements of the military. These are readily available, the first choice of civilian professionals, and stand to save the military the cost of independent development (re-inventing the wheel). This is in big part thanks to the insistence of subject-matter experts like the MWTC, NWTC and MWS training staffs.

Even though equipment used in mountain warfare receives considerably more use and abuse than its civilian counterparts, the following are just some of the commercial grade items that have proven themselves worthy and been accepted into the system: OR Pro gaiters and Ranger over-boots; OR Pro gloves and Mutant mitts; various Gore-tex, Polartec, fleece and Capilene products; Ridge-rests and Thermo-rests; Wilderness Engineering sleds; North Face tents; MSR and Coleman stoves; Petzl headlamps; Pelican flashlights; MSR water purifiers; Black Diamond Cams and ATCs; Omega carabiners; Gregory packs; Nalgene water bottles; Danner combat boots; Kolflach and Lowa plastic boots; Grivel crampons; Savage gaiters; Petzl and CMI ascenders; SMC and BD ice axes, BD and Grivel ice tools; Misty Mountain and BD harnesses; PMI, Sterling, Bluewater and New England ropes; Atlas snowshoes; Thommen altimeters; Life-Link and SOS Shovels; Ortovox and SOS avalanche beacons; Ascension and Colltex ski skins; Eagle and Blackhawk load bearing gear; Elan and TUA skis; Silvretta bindings; Ascension climbing skins; Leki poles; Leatherman, Gerber and Buck multi-tools; Suunto compasses; and Eagle, Blackhawk and Camelbak hydration systems.

CONCLUSION

Even though all equipment intended for use in a combat environment must be rugged and reliable, climbing and mountaineering equipment must still be as light as possible. The weight that a civilian mountaineer must carry is already significant, but when the additional combat load of a fighting man is considered, weapons, ammunition, demolitions, communications gear, etc, the weight becomes staggering.

If the military mountaineer is loaded down with too much weight, he will not be able to move efficiently, fight effectively or sustain himself for prolonged periods in the mountains. Excess loads will also cause increased fatigue, loss of stamina, decreased alertness and a higher probability of injury. This makes it that much more important to give the military mountaineer not only the best equipment available, but also the lightest.

(Photo—Sergeant Clearman)

27

MOBILITY IN THE MOUNTAINS

"Take advantage of the enemy's unpreparedness; travel by unexpected routes and strike him where he has taken no precautions."

—Sun Tzu, The Art of War.

The ability to out maneuver the enemy is pivotal to tactical planning and operations but mobility in the mountains is not a simple matter. Steep terrain, deep gorges, snow filled valleys, icy cliffs and fast flowing rivers are not only obstacles that slow movement, they also serve to channel troops into more navigable but predictable terrain. This channeling can work to a unit's advantage when trying to anticipate enemy movement, but also works to their detriment if facing a "mountain-smart" enemy.

The first priority for the military planner or intelligence officer will be to gain a detailed appreciation for the terrain in the planned area of operation (AO). Mountainous terrain can be first separated into two categories, Coastal or Interior, each with distinctly different weather patterns. The elevation on the mountains can then be divided in to Alpine and Sub-Alpine, usually delineated by the tree line. The terrain on the mountains can then be classified into one of three classes:

- Unrestricted—Going will be easy
- Restricted—Going will be slow
- Severely Restricted—No Go

With this information, the planner will be able to decide on the type of unit most suited to the operation; the clothing and equipment that will be needed; any specialized training that may be required; and the availability of appropriate logistical support.

Tactical considerations aside, command personnel must first consider the more practical aspects of mountain movement before committing their troops to alpine operations. The following are just some of the variables involved:

The Human Factor:

- Physical condition of troops
- Type of unit—Light infantry; heavy weapons; mortar teams; artillery; etc
- Training and experience in mountain warfare and military mountaineering
- Training and experience in snow travel with snowshoes and/or skis
- Training with crampons, ice axes and travel in rope teams
- Availability of qualified Assault Climbers and/or Scout Skiers
- Availability of Mountain Leaders and Mountain Guides

Logistical Considerations:

- Availability of suitable mountain clothing & related equipment
- Availability of suitable tents, stoves and sleeping bags

- Availability of ropes, climbing gear, crampons and ice axes
- Availability of mobility equipment—snowshoes, skis, sleds, snowmobiles
- Range and capability of radio communications
- Weight of equipment to be carried without mechanized support
- Availability of air assets / helicopters
- Altitude limitations of air assets
- Availability of re-supply once the unit leaves the road system

Terrain Appreciation:
- Availability of current, accurate maps or aerial imagery
- Type & Steepness of terrain
- Alpine or Sub-alpine
- Snow depth & current Snowline
- Probability of avalanches
- Ice conditions on trails, slopes and ridges
- River levels & Crossing points
- Maximum planned operational elevation
- Time available for acclimatization
- Suitability of camp sites

Environmental Variables:
- Season—Summer / Winter
- Climatic & Temperature range—Day / Night
- Weather patterns
- strength
- Anticipated snow fall
- Availability of accurate long range weather forecasting

All of the above will factor into the method of mobility, the distances that can be traveled and whether operations are even feasible with the personnel and assets available.

SNOW TRAVEL

One of the most basic considerations for movement in the mountains and on snow is the suitability of footwear, for without waterproof, insulted boots the troops will quickly become incapacitated with injuries such as frostbite and immersion foot.

Standard cold weather combat boots like the Danner and Matterhorn are both insulated and waterproof and suited to short to medium patrols in the snow. Their warmth and water resistance is improved with good socks and gaiters but they are not stiff enough to be considered crampon compatible.

The rubber VB boot is the next alternative for cold, wet conditions, with the black VB boot being good to -10 degrees F and the white "Mickey Mouse" boot to -20 degrees. There is no question that these boots are warm and dry and well suited to use in fixed positions and limited movements, however they are not well suited to extended marches and skiing and are particularly poor for climbing.

The best all around boot for winter mountain warfare and even summer alpine operations above the snowline is the double plastic boot (See Chapter 3). These must be sized larger than standard shoe size to accommodate the sock system without losing the airspace that is a critical part of the insulating process. Plastic boots should also be used in conjunction with a good gaiter.

The medium to high volume models that are best suited to broader American feet are the Lowa Civetta Extreme, the Scarpa Inverno, the Koflach Para for general use and Koflach Arctis for extreme or high altitude operations.

SPEED & EFFICIENCY ON SNOW

Apart from terrain limitations, the speed that a fighting force moves on foot is dictated by the size of the force and proximity of the enemy. Large armies, as were deployed in WWII, move very slowly, but there is little likelihood that troop concentrations of that magnitude would be committed to the mountains in modern warfare. However, there is still a vast difference between the speed that a reinforced rifle company can move in the mountains versus a smaller patrol element like a squad or platoon.

In closing with the enemy or even just moving through areas that may conceal guerilla forces, any unit, immaterial of size, must move more cautiously in a tactical mode with considerable time and energy given to advance and flanking security.

According to the military mountain operations manual, FM 90-6, a company of well-trained and

Steep terrain, deep snow and higher elevations drastically limit speed and mobility (Royal Marines)

well-equipped troops, traveling on foot, should be able to cover 4 kilometers per hour (KPH) on fairly hard, level terrain. An additional hour should be added for every 300m (1,000´) of elevation gain and an hour for every 600m (2,000´) of elevation loss. However, this is a very ambitious rate of movement for a large unit and does not take into account all the problems associated with high alpine snow travel.

Small groups of superbly conditioned Special Forces or mountain troops will be able to move faster than a company of infantry, but they may not be able to sustain that pace for more than a few hours at altitude or in deep snow.

When moving over snow, troops can move reasonably well breaking trail in 6″-12″ of snow provided they have waterproofed boots, good socks and gaiters. They could be expected to make 2-3 KPH under these conditions walking in single file, but slower if widely dispersed in tactical formations.

In snow deeper than 12″, movement will slow considerably and point men or trailbreakers will need to be rotated more frequently. Troops that would make 2-3 KPH on a broken trail will be slowed to 1 KPH or less when moving off the trail in deeper snow.

If snowshoes are available, and they may be essential when snow is deeper than 20″ considering the weight of packs, weapons and ammo, a unit can expect to make 1-2 KPH over new snow and 3-4 KPH on a broken trail or road. However, even though snowshoes give more float on deeper snow, they are quite slow and awkward to move in and expend a lot of energy. There is also a limit to just how much a man can carry and still derive an advantage from snowshoes. Even the biggest snowshoes like the Atlas 1044 are only recommended for an all-up load of 300 pounds and the smaller assault

shoes will lose efficiency at 200 pounds.

Apart from steep terrain and/or snow depth, there are two other factors that will slow mountain movement—the need for advanced reconnaissance and the need for force security, especially when moving into terrain already occupied by the enemy. Scouting teams must be sent forward to find trails, secure high ground and select suitable bivouac sites. They should be particularly alert for any areas that the main force may be channeled through that are devoid of cover or suited to an ambush.

Re-supply will also have a significant effect on how far and how fast a unit can move, since rations and ammunition are limited to what troops can carry in their packs and load-bearing harnesses. In general, most infantry units will deploy with about three days of rations and rarely will a man carry more than five or six days rations and ammunition, especially in the mountains where the capacity for work is reduced to less than 70% that at sea level.

Because of the lack of roads, mountain troops need to be re-supplied by helicopter on a regular basis, but this is not always possible at higher altitudes and in inclement weather. Even parachute air-drops are problematic and loads can be lost because of strong winds, steep terrain and often reduced visibility. A third option is mule trains provided these are available from the indigenous people and the trails are secure from enemy attack.

BRIDGING TECHNIQUES

Mountains catch rain and hold snow, which means run-off into rivers; and heavy snow melt means swift water that cannot be crossed by walking or swimming. So river crossing becomes an important skill for mountain mobility.

Rope bridging is taught in all mountain warfare programs and especially to pioneer engineering units. Under the supervision of an Assault Climber or Mountain Leader, the military mountaineer should be able to construct three different rope bridge systems with the standard issue ropes and carabiners carried by the platoon or company. The three types are:

- One Rope Bridge—which is a hasty bridge constructed by tensioning a static rope between two trees or boulders, and allows one man at a time to slide across clipped into the rope with a carabiner.
- Two Rope Bridge—built for follow on forces and utilizes a high rope to hold and a low rope to walk on, with the man clipped into the high rope with a sling rope and carabiner for safety.
- Three Rope Bridge—takes considerably more time to construct and can only be installed where the terrain and anchors allow. This bridge consists of two high ropes to hold like handrails and one low rope to walk on with all three laced together at the sides for added security. Three rope bridges are for heavier loads and semi-permanent installations.

All bridge and river crossings require security elements on both banks securing the movement and prepared to lay down cover fire.

SKI MOBILITY

Backcountry skiing is covered in Chapter 14 but there are a few points that relate specifically to military skiing that should be covered here.

Moving in the lowlands and on relatively level

ground the military skier (MS) needs only to maintain his balance and shuffle forward on skis, but he needs to be considerably more proficient to tackle the steeper and more challenging terrain found in the mountains.

There are also significant safety concerns involved in placing troops on steep terrain who lack the skill levels to turn and stop quickly when needed. In these situations it is preferable to have the troops remove their skis and walk, rather than risk casualties, who will then have to be carried or evacuated.

To move heavier items and group equipment, ski troops will use some form of sled like the ahkio or Norwegian pulk. Originally the US military used the heavy 10-man ahkio sled that some army units are still burdened with, but the Marine Corps and Special Forces have replaced this with the 4-man fire-team sled made by Wilderness Engineering.

The lightweight 4-man sled is much easier to handle in steep terrain and to deploy on and off of vehicles and helicopters. By going to the smaller sleds, the military was also able to do away with the heavy 10-man tents and Yukon stoves in favor of lightweight 4-man North Face tents and MSR or Coleman stoves. The 10-man tent still has some advantages for fixed positions and base camps where the additional size, room and warmth are a morale booster, but they are not suited to high alpine operations or rapid deployment.

Military units in countries that are heavily snowbound in winter, such Norway, Finland and Sweden, are very proficient on skis since most soldiers are taught to ski as small children and skiing is part of their lifestyle. This significantly reduces training time for ski troops, and as most military and Olympic biathlon teams have learned, it is easier to teach a champion skier to shoot than to teach a champion shooter to ski.

Since the US has never had to ski in combat, with the closest thing being NATO exercises in Norway, very few in our military become truly proficient skiers. This overall lack of proficiency is an important consideration for the field commander who is trying to calculate the speed and security with which his troops can move, knowing all the time that if he can out-maneuver his enemy he will be able to kick some serious ass.

The Royal Marines, who are the designated cold weather warfare troops of the British military, have a good grasp of their own limitations when it comes to moving a Commando Unit Group Formation over snow-covered terrain. The 3 Commando Brigade trains on skis for 2 weeks each year, in addition to 2 weeks of cold weather Arms Warfare Training followed by company and NATO exercises in northern Norway, for a total of 8-10 weeks of training. But even these professionals will limit movements to 8-9 kilometers per day on skis and as little as 4 km per day on snowshoes, illustrating the fact that large groups simply do not move quickly in deep snow and mountainous terrain.

Apart from just calculating the rate of movement

Under some conditions it is easier to carry the skis knowing they will give a significant advantage on the downhill runs (Royal Marines)

for a unit, days will be further shortened by the additional time that must be allowed for breaking camp in the morning and re-establishing camp in the evening. On a cold morning in the mountains, troops wake up slowly, take longer to get dressed, and require more time to make hot drinks and heat food—both of which are essential. Heavy sleeping bags and bivi sacks then need to be packed away; tents pulled and packed and sleds loaded. The commander must then plan on establishing camp early enough in the afternoon so that all the necessary work involved in digging fighting positions, placing tents and putting out listening posts can be completed before dark.

In winter warfare, additional time must be allocated each day to both make and break camp

SKIJORING

Skijoring is the use of a tracked vehicle such as a SUSV (Small Unit Support Vehicle) to tow up to a platoon of skiers at one time. This allows a group to cover more ground in less time with a reduced energy output.

The disadvantages are that movement is limited to roads and firm snow; it makes more noise; skiers can become colder on long movements; and it can be more hazardous to poorly trained skiers or if not done correctly.

To execute a skijoring movement correctly, requires a tracked vehicle, a driver, a safety spotter and a static rope.

Safety Rules:

1. Skiers must never tie themselves into the tow ropes
2. Maximum speed of 25mph for experienced skiers /15mph for less experienced
3. There must be at least a half ski length gap, tail to tip, between skiers
4. There must be a safety monitor on the vehicle with good communication with driver

Procedure:

1. Lay out and attach the middle of a static rope to the vehicle with a clove hitch (Army prefers 120´ ropes for safe spacing with a rifle squad)
2. Skiers assemble on the outside of the ropes—usually seven each side with the best skiers at the back
3. Starting from the front of the stick, skiers loop the towrope around the handles of their ski poles and place the poles under their armpit
4. When everyone is ready they give a thumbs-up to the safety monitor
5. The safety monitor gives the driver the signal to move forward and slowly increase speed
6. Skiers will keep their skis parallel on the flat and when going up hill. They will use a wedge for down hill movement to keep tension on the towrope and avoid bunching up
7. On turns, the last skier must try to ski the line wide to clear the corners
8. If one skier goes down, all those behind must release IMMEDIATELY and ski away to avoid running over him
9. The safety monitor signals the driver to stop immediately
10. The stick reforms and hooks up
11. Movement continues when everyone is ready

Note safetyman in tracked vehicle watching skijorers

Marines skijoring with ski poles wrapped into the tow line and baskets tucked under their arms

When one man goes down, everyone must release and ski clear to avoid further injuries

BV-206 over-snow tracked vehicles, also known as the M973 Small Unit Support Vehicles (SUSV) are a significant asset to mechanized infantry units but are limited to movement on roads and hard-packed snow. They can also be outfitted with a communications package to be used as a mobile Combat Operations Center.

Hummer with track modifications

Logan Manufacturing Corportation's LMC used at MWTC and by many mountain rescue teams

Military skiers skijoring behind a snowmobile

SNOWMOBILES

Another method of mechanized movement that is extremely well suited to small unit operations and border patrolling in deep snow country is the snowmobile. Light, fast and maneuverable in close terrain, these machines give a unit a level of mobility not found with any other system.

Norway's Finnmark Jaegers, who patrol the frozen forests and tundra of northern Norway, use primarily two forms of transportation—skis and snowmobiles. A company assigned to this region may have up to sixty-eight Yamaha Viking "skidoos" to support their long-range operations and quick strike capabilities.

The Viking is a 770-pound machine with a top speed of nearly 80 KPH that can carry two fully equipped soldiers along with 450 pounds of equipment (although this is not ideal). It can also tow a small trailer with an additional 420 pounds of gear. For the Norwegians, the whole package is air transportable and supported by a fleet of specially equipped Bell-412 helicopters.

A company of 124 Jaegers is composed of three rifle platoons along with a command and mortar platoon. Each rifle platoon consists of three 8-man teams equipped with 15 snowmobiles.

In each 8-man team, three men are equipped with G-3 (7.62mm) assault rifles complete with 40mm grenade launchers; three with MP-5 submachine guns; and two who function as snipers with Vapensmia 7.62mm sniper rifles (soon to get 12.7mm weapon systems). They are also equipped with the Carl Gustav 84mm antitank rocket, ERYX antitank system and Stingers for anti-aircraft capability. The unit also has three laser rangefinders and all drivers are equipped with night vision goggles (NVG).

To extend their range and capability, the Jaegers cache ammunition, fuel and supplies throughout the frozen wilderness in strategically placed dumps. Even with the constantly changing snow conditions, these well-hidden sites are easily located with GPS.

Even if snowmobiles are not fully integrated into a unit's capability on snow, one or two snowmobiles can be a tremendous asset. A snowmobile can be used to haul heavy items like tents, food, mortars, heavy machineguns, ammo and support equipment, leaving the troops free to ski or snowshoe with considerably lighter loads. A snowmobile can also tow several skiers in much the same manner as a SUSV or other tracked vehicle or be used to evacuate casualties. The three limiting factors with snowmobiles are:

- Range & fuel consumption
- Noise in otherwise quiet terrain
- Reliability under freezing conditions and rugged use

TACTICAL SNOW MOVEMENTS

Movement on snow leaves tracks that are easy to see and follow, especially when a large group has moved through an open area. To better conceal a movement, troops should move on bare, rocky ground and stay in the trees as much as possible to avoid aerial reconnaissance.

Moving at night is an option if night vision equipment is available and the unit has confidence in their night navigation ability. A unit can also move while it is snowing to take advantage of the fact that the fresh snow may fill their tracks (but not recommended on glaciers).

Since it is preferable to fight down hill, a unit should try to move as high on the mountain as possible without sky-lining themselves. They should also avoid valleys and gorges that may naturally channel the unit into an ideal ambush position for their enemy.

Noise can travel great distances in the mountains, particularly in valleys, so every effort should be made to silence equipment and move quietly.

A woodland camouflage uniform will standout in

snow just as over-whites standout in the trees, so the individual must be prepared to mix and match his uniform to best blend with the background.

When moving on snow, the weapon can be carried slung vertically behind the back, but closer to the objective it should be moved to the front by looping the strap around the neck. If there is any possibility of contact with the enemy then the weapon should be carried in one hand and both ski poles in the other.

Soldiers training and climbing at ANG-MWS with night vision devices

HELICOPTER INSERTION & EXTRACTION

Larger military helicopters such as the CH-46E and CH-53 can deploy troops directly into the mountains more efficiently than any other form of insertion. However, there are limits to the number of men, amount of equipment and maximum altitude that they are designed to handle. Each type and model of helicopter has specific altitude limitations called the Service Ceiling and specific altitudes where it can hover. When landing or coming close enough to the ground to allow troops to jump, the helicopter has the power advantage of being in ground effect (IGE), but if troops need to be rappelled in or hoisted out then the helicopter has to have sufficient power to hover out of ground effect (OGE).

The limits for hovering IGE or OGE are also affected by outside temperature, for example, the service ceiling of a CH-46E is 14,000´ and it can hover IGE at the same altitude, but on a +20 degree C day, it can only hover at 12,580´. Hovering OGE the Chinook can only handle 11,970´ but at +20°C this is pushed down to 8,600´.

Similarly, the UH-60L Black Hawk can hover OGE at 11,125´, but at 70 degrees F, 9,375´ and at 90 degrees F only 7,650´.

	CH-46E	CH-53	UH-1N	UH-60L
Service Ceiling	14,000´	18,500´	15,000´	19,150´
Hovering IGE	11,970´	11,550´	12,900´	
Hovering OGE	8,600´	9,500´	4,900´	11,125´

CH-53 deploying Marines at over 8,000´ for a mountain movement exercise in the Eastern Sierras

Helicopters should not be used to deploy troops from sea level to altitudes above nine or ten thousand feet without giving them time to acclimatize. But helicopters can be used to move and re-deploy troops already acclimatized at altitude, bypassing obstacles such as rivers, gorges, valleys and ridgelines that could otherwise slow progress for days.

Helicopters can be used to re-supply troops in the mountains, bring in replacements, evacuate casualties and for additional firepower. They can also fly aerial reconnaissance missions, be used to photograph the terrain for route planning and even drop explosives onto snow-covered slopes to clear avalanche hazards.

In addition to altitude limitations, a helicopter loses a considerable amount of power and lifting capacity at higher elevations (30%-50%), and even though colder air normally helps performance, it is not sufficient to offset the loss caused by the thinner air in the mountains.

Because of the added equipment load, mountain troops will require at least one and a half seats per man, plus additional room for sleds and skis. This reduces the number of troops that can be carried on each run. For example, the CH-53 can usually carry 24 troops at sea level but only 10 with full combat gear when operating at 10,000´ (MSL).

The CH-46 will take a full squad with equipment at sea level but with the added weight of a full fuel load they are limited to 6 men at altitude and 8 when they have burned off fuel. The bigger more powerful CH-47 can handle twice that but they will also sink deeper into the snow on the LZ.

LANDING ZONES

When selecting and preparing LZs for military helicopters, it is necessary to have an area 150 meters by 150 meters for CH-46s and CH-53s and 100 meters square for the UH-1N. For multiple landing points, 150-meter spacing is required between ships and the ground slope cannot be more than 14 percent (8 degrees). It is also important that helicopters can easily clear any obstacles such as trees on the approach and take-off.

The minimum information required by the pilots for a landing zone brief is:

- Your Call Sign
- LZ Location
- Radio Frequency
- LZ Marking
- Wind Direction & Velocity
- LZ Size & Elevation
- Obstacles & Snow Conditions
- Visibility
- Approach / Retirement Direction (recommended)

The procedure for extracting or re-deploying troops in a snow environment is similar to that which is practiced at sea level except the blowing snow and colder temperatures can create additional problems.

In brief, when a sled team is preparing for a helicopter extract they will first secure all their personal clothing and equipment and then bundle all their skis onto the sled. They will also remove their overwhites to give better contrast on the snow for the pilot.

When ready, the team leader will lead the squad or fire-team out onto the LZ to the right-front of the anticipated landing site for the helicopter, with the assistant leader bringing up the rear.

The team will huddle on the LZ at the 2 o'clock position, all loose gear secured, hoods up and goggles on since the blown snow can be not only cold and painful, it will cause near whiteout conditions until the ship has settled and the pilot rolls off the power.

Once the crew chief or pilot is ready to board the sled team, they will signal to the team leader to lead his group directly to the ramp. The sled team leader enters first, followed by the team and designated sled handlers. Each man hand carries his snowshoes and weapon and places them between his legs when seated. Skis are carried parallel to the deck, placed beneath the feet and far enough forward to allow the sled to be loaded.

Establishing communications in the mountains can often be a problem, especially in valleys and when transmission is blocked by high mountain ranges

CH-46 flaring at 8,000 feet (MSL) to pick-up Marines for re-deployment to another ridgeline during winter warfare training. The sled team should already be huddled on the LZ at the pilotís 2 o'clock position.

Sled teams moving out onto the LZ—all gear secured, goggles on, hoods up and long items kept parallel to the ground

Marines huddled at the pilot's 2 o'clock position and wearing camo ECWCS to contrast with the snow and be easily visible in the sun's reflected glare

On signal from the crew chief the sled team moves to the rear ramp to load. They stay low just in case the heavy CH-46 should sink into the snow and cause the rotor blades to come closer to the ground.

HELICOPTER SAFETY

If the snow is deep, a helicopter will sink down bringing the rotors closer to the snow than usual. It is therefore important to keep your head down and any long pieces of equipment like antennas, skis, poles and weapons should be carried parallel to the deck.

Under various conditions, ice can build up on the rotor blades and when it sheds, it comes off in a myriad of flying projectiles. The safest place to be is on the ground, hood up and goggles on with the face covered.

Note: Chapter 21 has additional information on helicopter operations

28

ALPINE & COLD WEATHER WEAPONS TRAINING

Virtually any modern weapons system should function in alpine conditions provided it is correctly lubricated and maintained. However, mountain and wilderness operations require weapons with a greater range capability than urban assault weapons such as the MP-5 and handgun. The average engagement distance in an urban environment is less than 50 meters and most often less than 10 meters. Even SWAT snipers will seldom get the opportunity to engage a target at over 100 meters which would be considered extreme close quarters for a military sniper in the mountains.

Selecting the ideal weapon system for alpine operations could be dictated by the nature of the mission, the type of terrain, anticipated engagement ranges and the capabilities of the enemy but most often the selection will be made on simply what is available and with which weapons the troops are most familiar. Another consideration could be the tactical flexibility needed in a weapons package required to handle the fast paced, rapidly changing, unpredictable nature of combat.

It would seem that the standard .308 (7.62mm) assault rifle would be a good choice for mountain warfare, having better long range capabilities and terminal energy, but .223 (5.56mm) weapons are lighter and allow the operator to carry more ammunition. Both important considerations when weapons must be carried long distances and re-supply is not readily available.

The solution is a mixed bag of weapons to give the commander or team leader the maximum flexibility in deploying his troops. Light assault rifles like the M16A2, M4/CAR-15 and HK 33/53 are all excellent general issue weapon systems, provided they are augmented by more powerful M14, M60 or HK G3/91 type weapon systems carried by selected members of the unit.

Close quarter weapons like the MP-5 and shotgun would only have application in clearing caves or heavily wooded areas, but the suppressed MP-5 is excellent for silent takedowns. Handguns should be issued to snipers as a back-up weapon and are also useful for assault climbers who must scale cliffs and set ropes for follow-on forces without the ability to carry a primary weapon.

Primary snipers should be armed with their issue M24 or M40A1 .308 (7.62mm) sniper rifles, while the number two sniper should have more firepower in the form of a scoped M16/M203, scoped M-14 or G3/SG1.

On the subject of sniper weapons, for the last few years we have been doing considerable test and evaluation work at STTU to find the optimum long-range sniper cartridge. To facilitate this program we had Robar build a set of rifles in .308, .243, 300 WinMag and 7mm Magnum, with various barrel twists and all topped with Leupold Mark 4 long range scopes to run comparisons with both conventional and VLD (very low drag) projectiles. Most

Captain Lord of the 3rd Battalion / 172nd Infantry with the Army M24 sniper rifle

Author with Robar SR60D

Seal equipped for cold-weather/arctic warfare with an M14

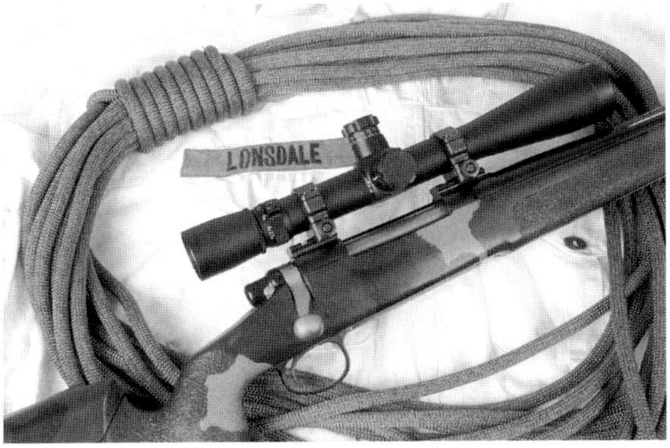

Robar 7mm magnum built on a McMillan A3 stock and topped with the Leupold's 3.5X - 10X Long Range M1 scope

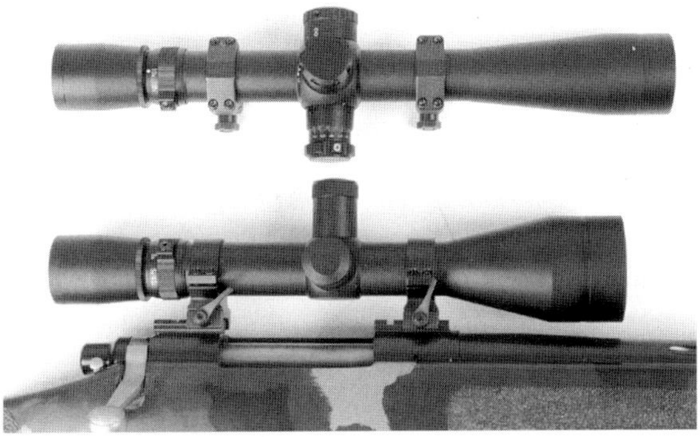

Leupold variables: 3.5X-10X Long Range M1 with 40mm objective (top) and 4.5X - 14X Long Range Tactical with 50mm objective (bottom)

SR60 with Mark 4-M1 10X fitted with an extended lens shade

testing is done at 1000 yards and is still on going as of this date, but I can give a few preliminary non-scientific observations.

The SR90 300 WinMag with a 190-grain Sierra Match King projectile has been an excellent performer at 1000 yards hitting the steel targets noticeably quicker and harder than the .308 sniper rifles shooting 168-grain BTHPs. The 190-grain in 300, as with most heavier and faster projectiles with a good ballistic coefficient, also does an excellent job of bucking the wind giving a higher first round hit probability.

Even though the long action SR90 is heavier and has more recoil than the M40A1s and SR60Ds, every military sniper who shoots this weapon is so impressed with the terminal energy dumped into the targets that he wants to go out and buy one for him self. There is something very satisfying about consistently center-punching a man-sized steel target at 1000 yards in variable crosswinds.

In 7mm Mag, we have set up one SR90 with a fast-twist barrel to handle 179-grain and 182-grain VLDs and another on a McMillan A3 stock with a conventional twist barrel to handle the 168-grain Match King projectile. Each rifle is topped with the newer Leupold Long Range M1 3.5X-10X or 4.5X-14X Variable scopes.

Even though the heavier VLD projectiles handle the wind better at longer ranges and have proven to be extremely accurate, they are too long to load in the magazine and not suited to military applications. The fast-twist barrel and higher chamber pressures also erode the throat of the barrel more than would be acceptable for an issue weapon system that would be required to have a longer work life.

The Robar/A3 7mm Mag shooting the 168-grain BTHP is very similar to the Secret Service CST rifles and is proving to be a real winner with less recoil and bark but all the long range performance of the 300 WinMag. We have also built this rifle a little lighter than the SR90 making it more suited to long carries in the mountains.

However, here is the big revelation. Although the Leupold Long Range variables have become our first choice in scopes, and the quality of the Robar rifles is flawless, exotic calibers are not suited for military issue at this point in time. Even though the magnums hit harder and faster, they also kick and bark more, are more difficult to master and maintain, the ammunition is more expensive and barrels wear out faster.

An Army, Marine or Seal sniper spends a tremendous amount of training time developing data on his issue rifle with the 308 Win/7.62mm round. To switch to a new caliber would require that he go out and develop all new data on both the round and the rifle, so he is better to stay with what he knows best and take full advantage of the accuracy and consistency that match grade ammo has proven itself capable of producing.

The most important skill of the sniper is to consistently make first-round hits at long range and this can only be achieved with total familiarity and confidence in the weapon system.

For the individual looking for a good long range rifle, and has the additional budget for the more expensive ammunition and training, the 7mm Mag and 300 WinMag are hard to beat. But more 1000-yard matches have been won with the .308 than any other caliber, proving that intimate familiarity with a particular rifle and caliber is more important for accurate shooting than just owning a more powerful bullet thrower. With all the precision long-range rifles we have at STTU, and even though I enjoy shooting the magnums, I still revert back to my 14 year old SR60D and Federal Match .308 ammo when hitting what I shoot at is REALLY important.

FACTORS AFFECTING WEAPONS FUNCTION

Rust is the perennial enemy of all things metal and any fighting man who has served in Asia or the tropics will know this all too well, as will Seals who must operate in salt water on a daily basis. Left wet and unattended, a weapon can turn to a mass of red ferrous oxide in as little as 12 hours.

The fighting man's primary responsibility is to maintain his weapon system in a constant state of readiness, for without it he is just another walking casualty waiting to be fitted for a body-bag. If his weapon is not in perfect working order he is a liability to both himself and his teammates, so this means it must be cleaned, dried, lubricated and maintained—sights zeroed, screws tightened, moving parts oiled and the bore and ammo kept dry.

Soldier armed with a shotgun for protection against polar bears in Greenlnd (SFC Gordon)

Operating in arctic or sub-arctic conditions presents the shooter with many of the same maintenance problems encountered in jungle warfare and maritime operations—an over abundance of water, but in this case in the form of snow and ice.

It is virtually impossible to keep the outside of a weapon dry when operating in the snow, particularly when climbing, crawling, skiing (crashing) and shooting from a prone position. Even deploying in and out of helicopters puts the fighting man in a maelstrom of flying snow whipped up by the powerful rotor-wash, so every attempt must be made to keep snow out of the barrel and action.

Even taking a weapon into a warm environment can create problems. Moving with an apparently dry weapon from near freezing temperatures into a heated tent or building can create condensation on a weapon in much the same way fog forms on a pair of glasses or goggles. This condensation will eventually rust and disable a weapon if not dried off and replaced with some form of protective coating.

Weapons maintenance is important in all environments. Apart from using the correct lubricants, the bore must be periodically checked for snow and obstructions

LUBRICATION & MAINTENANCE

Lubricants do not get a lot of consideration in warmer climates except when they become contaminated with dust and sand, but in colder conditions lubricants stiffen up causing the action of a weapon to become sluggish. This is a result of the normal thickening of the lubricant and corresponding loss of viscosity.

The military recommends CLP (Cleaner, Lubricant and Preserver) in cold climates down to –35 degrees F. Below –35 degrees CLP will freeze so for temperatures from zero down to –65 degrees, the recommended lubricant is LAW (Lubricant Arctic Weapon). If LAW is not available, the military recommends firing the weapon dry rather than risking the CLP freezing up.

The lubricants that we use at STTU year round are also well suited to sub-zero conditions. These are Shooters' Choice All Weather High-Tech Grease which has a temperature range of –65 F to +350 F and doesn't stiffen when cold. It also comes in a convenient syringe applicator. For a general-purpose lubricant we use FP-10 Lubricant Elite, which has a temperature range of –76 F to +500 F, and another useful product is their Rust Prevent, which is an aerosol rust preservative lubricant with a superior anti-oxidation and moisture displacing agent. For barrel/bore cleaning and conditioning, Shooter's Choice MC#7 Bore Cleaner works well.

In addition to lubrication and routine maintenance, there are protective finishes that can be applied to weapons to make them more resilient to moisture. Virtually all our weapons have some form of protective finish since we operate so much around salt water and in hot, humid climates. The three that we recommend are Roguard, Armor-Tuff and NP3.

Wilson CQB 45 treated with Armor-Tuff and Sig 226 coated with Roguard and NP3

Robar Industries' Roguard is polymer technology at its best and is used on all our sniper weapons. Twice as hard as Poly T and twice as corrosion-resistant, Roguard will take far more punishment than any other black finish available—tested both in the laboratory and in the field. Robar is so confident about Roguard that they will give a lifetime guarantee against peeling and corrosion! This guarantee does not cover wear and tear, abuses or alteration to the firearm, or corrosion in abraded areas due to wear but it still says a lot for their faith in the finish.

The second finish that we recommend is Wilson Combat's Armor-Tuff, a Wilson in-house process consisting of a Phenoloc resin that is heat cured onto the weapon. It is not a soft Teflon finish and is harder and more abrasion resistant than Teflon based products, which tend to give a slick feel that is not altogether good on a weapon. Armor-Tuff will withstand extreme temperatures from –250 to +500 degrees F and saltwater spray or immersion for over 1000 hours. It can be done in a matte black or OD green making it ideally suited to tactical weapon systems.

For most internal surfaces that operate with closer tolerances, we prefer Robar's NP3 which is a satin gray, non-reflective surface treatment for steel and metal alloys that combines sub-micron particles of P.T.F.E. (polytetrafluoroethylene) with electroless nickel. NP3 is a very accurate and even coating that can be plated to all internal parts giving a smoothness to the action not found with other coatings; and has a high lubricity and low friction coefficient, therefore the life expectancy of a weapon will be greatly increased due to less frictional wear.

One of the best qualities of NP3 for all operations is that it is very corrosion resistant, (a 1 mil. (.001) coating exceeding 240 hours in a salt-spray test), and in cases where the NP3 has been perforated, the corrosion shows no tendency to spread or migrate under the coating.

Breakage is also a problem in the extreme cold as some plastics and metals become brittle at freezing temperatures. Banging a cold weapon against anything hard could cause hand-guard breakage that under normal conditions would not have occurred. Internal metal parts in a weapon striking against each other in the firing sequence are also prone to breakage, especially sears, firing pins and operating rods.

When a weapon is being fired in sub-zero temperatures, the barrel will quickly heat to between 200 and 750 degrees Fahrenheit depending on the number of rounds fired and the rate of fire. This constant cooling and reheating can alter the temper of the metals. Similarly, laying a hot weapon down on the cold snow can cause warping of the barrel.

Snow can also block an action or barrel and cause malfunctions and stoppages. Any snow in extra magazines can also create problems. Snow that has melted in the warmth of a tent will turn to water and then re-freeze to ice when taken outside. This can freeze an action or magazine solid so every opportunity should be taken to wipe away snow and visible moisture.

FIRING CONSIDERATIONS

Operating in cold conditions and on snow creates some unique problems for the SWAT team or rifle squad that they would normally not encounter in more temperate climates.

At temperatures below -20 degrees F, visibility may become obscured while firing. This is a result of the water vapor in the air crystallizing creating minute ice particles and fog that can hang in the gunner's line of sight for several minutes. This is not too much of a problem for the individual rifleman who can move about but for crew served weapons and snipers in fixed hides and concealed positions it could be more of a problem.

If a weapon has a bi-pod or tri-pod, particularly crew served weapons, machine guns and mortars, the adjustments on the legs and various wing-nuts and locking devices can become frozen so should be treated with CLP in anticipation of this problem.

If sights become filled with snow or ice over, this can also interfere with the shooters ability to aim or shoot accurately. Even optical sights can become fogged in freezing temperatures if they are not dry nitrogen filled or get any moisture inside them at all.

Standing on skis using poles for support

Standing using poles for support

Kneeling on skis

Kneeling on skis

Prone with skis to one side and using pack for support

Prone with skis splayed

Modern high quality optics like those made by Leupold and Aimpoint should not have these problems, but inferior optics can be a nightmare in cold weather, especially when taken from inside a warm tent or shelter out into the colder air.

Identification or marking of a shooting position to the enemy is another problem. Just as in the desert, where a muzzle blast from the prone position kicks up a cloud of dust marking a position; in freezing temperatures a discharge may leave fog hanging in the air much the same as when someone exhales their warm breath into cold air.

Finding an ideal shooting position in the snow can also be a problem. Apart from the thermal considerations for the shooter, the shape of the shooting position will be constantly changing as the snow compresses or melts as the shooter readjusts position and especially under recoil.

On firm dry ground the ground itself makes for a stable weapons platform that helps absorb some of the recoil and vibration of firing, but on snow, the weapon will sink into the snow or melt it if it becomes too hot. On ice there is no absorption of recoil causing more vibration and muzzle jump plus the weapon and shooter may slide around after each discharge.

When engaged in firefights, it should be realized that snow is very effective at stopping bullets and smothering fragmentation munitions so indirect fire may prove more effective than direct fire. Although 40mm grenade launchers like the M203, older M79

Dust generated by the muzzle blast of the Barrett M82A1 50 cal

and crew-served Mk19 work well in the mountains, grenades in particular lose much of their effectiveness when they land in snow along with having a reduced casualty radius. Therefore, maximum use should be made of airbursts with time and proximity fuses.

Because of their high trajectory, mortars also make for very effective indirect fire support in mountainous terrain.

For direct fire missions, it is obviously advantageous to control the high ground and shoot down into the enemy positions. However, the cold also slows the burn rate of the propellant in a cartridge, which reduces the velocity and range of projectiles. This will result in low bullet strikes so it is recommended to re-zero weapons for local conditions and re-do range cards.

BALLISTIC PROTECTION

A standard segment of the MWTC winter warfare package is a demonstration of the penetration of various types of ammunition on snow banks and manmade barricades. Lighter rounds like the 5.56 tend to skid all over the place with unpredictable results but heavier rounds like 7.62 penetrate better because of their weight and kinetic energy.

One lesson learned from this exercise is that the rifleman can improve his fighting position with better ballistic protection by building blocks of icecrete. This is done by filling any available bags or boxes, such as MRE boxes or ammo crates, with a

Mountain warfare instructor runs penetration tests on boxes of improvised ice-crete

mixture of dirt, rocks, sticks, snow and water and then allowing it to freeze over night. By then arranging these into multiple barriers they seem to work better than one thicker single barrier.

EXPLOSIVES

Explosives are more difficult to deploy in the cold because of the need to wear gloves and the fact that most explosives become very hard. Plastic explosive (C-4) becomes so hard that it is all but impossible to insert a blasting cap unless it was done earlier and indoors when the plastic was still warm. In extreme cold C-4 has been known to simply shatter when the blasting cap fires rather than detonating the explosive.

Detonating cord also becomes stiff and difficult to tie in cold weather and will break easily in extreme cold. Time fuse also remains curled in cold weather and will usually break if uncurled.

Condensation and moisture contribute to the increased probability of misfires so hang-fire waiting time should be doubled in colder environments. However, explosives like TNT are very useful for blasting fighting positions in ground frozen too hard to dig.

LONG RANGE SHOOTING

Long range shooting is covered in more detail in the STTU SNIPER II manual but there are several topics related to mountain warfare that warrant mention here.

Wilderness and mountain operations will put the shooter in positions where he may have to engage targets at greater distances than those associated with urban operations and police actions. To operate in the mountains, a rifleman should be able to effectively engage targets out to 400-600 meters and snipers with scoped rifles at longer ranges out passed 600-800 meters. However, at these distances, various environmental conditions can significantly affect the path and strike of a bullet, particularly

www.sttu.com

lighter projectiles.

Under different atmospheric conditions, air resistance can vary with temperature, humidity, and altitude. Hot air is thinner than cold air so the bullet moves easier; humid air has more moisture and is therefore denser, slowing the bullet; higher altitudes have thinner, less dense air and therefore less resistance against the bullet.

Wind however is the most significant variable since it can vary in velocity, be constant or gusting, go with or against the bullet, or cross its path at varying angles. Accurately calculating winds in the mountains is very difficult but a task made considerably easier with the use of a wind meter like the Kestrel 2000. But even this is of limited value since it can only give the wind at the shooter's location.

It is only through practice and adequate amounts of long range shooting under a variety of conditions that a shooter will develop the skill and experience to dope the wind with reasonable accuracy.

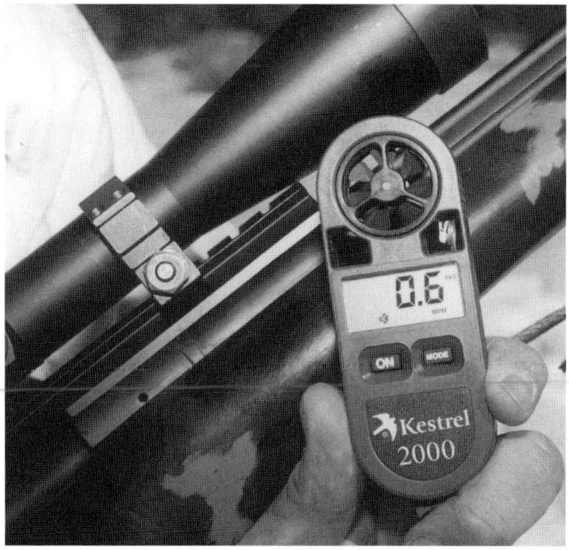

The Ultralight Kestrel 2000 wind meter measures wind speed from 0.7 -89 MPH, along with temperture and wind chill. They are ruggedly reliable and run 400 hours on a CR2032 watch battery

RANGE ESTIMATION

In modern warfare, the most accurate method to measure range is with a laser rangefinder, but snipers can also become quite proficient at "milling" targets with the mil-dot reticles in their scopes. There are also several other accepted but less accurate methods used to estimate range, some of which require extensive training and experience to attain proficiency.

Map Method—The map method is a simple matter of measuring the distance between the shooters location and the target on a topographical map. For example, if the map scale is 1″ to 1000 meters and the measured distance is six tenths of an inch on the map, then we have a distance of 600 meters. If the map scale was 1″ to 1 mile, then six tenths of an inch would equal 6/10ths of 1 mile or 1056 yards.

100-Meter Unit Measure Method—Most sniper or competition shooters use this method almost subconsciously. The 100-meter (yard) method simply entails dividing the distance between your position and the target into 100-meter increments. Shooters are used to practicing over formal rifle ranges that are already divided into measured 100-meter lines, so the relative appearance of this distance is a familiar one.

Averaging Method—To increase the accuracy of range estimations, a team leader can take the estimates of each member of the team and average them.

FACTORS EFFECTING RANGE ESTIMATION

When using any of the above methods for range estimation, there are several factors that can have an adverse effect on the final estimate. If these factors are observed and understood then appropriate compensations can be made.

Nature of the Target—Objects of a regular shape or high contrast, such as a building, will appear closer than irregularly shaped more obscure objects. In addition, a partially exposed object will appear

Similar to the Mark 4 and earlier Ultra, the Leupold variable 3.5X-10X M1 Long Range scope has the option of a Mil-Dot range estimation reticle

more distant than a fully visible object.

Terrain—Over irregular terrain, the tendency is for the eye to follow the irregularities and overestimate the distance. Over smooth terrain like sand, snow or water, the tendency is to underestimate.

Lighting—The more clearly a target can be seen, the closer it appears. A target in full sunlight will appear closer than the same target at dusk, dawn or through rain or smoke. Also, if the sun is behind the shooter the target will appear closer, and if the sun is behind the target then it will be harder to see and appear more distant.

RANGE ESTIMATION RETICLES

Scope reticles that are advertised as "range-finding" are range estimation aids at best. The Police model Leupold Vari-X III makes use of a known distance of 16″ between the center crosshair of the scope and the thicker duplex line. When laid over a target this is designed to represent the width of a man's body. When the power setting is adjusted to match, then the range can be read off of the back of the adjustment ring.

Similar to the USMC Unertl 10X scope, the Leupold Ultra and Mark 4 scopes are available with a Mil-dot reticle configuration. With the Mil-dot scale, the shooter can use the size comparison Range Estimation Formula to calculate the range.

METRIC / ENGLISH CONVERSION:

Yards	Yards or Meters	Meters
109	100	91
219	200	183
328	300	274
437	400	365
546	500	457
656	600	549
766	700	640
875	800	731
984	900	823
1094	1000	914
1203	1100	1005

MWTC instructor Sgt Clearman supervises training in long range slope angle shooting

Sniper teams must convert line-of-sight range estimations to flat-line ranges by use of the cosine formula

Snipers calculate slope angle with the aid of a protractor and piece of string used as a plumb line

RANGE ESTIMATION FORMULA

To estimate range with a mil scale or mil-dot reticle requires that the shooter know the size of a specific object, usually a man, but it can also be any other piece of equipment that is of a known dimension.

$$\text{The formula is: } \frac{\text{Size of Object (in YARDS) X 1000}}{\text{Size of Object (in Mils)}} = \text{Range to Target}$$

In practical terms, a sniper sees that a standing man is 4 mils in his scope. The man is approximately 2 yards (6ft) tall.

$$\frac{\text{2yds X 1000}}{4} = \frac{2000}{4} = \text{500 yards}$$

RANGE ESTIMATION TABLE

The following table has been developed as a quick reference for snipers using a mil scale or mil-dot scope. The first range estimation is for a standing man 6 feet tall, then a sitting or kneeling man 3 feet tall, and finally a standing man who is 5 feet 6 inches. A 6′ average would be used for a European or American foe and 5′ 6″ for possibly smaller Asian or Middle Eastern terrorists.

HEIGHT IN MILS	STANDING 6′ 0″	SITTING/KNEELING	STANDING 5′ 6″
1	2000	1000	1800
1.5	1333	666	1200
2	1000	500	900
2.5	800	400	720
3	666	333	600
3.5	571	286	514
4	500	250	450
4.5	444	222	400
5	400	200	360
5.5	364	182	327
6	333	167	300
7	286	143	257
8	250	125	225
9	222	111	200
10	200	100	180

SLOPE ANGLE SHOOTING

Most hunters know to compensate for bullet impact when shooting up hill or down hill but this is usually done by "guestimation". The alpine sniper needs a little better appreciation of this potential error, since he will invariably be engaging targets at longer ranges and usually from higher ground.

Shooting on an upward or downward slope requires the same amount of correction, but the sniper does not use the line of sight distance to the target but the flat line distance—In other words, the base of a right-angle triangle, not the hypotenuse.

The following is a list of compensation factors based on the Cosine formula to be used in setting the scope sight. To use this table, find the angle and then multiple the estimated range by the decimal factor on the right. For example, the line-of-sight range to the target is 800 meters, with a down angle of fire of 30 degrees. The scope should be set for:

800m X .87 = 522 meters

Degrees of Slope Angle Up or Down	Multiply Range By
05	.99
10	.98
20	.94
30	.87
40	.77
50	.64
60	.50
70	.34
80	.17
90	.00

Sniper observer teams working with the SASR M82A1

Suunto compass with inclinometer being used to measure slope angle

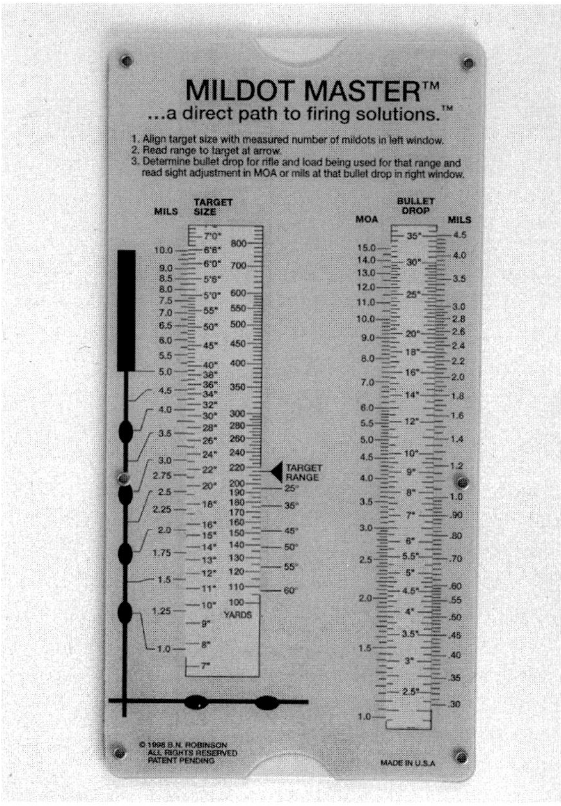

Slide rule type mil-dot calculator

EFFECTS OF ELEVATION

The air at higher elevations is less dense so produces less drag on the flight of the projectile. This in turn causes a higher impact on the target.

Even though results will vary with caliber and muzzle velocity, and we have not had the opportunity to verify the results, the following Marine Corps sniper chart shows the approximate effect in impact of changes in elevation from sea level to 10,000´. Elevation adjustments are in minutes of angle.

Range (yds)	2,500´	5,000´	10,000´
100	.05	.08	.13
200	.1	.2	.34
200	.2	.4	.6
400	.4	.5	.9
500	.5	.9	1.4
600	.6	1.0	1.8
700	1.0	1.6	2.4
800	1.3	1.9	3.4
900	1.6	2.8	4.8
1000	1.8	3.7	6.0

Charts such as this should only be used as a rough guide so the sniper is always encouraged to go out in the field and develop and log his own data.

WEAPONS TRAINING

At STTU, for over two decades we have been preaching to both law enforcement and military the need to increase ammunition allocations for training. In some cases we have had an affect but in others the agencies are still woefully deficient which is reflected in the poor shooting ability of their personnel.

To be proficient a shooter must have both well-structured training and sufficient ammunition to develop the muscle memory and coordination necessary to become an effective operator. For soldiers, Marines or tactical teams required to go in harms way, this should be a minimum of 5,000 rounds per man per year just for their primary weapons. Agencies that allocate as little as 300 to 400 hundreds rounds a year for sustainment training are bordering on criminal negligence in their training policies.

500 rounds per year is less than 10 rounds per week, and no one can maintain proficiency with that pitiful allocation. Even 5,000 rounds per year is only 100 rounds per week, which is not much considering each operator shoots over 300 rounds per day in most STTU training programs.

"It is not sufficient that the soldier must shoot, he must shoot well" —Napoleon (1769-1821)

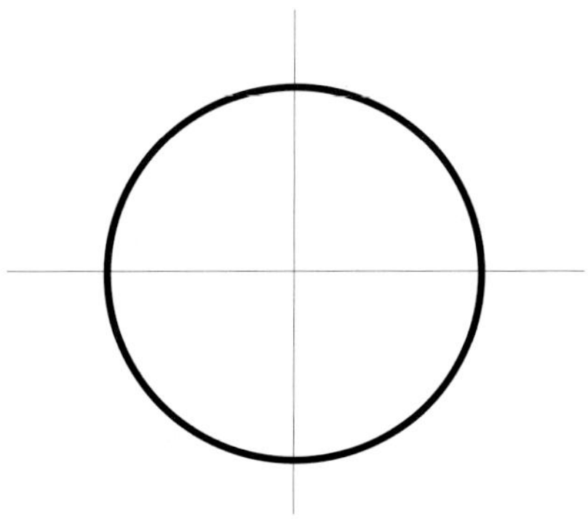

(Lonsdale)

29

BASIC TACTICAL PRINCIPLES

Small unit tactics, whether for jungle, desert, urban or mountain warfare, have many similarities derived from the fact that the foundation of tactical doctrine has not changed much in over two thousand years. The lessons taught by Sun Tzu in "The Art of War" (500 BC) can be applied to modern warfare just as they were to ancient battles.

Military and law enforcement tactical operations are a team effort, requiring coordination of several elements in a carefully choreographed sequence of events called tactics. It is these tactics, learned and perfected through constant repetition, that give a small well-trained cohesive group the advantage over a much larger but less organized adversary.

In mountain warfare, the terrain, remoteness of location and weather must also be factored into the tactical equation. In fact these are all double-edged swords that can both help defeat an enemy or immobilize friendly forces. The unit that is most comfortable in a winter environment and best able to utilize the terrain will have a distinct advantage over an enemy that is ill prepared to operate under those conditions.

The foundation of infantry tactics is the ability to shoot, move and communicate—all of which are considerably more difficult in the mountains. Shooting can be over longer distances, up or down slopes, with strong cross winds and in freezing conditions. Movement is restricted by steep rocky terrain, deep snow, hard ice, poor visibility, along with the need for specialized clothing and equipment. Communications are always problematic in the mountains since one can seldom get a good line-of-sight transmission and the distances involved can be significant. Radio batteries are also adversely affected by cold conditions.

But these three principles are a little too broad and somewhat over simplified, so the following points have been included to give the team leader or operator additional food for thought.

KNOW YOUR ENEMY—It is important to know more about your enemy than he knows about you—specifically, size and strength of their forces, type of weapons, level of training, morale, tactical procedures, and competency in the mountains. Are they indigenous to the region or just a guerilla unit using the mountains for refuge? Do they have sympathizers in the area? How good is their equipment and how long can they sustain themselves without re-supply? Do they get supplies given to them by the local populace or by raiding local villages and settlements?

KNOW YOURSELF—Know the capabilities of your own team, their fitness, endurance, training levels, capabilities, morale and weaknesses. Also

support equipment, ammunition and rations on hand, proximity to resupply, and availability of air support and casualty evacuation. Only by knowing all of these will a team leader be able to make sound tactical decisions.

KNOW YOUR TERRAIN—A unit should not only know how to live and travel in the mountains, but also how to best use the terrain to gain a tactical advantage over the enemy. Use of the high ground gives a unit better observation and fields of fire; but it is also more exposed and more difficult to conceal troop movements. The fact that valleys channel movement makes the ambush an excellent tactic for small unit operations and it is always easier to fight from the higher position.

However, units should still avoid fighting in the mountains whenever possible, since the nature of the terrain limits speed and maneuverability and makes it difficult to observe all approaches. It is also a dangerous environment even without an adversary.

DEFENSIVE POSITIONS—Mountains are not the best place to establish a defensive position since the terrain blocks any long distance observation making it easier for an enemy to approach. Troops also become very passive in a cold alpine environment, more interested in creature comforts like shelter and warmth than aggressive patrolling and being alert in an OP. The successful unit is usually the one that takes the offensive in the mountains and this requires the freedom to maneuver.

FIRE & MOVEMENT—Cover fire should always be available when moving across open terrain or involved in climbing or river crossings. Since climbing and rope installations require both hands and considerable concentration, the assault climbers must have a security team to protect their movements. Usually the heavier automatic weapons are designated to cover movement since they can lay down a significant amount of suppressing fire when needed.

MAXIMUM USE OF COVER—Concealment may hide an operator, but cover will stop bullets. In all aspects of special operations, especially when approaching or scouting, there should be a constant awareness of the nearest available cover, should gunfire erupt from a suspect location or stronghold. Deep snow, large rocks and thick trees are all good bullet stoppers.

NUMERIC SUPERIORITY—It is tactically sound to have a larger and stronger force before engaging the enemy. In many cases this show of force may well intimidate the enemy and deter any resistance. However, a smaller well-trained unit is usually more effective in the mountains than a large poorly trained rabble. And even one well trained man in a position of advantage can delay a large force for several hours.

FIREPOWER SUPERIORITY—Never intentionally enter into a confrontation out-gunned. The Seal Teams are a good example of relatively small units carrying tremendous firepower to bring a quick and efficient solution to most problems, or at least lay down enough lead for a tactical withdrawal. For law enforcement, the weapon most commonly confronted in the wilderness is the scoped hunting rifle, which has more range, power, accuracy and penetration than most police weapons and will punch through most body armor.

SPEED—The speed with which a unit can move in the mountains is critical to out maneuvering an enemy. Speed also reduces the exposure time to hostile fire; surprises and confuses the enemy and increases the chance of success by striking where least expected. Speed in the mountains is a combination of physical fitness, training, experience, lightweight gear and favorable terrain.

SURPRISE—Surprise is probably the single biggest advantage that a unit can have entering into a confrontation. With surprise it is possible for a small aggressive force to overcome a numerically superior group. But it should also be remembered that surprise is another of those double-edged swords in that a good team leader should never allow his troops to be surprised in the field. This can only be avoided by developing good intelligence, frequent scouting missions, aggressive patrolling and the necessary security precautions when moving or encamped.

SHOCK ACTION—When over-whelming firepower, noise and confusion are combined with speed and surprise then an assault team has achieved shock action. Shock action is intended to temporarily paralyze an enemy in the first few critical seconds of a confrontation giving the aggressor time to totally dominate the situation. A well-prepared ambush is a good example of both surprise and shock action, and an enemy encampment can be raided with the same level of shock action in the early hours of the morning.

Soldiers use a frozen gully as an unexpected line of approach

DISTRACTION—Any time an enemy can be distracted, by subterfuge or loud noise, from the primary direction of assault, the team will have gained precious seconds to close in and dominate the situation. Diversionary fire, stun munitions, concussion grenades and even explosive breaching are all excellent forms of distraction and diversion.

ACCURACY—Firepower is wasted if it only tears up the real estate, and speed without accuracy is also ineffective. To be effective, weapons-fire must neutralize the threat in the first few seconds of an engagement and before the enemy can seek cover, go to ground, reorganize or return fire. Quick target selection and identification, combined with the fast, accurate use of force, will reduce the probability of accidental shootings and improve safety for the assault team.

Lt.General Lewis ‘Chesty’ Puller summed it up when he said, “You don’t hurt ’em if you don’t hit ’em”.

SAFETY—Safety is a relative term in combat since war is not safe, but professional planning can eliminate much of the confusion that causes accidents. Individual and team safety depends on common sense and good judgment, but it must also be a formal part of the planning and briefing process. Employing good tactics, combined with speed, surprise, shock action, accurate fire and safe weapons handling all contribute to the safety of a team in a firefight.

MORALE—Troops routinely grumble about the conditions in which they are required to operate, but this is not necessarily a sign of low morale. As long as the commander gives consideration to the creature comforts of his troops - warmth, food, shelter and rest—morale will usually be good. Knowing that your training and equipment is superior to the enemy's, for example having night vision capability, is also a tremendous morale booster.

The best way to break the morale of the enemy, is to create discomfort by denying him sleep and resupply. This can also be achieved by pursuing or drawing him into terrain and conditions that he is neither trained nor equipped to handle. Even when he has good equipment and feels safe in his camps, snipers can be used to prove otherwise.

OPERATIONAL SECURITY—Op-Sec begins in the planning stages of an operation when the dissemination of information, particularly troop movements must be carefully controlled. A mountain unit is most vulnerable when it is being initially inserted into the mountains and the noise of helicopters draws attention to this activity. Multiple landings and take-offs from a number of different sites can confuse the enemy as to the exact location of the actual insertion.

Operational security in the field is the product of secure communications procedures, unobserved movement, diversionary tactics, and the use of scouting patrols to establish enemy locations. When making camp, it is better for a small unit to use stealth and literally disappear into the snow rather than trying to dominate and defend the high ground.

Like mobility and resupply, communications are always a problem in the mountains

TACTICAL MOVEMENT

As any successful deer hunter or veteran of long-range reconnaissance missions can attest, there are several ways to alert the enemy or scare game. These include unnecessary movement, noise, light, smell, shape, shine or color—some being more significant than others.

Movement—Movement attracts attention, especially fast, sudden or jerky movements. If necessary, move slowly picking the path carefully. It is important to move a short distance, stop, listen and then move again. Speed will be dictated by terrain, available cover, light conditions and proximity to the enemy.

Noise—If within earshot of the enemy, any sound can be a dead give-away. This is especially true of metallic noises and those that do not blend with the ambient noise level. Although an accidental discharge could prove catastrophic, the most common errors are failure to turn the radio volume down and banging the primary weapon against unmuffled load-bearing equipment, pack frames, skis or ski poles. In the mountains, the noise of the wind, rushing rivers, rock slides and snow falling from the trees can be used to mask the sound of movement. Moving quietly on snow is not difficult but concealing tracks is.

Light—Poor light discipline at night, non-essential use of flashlights, and 'stressing' on weapon's light-mount switches will all mark the operators location. One must also be careful with stoves, fires and cigarettes.

Shine—Reflection and shine, like light, can be a problem during the hours of daylight or at night. All equipment should be selected or modified so that it has a dull, non-reflective, matte finish. The snipers and observers must constantly be aware of the potential for reflected light flashing off of binoculars, glasses, scope lenses and light mounts. This can be from the sun or artificial light sources.

Shape & Color—Just like moving on a ridgeline, wearing full woodland camouflage on a snow backdrop or over-whites in the trees will silhouette an operator. By using over-whites and a pack cover, the sniper or rifleman should be able to drop and all but disappear into the snow. Another shape that can give away a position is the hard outlines of fighting positions dug in the snow, so it is important to round all

Note how the snow suit stands out against the trees making over-whites more practical for moving in and out of the tree-line

Note how the camouflage fatigues or ECWCS stands out on the white snow background

the square edges that do not appear naturally in nature. Remember, stay low, move slow and use anything that will disrupt your outline.

Smell—When moving in very close proximity to the enemy location, especially if dogs are present, the operator must eliminate strong personal smells such as cologne, after-shave and soap, and the cleaning solvent on weapons. Personal waste should also be sealed in plastic bags if staying in one position for any length of time.

Animals—Just as watchdogs in the city will alert to an intruder, the finely tuned senses of animals in the wilderness can detect smell, noise or movement at greater distances than any human. This can work both for and against an operator in the mountains in that any observed changes in animal behavior must be interpreted as a warning of approaching predators or humans.

In addition to these basics, moving in snow and at altitude can also have an adverse effect on an individual's movement. A soldier who has been breaking trail through deep snow all day will become tired and clumsy, which in turn will cause him to be less conscious of the noise he may be making.

At higher altitudes and with reduced oxygen to the brain it becomes even worse. Individuals who have not adapted well to altitude (above 14,000´´), combined with a loss of appetite and inability to sleep will quickly become fatigued and apathetic. They will be less alert, stumble more often, make noise, drop or lose equipment and deteriorate to a point where further movement would be not only tactically unsound but also unsafe.

Troops that are pushed too long and too hard under cold, adverse conditions will gradually lose their tactical mindset and begin to think more about their own misery than the mission. This makes regular rest, re-hydration and re-warming a tactical necessity above the snowline.

Marine Sniper suitably equipped for summer alpine operations below the snowline

OBSERVATION POSTS & SNIPER HIDES

A big part of mountain warfare is reconnaissance work, which is a combination of scouting and static observation, which is often best done from a position of concealment. For prolonged information gathering missions or sniper operations, the selection of this hide is critical to both the security of the mission and personal comfort.

In selecting a suitable sniper hide or OP, there are two primary considerations:

1. To provide maximum fields of observation and fire
2. To provide maximum concealment from enemy observation

When selecting a suitable hide location, there is a tendency to go for the high ground and although this may give good all around observation it does not offer good concealment. The base of rocks, trees

and natural depressions are more appropriate especially if they are less than obvious to the trained observer.

In selecting an OP or final firing point (FFP), the rifleman should not position himself in or around significant terrain features and should avoid locations that at first look appear ideal, since they will appear equally obvious to the enemy.

When selecting a hide, also consider the following:

a. A secure approach route
b. Ease of entry or exit
c. Suitable line of fire
d. Good arcs of observation
e. Security while in the hide
f. Sufficient room for two
g. Comfort

Secondary considerations:

a. Expected length of stay
b. Activity & Proximity of Indigenous Personnel
c. Emergency evacuation procedure
d. Radio use and reporting
e. Movement of friendly forces
f. Special hide construction materials required
g. Disposal of human waste

Equipment considerations for the OP in winter warfare are:

- Warm, waterproof clothing & over-whites
- A waterproof ground sheet to line the position
- An insulated iso mat to sit or lie on
- A waterproof notebook
- A thermos
- High energy snacks
- A waterproof bivi sack
- Plastic bags for human waste

APPROACH CONSIDERATIONS

The following should be considered in approaching the hide, OP or FFP.

- Avoid equipment that rattles or shines
- Do not talk during the approach except in close whispers
- Turn radio volumes down
- Night vision devices are preferable to flashlights
- Place your feet with care and avoid leaving tracks in open areas
- Do not lean on or brush up against snow covered trees
- Make maximum use of natural depressions in the snow for concealment
- Rocks, large trees and snow drifts all make good cover
- Practice good weapons discipline and safety
- Keep finger off the trigger and control the muzzle
- Check the bore for snow or debris after a movement

STEALTH MOVEMENT ON SNOW

Moving on snow with minimum disruption to the surface of the snow or without leaving a trail is difficult, but the following is one technique that is taught to snipers.

After camouflaging the weapon with white tape and not paint, (as the sniper may have to move back into the tree line or out of the snow,) the sniper will then put on warm clothing and over-whites for both thermal protection and concealment when crawling on the cold wet snow.

All additional equipment will be secured in the packed, which is then placed on an iso mat so that it can be dragged behind the sniper with a piece of 550 cord.

The sniper then lays down with his skis under each arm and begins to slide across the snow on his belly towing the pack and iso mat. In this way the skis form a larger surface area so that the sniper does not sink into the snow and the iso mat smoothes the trail left behind by the skis.

Once the sniper finds a suitable hide location at the base of a tree, he will dig into the snow and create a suitable hide. The lower branches on the tree will add to the concealment and the tree itself will help protect the sniper from the elements.

Use of natural terrain features

Sniper prepares his gear for the approach

1. Sniper sets pack on iso-mat to be dragged behind

2. Use of skis to float over the snow and dragging pack to flatten tracks

3. Sniper begins to dig a hide at the base of a tree

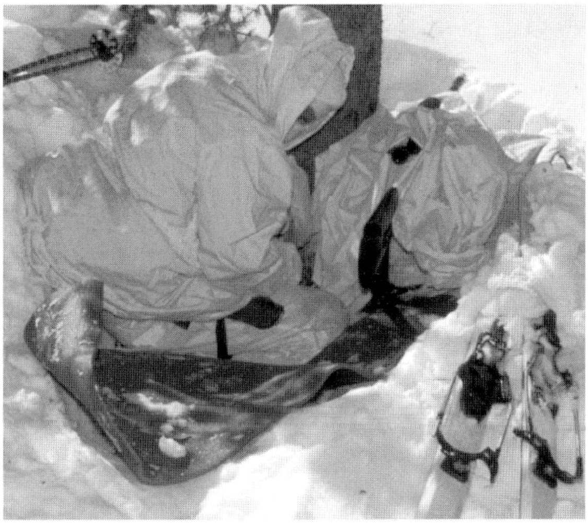

4. Iso-mat used in bottom of pit for thermal protection

5. In position but skis at hand ready for a quick bug-out

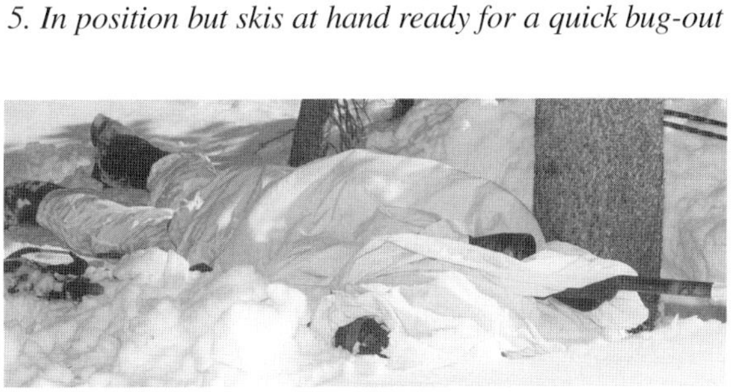

FIELD CAMOUFLAGE

The amount of time a team dedicates to field camouflage will be dictated by their particular mission and area of operation (AO). Obviously, Marine Scout/Snipers spend many weeks perfecting their ability to move undetected behind enemy lines. Police snipers would seldom need that level of proficiency for urban operations, but in the mountains it is a different story.

The single most important piece of equipment to the scout/sniper in a wooded environment is his ghillie suit. This hand-made suit is designed to break-up his outline, blend in color and texture to the AO, and muffle his movements. However, in the mountains, this is a heavy piece of gear that is not suited to rocky terrain and once above the snowline would be replaced by over-whites.

The over-whites work well, but they do not make the sniper invisible. He must still make maximum use of natural concealment; terrain features, depressions, shadows, rocks, trees and vegetation, when moving in "Indian country".

Unlike military snipers, many police teams neglect their weapon systems when it comes to camouflage. That black shiny barrel, new scope and lacquered wood stock can easily give away a sniper position through shape, color and shine.

When ordering a new weapon, it is recommended to request a woodland camo stock and matte finish on the barrel and action. Then mask off the scope lens and dials, bolt and chamber, safety, trigger and muzzle and spray with matte colored paint that matches the local environment. This could be woodland green or desert brown with additional gray or black accents to break-up the outline. Additional similar colors can then be added to complete the mottled camo appearance.

The alpine sniper will usually begin with a green/brown camo weapon and should not paint it white unless the planned AO is totally snow bound for the duration of his mission. In most cases, the sniper will be moving in a combination of snow, trees and rocky terrain so the woodland camouflage is the best with white tape and tube gauze being used for the snow camo when necessary.

NIGHT OPERATIONS

Even with night vision devices, there are many problems associated with night operations in the mountains. Apart from being more difficult to navigate and see significant landmarks, it is also difficult to see hazards such as cliffs, crevasses and avalanche prone slopes. The batteries in the NVDs will also have a reduced life in colder conditions so spares should be kept warm and carried close to the body.

The advantage of moving at night is that it conceals movement, the snow is firmer and more stable and it is easier to see an undisciplined enemy's lights, fires or cigarettes. With a full moon on a clear night, moving on snow is almost as easy as moving in daylight.

However, it should also be realized that natural night vision is not as good at altitude as it is in the low lands, being reduced to half that at sea level when above 12,000 feet.

SETTING AN AMBUSH

As in most conflicts, the ambush is one of the most effective tactics in the infantryman's bag of tricks, but they require that the ambush team wait for several hours or days for the enemy to walk into the kill zone. In cold, snow-covered terrain, this creates not only problems with concealment but also exposure, hypothermia and even frostbite.

Once a probable ambush site has been selected, the ambush team should ski into the area through the trees being careful to stay well clear of the road or any local populace. A safe distance to the rear of the ambush site they can set-up a well-concealed warming tent hidden in the trees and snow, complete with flanking security posts. The tent should be far enough back that any noise of stoves, etc, could not be heard from the ambush site.

The ambush team should then dress as warmly as possible in pile, Gortex and over-whites in anticipation of the long hours lying in the snow. Weapons should be loaded in the rear if they are not already and all loose equipment muffled. Skis are exchanged for small assault snowshoes for the

approach to the ambush site but since they do not give a lot of float on deep snow, heavy packs should be left in the rear.

Remaining concealed and moving slowly, the scout should lead the team to within view of the ambush site where the team leader will usually take the center position covering the kill zone. The rifle team will then spread out to either side at appropriate spacing with the machine gunners covering each end of the kill zone.

Each man or 2-man fire team will then prepare hides. The base of trees make excellent individual fighting positions, especially if the soldier can dig in from behind a snow drift creating a small snow cave with a firing hole in the front. If digging a fighting position in a more open area, the edges of the pit should be rounded to break-up hard outlines and shadows that would look unnatural in the snow. An iso-mat should then be placed on the bottom of the pit to further insulate the shooter from the cold wet snow.

Depending on temperature and activity on the trail, individuals can then rotate back to the warming tent to eat, drink, pee and rest. In this manner the ambush site can be maintained for as long as rations are available and the security of the unit remains intact.

Note: Additional notes on tactics and movement in the field can be found in the other STTU training texts listed in Appendix F.

POLICE

30

TACTICAL RAPPELLING

Rappelling is a fundamental skill in climbing, mountaineering and rescue work and is used as a confidence builder by Outward Bound schools, police departments, fire departments and the military. It also has numerous applications in military and law enforcement tactical operations.

The following are some of the more obvious applications for rappelling:

Mountain Rappelling:

- Descend cliffs and steep terrain
- Retreat off of a mountain or ridgeline
- Access victims on cliffs and mountain sides
- Vehicle over the side accident access
- Insert rescue teams from helicopters in rough terrain
- Military helicopter insertion method

Urban Rappelling:

- High-rise building & bridge rescue work
- Escape from burning buildings
- Roof access from helicopters
- Tactical scouting missions
- Window and balcony access and entry
- Access to windows for mechanical & explosive breaching
- Deployment of distraction devices and stun munitions
- Diversionary movement
- Ship and oil rig boarding

With all these applications for rappelling, mountain guides and mountain warfare instructors, as subject-matter experts on rope access techniques, will be called upon to host and teach a broad range of rappelling programs. Offering these programs is also another way in which a mountain warfare school can expand their capabilities and overall usefulness to the military.

EQUIPMENT

Much of the equipment used for mountaineering can be used for tactical rappelling, but there are also several items that are more specifically suited to this mission.

Rappelling Ropes:

Rope is well covered in Chapter 6 but there are some additional points related to tactical rappelling that are covered here. Although mountaineers routinely rappel on dynamic rope, the additional stretch makes it far less suited to tactical rappelling than static kernmantel. Static rescue ropes and rappel lines are also designed to be harder wearing and have thicker sheaths than the softer, more manageable dynamic climbing ropes.

There are three different sizes of static line commonly used for rappelling, each with its own advantages and disadvantages.

The lightest general-use static rope is 3/8″ (9.5mm) line, which is rated for approximately

British SAS used rappel tactics in assaulting and breaching the Iranian Embassy in London, May 1980. Note in the upper right photo that one trooper became jammed up on his rope and was burned before dropping to the balcony to continue the assault (SAS)

5,000 pounds and can be used as a one-man tactical rappel line, fixed line or hauling line. This smaller diameter rope is well suited to use in leg bags but does not generate much friction within a figure 8 rappel device so is not usually the first choice for most teams.

7/16″ (11.1mm) rope is rated for 6,500 pounds and is commonly referred to as tactical assault line. This is a good balance between size and weight and the most popular for general rappelling applications. It is also NFPA rated for One-Person rescue applications.

1/2″ (12.7mm) CMC Rescue Lifeline and KM III are rated for 9,500 pounds and in black make excellent rappel lines for operators wanting a more controlled descent or carrying heavy loads. They are also NFPA rated for Two-Person rescue applications such as pick-offs.

Static rope and rappel line is generally purchased in black to differentiate it from OD green dynamic climbing ropes in the military or the more colorful ropes used for civilian applications.

The length of a rappel line will vary with the tactical application, but in general, tactical rappels are not long rappels since operators are potentially exposed to hostile fire when rappelling.

For the typical operation where an operator is rappelling from the roof of a building to make entry on a balcony or window below, the rope should be first and foremost long enough to reach the objective, but secondly, long enough for the operator to continue to the ground if compromised. A third and recommended consideration, is that there should be sufficient rope coiled above the anchor to lower the operator to the ground if he should become injured or entangled. So the general rule would be to deploy rappel lines that are twice the length needed, or twice the height of the building.

The way in which this works is to place an additional large Rescue 8 or brake-bar rack at the point where the rappel line connects into the anchor strap(s). Then sufficient rope is measured out to reach the ground and that becomes the point where the rope is attached to the Rescue 8 just as if it were a lowering system. The difference is that the rappel rope is laced into the rescue 8 and secured, ready to be released and used to lower the operator if he should need it.

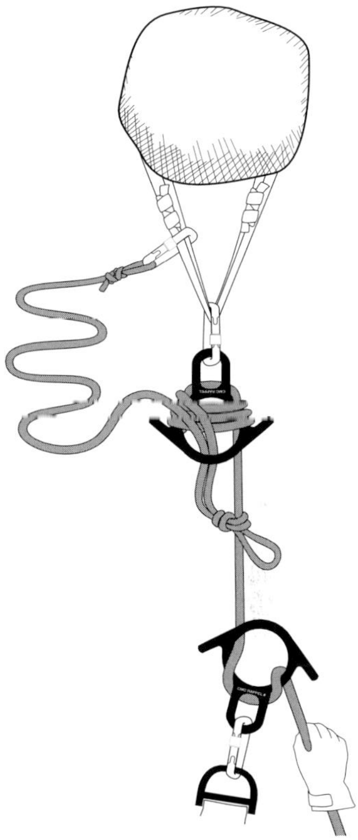

Contingency rappel anchor that can be released to lower a tangled rappeller (CMC Rescue)

The first part of the rope to be used for the rappel can then be stuffed into a rope bag in preparation for deployment.

Rope Bags:

There are primarily two types of rope bags used in tactical rappelling; the standard rope bag used to store, protect and deploy ropes, and tactical leg bags designed to be attached to the operator's leg.

The leg bag allows the operator to move to the helicopter or rappel station already rigged for the rappel and only needing to snap into a suitable anchor point. The leg bag also eliminates the need

French GIGN training on the tower at their headquarters just south of Paris

for the operator to pre-deploy the rope down the side of a building and possibly warning the enemy of his approach.

Leg bags do not hold a lot of rope, so are usually loaded with 50′ of 7/16″ or 75′ of 3/8″ rope. They are not suitable for long heavy ropes, but a conventional rope bag can be clipped to the rappel harness for that application.

Edge Protection:

Edge protection is critical for urban rappelling where most buildings have numerous sharp edges, rusting metal, roofing iron and squared concrete edges that can abrade ropes.

Rappel Harness:

The selection of a harness will depend on the specific mission and how much time one expects to be hanging in that harness. For a SWAT team that uses rappelling to drop from a helicopter or access a window, a strong but simple tactical rappelling harness will be sufficient. A tactical rappel master, who will be spending considerably more time on rope, would be more comfortable in a more substantially padded Pro series harness.

A good harness should hold the rappeller securely even when inverted

Rappel Device:

Most teams use some form of 8-plate rappel device for their simplicity and versatility. The two that we use at STTU are the CMC Rappel 8 for individual rappelling and small ropes, and the CMC Rescue 8 for heavier ½″ ropes and as adjustable anchor points. The ears on these devices make it easier to lock off and prevent the rope from slipping around to jam the device in a girth hitch.

CMC Rescue 8 and PMI double-palmed rappel gloves

Carabiners:

All operators should have four aluminum locking Ds for personal use, while team carabiners for anchor points should be steel locking Ds.

Prusik Loop:

Each operator should have at least one prusik loop set-up to be used as a safety, self-belay device.

Rappel Gloves:

Since tactical rappels are neither fast nor long, rappelling gloves should be thick enough to protect the hands but thin enough to give good dexterity and easy weapons manipulation.

Rope Cutter:

Every rappeller should be equipped with some form of rope cutter so as to be able to cut free from entanglements and curtain cords. A conventional knife blade is dangerous to use around ropes under tension so a protected hook-knife, line-cutter or EMT shears are preferable.

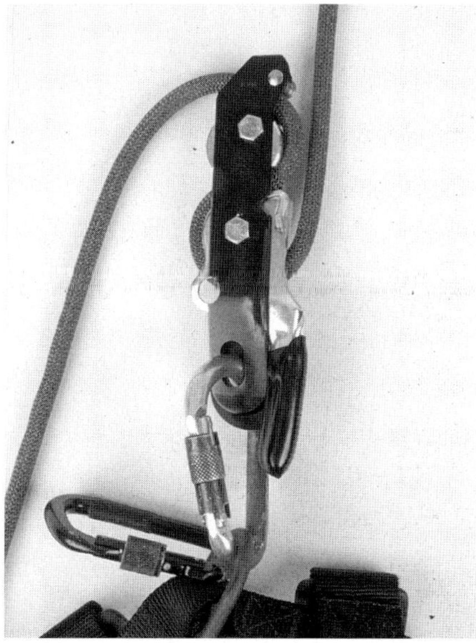

The Petzl Stop is popular with some teams for its controlled braking and hands-free auto-stop feature

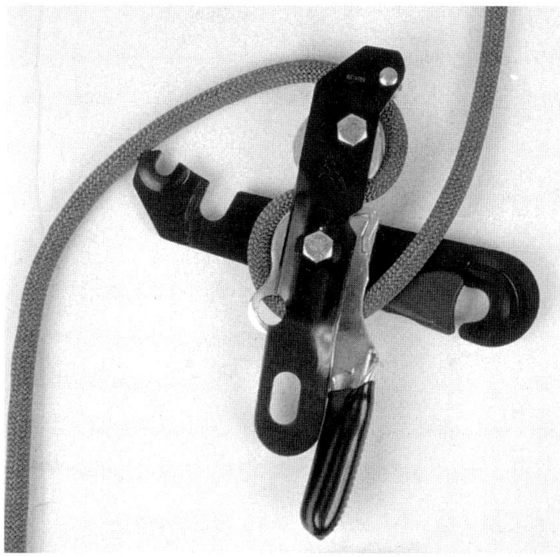

Personal Tactical Rappel Kit with anchor straps, rope, rope bag, harness, Rappel 8 and essential hardware (CMC Rescue)

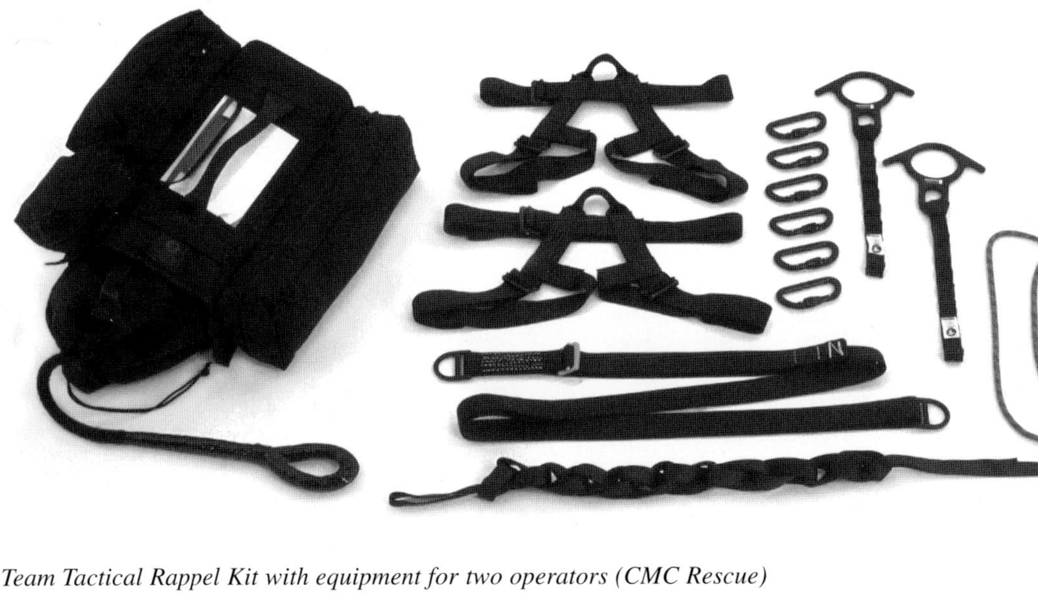

Team Tactical Rappel Kit with equipment for two operators (CMC Rescue)

Tactical Rappel Systems:

CMC offers pre-configured tactical rappel systems for law enforcement and military operations. These contain everything required for not only the rappel but also to set-up a simble anchor system. Their Personal Tactical Rappel Kit contains:

1 75´ x 3/8˝ CMC Lifeline, Black
1 CMC Stealth Rope Bag
1 CMC Rappel 8 & Link Strap
1 Tactical Rappel Harness
1 Pro Series Variable Anchor Strap, Black
1 20´ x 1" Tubular Web, Black
4 CMC Locking Carabiners, Black
1 5´ x 7mm Prusik Cord, Blue

CMC's Tactical Team Rappel Kit has 150´ of black 7/16˝ Rescue Lifeline and everything needed for two tactical officers.

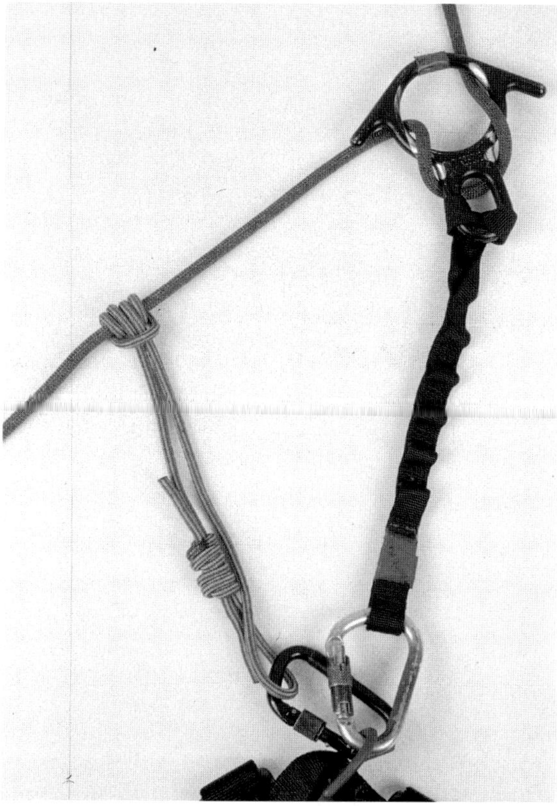

With the CMC Link strap, the prusik can be positioned blow the 8 so that both the descent and the prusik can be controlled with the left hand leaving the right hand free to shoot

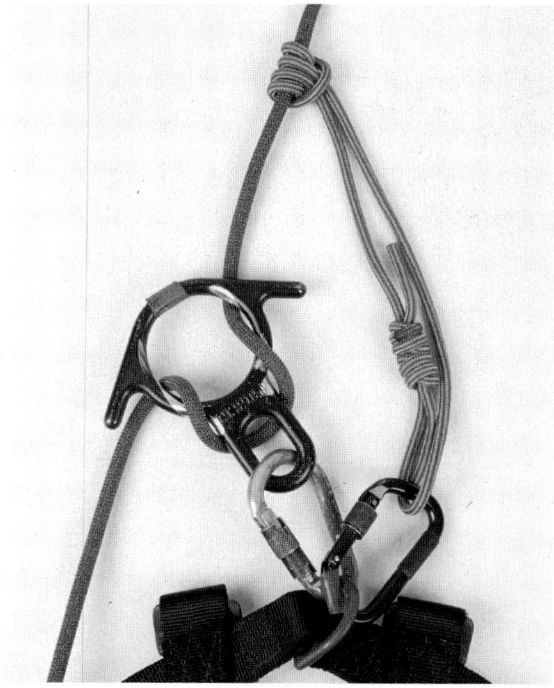

When using a safety prusik in this configuration, the rappeller must control the descent with his left and the prusik with his right. This does not leave a hand free to handle a weapon efficiently

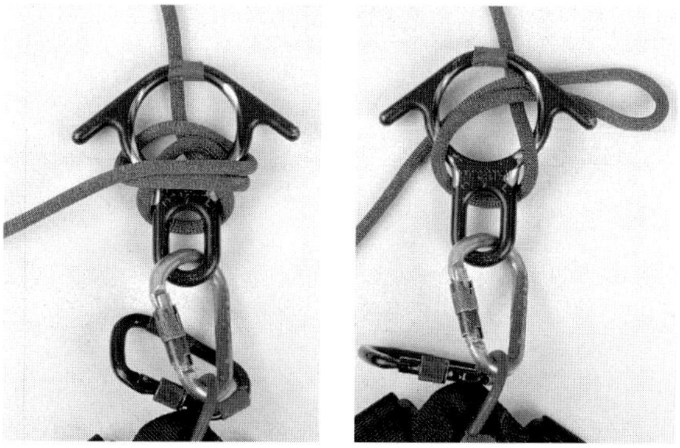

Rescue 8 locked off on left and set for a quick-release on right

Micro Rappel Kit (CMC Rescue)

MICRO RAPPEL SYSTEM

The Micro Rappel System was developed by New England Ropes for military special operations teams in cooperation with the US Army research and development labs. It derives its small size by using 25 meters of New England Ropes' Tech Cord (5mm Technora) packed into a small belt pack with integrated rappel belt and harness. In place of a large conventional 8 plate, this system utilizes the compact steel Micro Rappel Descender which offers multiple rigging configurations to control descent rate according to the weight of the operator and his equipment load. It is not intended as a rappel system for routine tactical operations or regular use, but more an emergency escape device or rappel system for unanticipated tactical options.

Components of the micro rappel system; note the leg straps stowed inside the waist belt

Safariland 3004 Special Ops tactical leg holster for the SIG 226

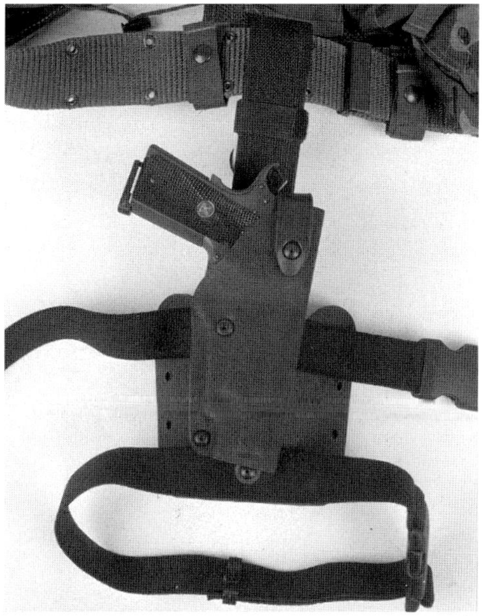

Safariland 6004 Tactical system for the Wilson CQB .45ACP Colt Government

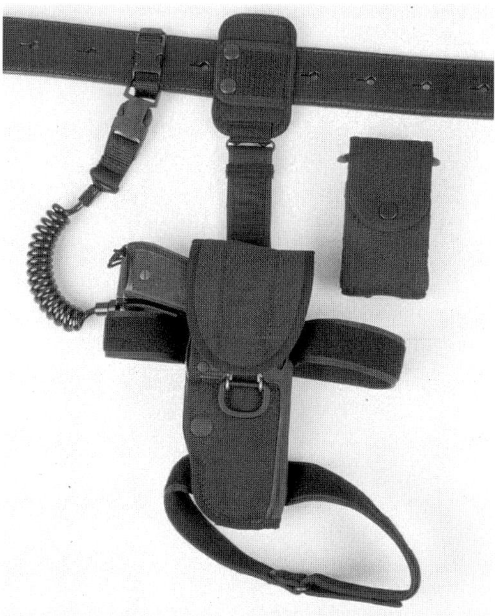

Bianchi T1500 Tactical holster with the M9/Beretta 92. Also fits the SIG 226, 228, Glocks and H&K USP

Paul Evers holster with light-mounted SIG 226, built for the British SAS

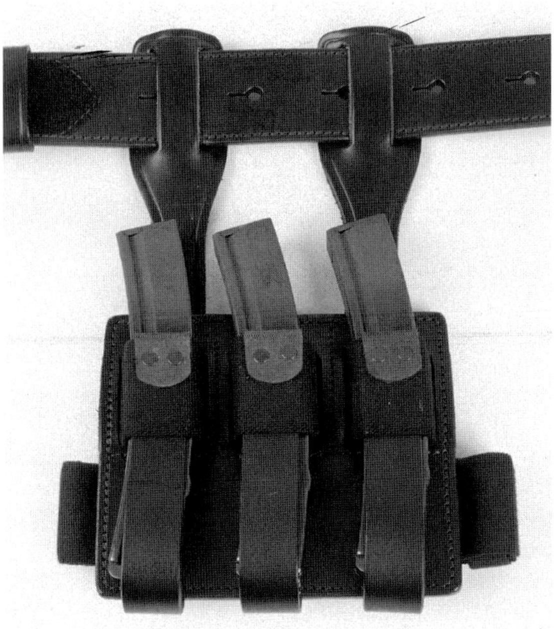

Evers triple magazine pouch for the MP-5 submachine gun

The rappelling harness, pistol belt and load-bearing equipment must be compatible allowing the operator to rappel with his weak hand and shoot, deploy stun munitions or handle breaching tools with his strong hand

Tac-Ord retaining leash recommended for all vigorous activity, rappelling and climbing

WCSO deputies hook one boot into the window frame to stabilize themselves as they practice weapons handling during an STTU tactical rappel program

The Chalker chest sling is ideally suited to tactical rappel operations

Basic rappel system with no belay safety

Basic rappel system with no safety prusik

Rigging of the 8 Link and prusik loop

Rigged with CMC 8 Link and safety prusik loop

The prusik hitch is slid down by the rappel-hand leaving the gun-hand free

Marines going through basic rappel training. This tower also permits simulated helicopter deployment from both the skids and hell hole.

Simulation of an injured or unconscious rappeller with a bottom belay

TACTICAL RAPPEL PROGRAM DEVELOPMENT

Tactical Rappel training is far more than just sliding down ropes as illustrated by the following overview of an STTU tactical rappelling program. This is an intense three or four-day program designed to prepare law enforcement and military special operations personnel to utilize rappelling as a tactical option. The first three days are for operator training and the fourth day is for the training and evaluation of potential Rappel Masters.

Training Requirements:

- Classroom with whiteboard and carousel slide projector
- Rappel Tower / Multi level with windows and balconies
- Access to high-rise building (if available)
- Tactical Shooting Range

Individual Equipment:

- Full Tactical Equipment and body armor as worn on call-outs
- Belt & Handgun Holster that will accommodate a Rappel Harness
- Rappel Harness and Leather Rappel Gloves (not too thick)
- One Rappel 8 with ears and four Locking Carabiners
- Helmet—Lightweight Protec, climbing or ballistic
- Radio with tactical ear-piece, throat mic or boom mic (optional)

Day One—Classroom

- Introduction to Basic Rappelling
- Safety Lecture
- Rappelling Ropes & Rope Management
- Essential Knots & Hitches
- Hardware Selection & Rappel Devices
- Anchors & Anchor Systems

ALPINE OPERATIONS

Rappel weapons training in STTU training programs

Day One—Rappel Tower
- Safety Brief
- Setting Anchors
- Back-up Anchor Systems
- Basic Rappelling Drills w/ Bottom Belay
- Rappelling w/ Self-Belay

Day Two—Classroom
- Applications & Limitations of Tactical Rappelling
- Tactical Rappel Systems
- Risk Analysis & Accident Management
- Weapons Handling

Day Two—Shooting Range
- Safety Brief
- Weapons Handling
- Simulated Shooting on Rappel
- Shooting on Rappel
- Stun Grenade Deployment

Day Two—Rappel Tower
- Advanced Rappel Techniques
- Bottom Belay & Top Lowering Systems
- Self Rescue & Belay
- Rescues & Pick-Offs
- Introduction to Window Entry
- Coordinated Rappels
- Basic Weapons Handling
- Rappel / Rope Management Practical & Written Tests

Day Three—Tower
- Rope Deployment—Overt / Covert
- Lateral Movement on Rope
- Weapons Handling
- Scouting & Window Entry
- Diversions, Distractions & Breaching
- Coordinated Cover Fire
- Simultaneous Entry Drills
- Immediate Action Drills

Night Three—High-rise Building
- Tactical FTX
- Response, Planning, Deployment & Execution

Deputies prepare a simultaneous assault through three windows with all movement covered by the sniper team

Note poor weapons handling by the rappeller about to make entry

Note the considerable amount of smoke from stun munitions obscuring the operators' view into the stronghold

Day Four—Classroom & Tower

- Role of the Rappel Master
- Rappel Program Development
- Risk Analysis & Accident Management
- Tactical Rappel Operations Planning & Briefing
- Evaluation of individual Rope Management & Rappel Skills
- Setting & Supervising Anchors
- Safety Checks & Dispatching Rappel Teams
- Emergency Rescue Systems
- Final Practical Evaluation of Supervisory Skills

Helicopter rappelling and fast roping would each require an additional day of training.

Sheriff's SEB fast-roping onto the deck of a boat during a drug interdiction simulation

FBI-HRT fast-roping from a Huey, training for the '84 LA Olympics

French GIGN

French GIGN counter terrorist team rappeling and fast roping onto the top of a hijacked aircraft

SEALs practice contact drills while wearing snowshoes (NSWG)

31

LAW ENFORCEMENT TACTICAL OPERATIONS

In light of the casualties that the military has sustained in mountain warfare over the years, a law enforcement agency needs to seriously consider the possible consequences of launching a tactical operation in a mountainous or cold weather environment. Very few law enforcement agencies are able to maintain even the minimum standards for basic urban SWAT operations, let alone take on the complexities of a wilderness operation. In addition, very few teams have the clothing or equipment necessary to sustain themselves on a winter call-out, especially miles from their warm vehicles, fast food restaurants and accustomed logistical support.

In the absence of a significant budgetary and training commitment, there are five obvious reasons why civilian agencies should avoid extreme cold weather or mountain operations:

1. Fitness: There is a big difference between sprinting across a street in SWAT gear and humping a pack all day in cold mountainous terrain. The general physical condition of most law enforcement officers deteriorates from the time they leave the academy (or go through their first divorce). Only professional full-time tactical teams such as LA Sheriff's SEB and the FBI-HRT have the time, budget and motivation to maintain all their personnel at the necessary peak levels of fitness.
2, Training: The time required to become proficient in all aspects of wilderness living, traveling and patrolling is usually well beyond the time availability and budget of most tactical teams. Those that do receive this type of training often have difficulty maintaining those perishable skills.
3. Clothing: Most agencies do not have the technical clothing systems and related training needed to sustain men in the field for any length of time without succumbing to cold related injuries. There are agencies, that because of their location, perform routine law enforcement duties in extremely cold conditions but there is a big difference between standing around in the cold for a few hours and moving into the mountains for several days.
4. Equipment: In addition to all the equipment listed in Chapters 4 & 5, the team would need ECWCS, tactical load bearing equipment, communications gear and weapons more suited to mountain warfare than the limited ranges encountered on most urban operations.
5. Experience: Comfort, safety and proficiency in the mountains can only come with experience, and few tactical teams would ever get enough training time to become proficient in those potentially harsh conditions. The only exception would be teams with members who have been formerly trained in mountain warfare or are already avid mountaineers and backwoodsmen.

Many law enforcement agencies in the northern and northeastern United States, Canada and Alaska

Gore-Tex outer wear, longer ranged weapon systems and military type load-bearing systems (Eagle or Black Hawk) will be required for law enforcement operations in the wilderness

already have cold weather clothing systems designed for routine duties, but the bulky thermal suits and heavy coats that they generally issued are designed more for warmth alone and not to handle the high energy heat output or freedom of movement required for alpine or tactical operations.

However, apart from county rescue teams, there are many US law enforcement agencies that have taken the technical clothing systems used for mountaineering and alpine skiing and adapted them to suit their needs. Initially, these clothing systems were only used by park rangers and US Fish & Wildlife agents, but most agencies that are located near ski resorts and winter recreational areas have now made the move to the three layer concept of wick, warmth & wind. This is most noticeable with those working in states like Colorado and Utah that have numerous ski areas that are small towns within themselves.

With adequate clothing and thermal protection, a metropolitan tactical team is able to deploy in snow and winter conditions within the city or developed suburban areas, provided it is for only limited amounts of time. This is in part due to the fact that they can rotate personnel as they become cold, and also because they have ready access to vehicles, warm food and logistical support. However, once that team moves into a rural or wilderness environment, the deficiencies in their equipment, training and communications will become immediately evident.

WILDERNESS TRAINING

If an agency has a real need for a wilderness or alpine operations capability, then it is initially a matter of allocating the funds, making the training time available and selecting suitable personnel. But before venturing into the mountains, a special weapons team that already has a high level of proficiency in urban swat operations would require additional training in several wilderness related specialties. The actual training modules needed by a specific team would be driven by the mission capability that the team desires, anticipated area of operation, duration and local weather conditions.

With six external pockets, the ruggedly constructed Becker patrol pack is ideal for both military and law enforcement tactical operations

The following is a list of training programs recommended for teams or agencies involved in or anticipating wilderness operations. In some cases, this type of training can be obtained from the military or civilian mountaineering schools but these will not be tailored to law enforcement. For programs specifically designed to meet the needs of the law enforcement community, agencies should look to organizations such as NTOA, STTU and the National Wilderness Training Center for Law Enforcement (NWTC).

The following courses can be taught individually or grouped into more comprehensive training packages, varying in length from 2 days to 2 weeks:

- Wilderness Basics
- Tactical Wilderness Basics
- Land Navigation & Tactical Patrolling
- Emergency Wilderness Survival
- Extreme Weather Survival
- Tactical Tracking
- Booby-Trap Identification & Avoidance
- Helicopter Insertion & Extraction Techniques
- Mountaineering—Basic
- Backcountry Snow Travel
- Backcountry Planning & Deployment (Command & Control)
- Mountain Rescue Techniques—Basic
- Wilderness First Responder medical package

Even with the above types of training, a team that has little actual experience in the mountains or working in snow-covered terrain, should not go operational without the supervision and guidance of professional mountain guides or qualified mountain leaders.

TACTICAL OPERATIONS

For the swat team that has the suitable equipment and training, the execution of tactical operations in a wilderness environment still requires considerably more planning and logistical support than an urban operation. The team cannot simply respond to a location in their vehicles, secure the perimeter, collect intelligence and launch an assault all in a matter of hours. They will not be operating in a familiar neighborhood that they have driven through on a daily basis. Because of the remoteness of most objectives, successful wilderness operations can take days or even weeks to plan and execute.

The Shomer-Tec Raid pack has many climbing features making it suited to both conventional and alpine tactical operations

Some of the problems to be considered are:

- Remote locations that are difficult to access
- Lack of familiarity with the area and terrain
- Increased distances to the target that may have to be covered on foot
- Much larger areas to observe and/or contain
- Significant distance from re-supply & medical support
- Lack of an effective communications net

The following are lists of some of the basic planning considerations for intelligence collection, logistical support and execution of a wilderness operation.

INTELLIGENCE COLLECTION

Accurate information collection and intelligence development is critical to any operation, and even more so in the mountains where the terrain and weather can be as dangerous as the enemy. The information flow leading up to an operation must be **current**, **confirmed** and preferably **first hand**—and this means putting men on the ground.

There is a saying within the military community, "Assumption is the mother of all fuck-ups". Translation: Assume nothing, not even that maps are accurate or aerial photographs are current, and especially don't believe informants until their information has been corroborated and verified. It is important to get "eyes on" the target location as soon as possible to begin the flow of tactical data back to the command post.

To collect the initial pieces of the puzzle, the military acronym is useful—S.A.L.U.T.E.

Size of the suspect group
Activity they are involved in
Location of activity
Unit if they are part of a larger group
Time of observations
Equipment and weapons seen

The following is additional information that should be collected or considered in an intelligence build-up:

- ❑ Target location with map grid references and identifiable landmarks
- ❑ Nature and scope of the suspect operation / illegal activity
- ❑ Techniques being used to disguise the operation
- ❑ Ingress and Egress routes being used by the suspects
- ❑ Noise signature of the operation, generators, vehicles, etc
- ❑ Security measures visible—Guards, alarms, booby-traps, fences, etc
- ❑ Radio antennas visible
- ❑ Number of occupants and their Nationalities
- ❑ Armament and apparent level of training
- ❑ Propensity for, or history of violence
- ❑ Daily & Nightly routines observed
- ❑ Movement of illicit product / timings / routes / security

A camouflaged or white suit is not a substitute for good concealment and better use of trees, shadows and terrain features

LOGISTICAL CHALLENGE

One of the greatest challenges in wilderness operations is the magnitude of the logistics effort to support even a small unit in the field. Consideration must be given to:

- ❑ Command Post location, manning and set-up
- ❑ Suitable Transportation—Helicopters, four wheel drives, trucks, snowmobiles.
- ❑ Communications—More powerful radios, hilltop repeaters, airborne command and coordination.
- ❑ Longer Range Weapons—Assault rifles, sniper rifles, ammunition, scopes, night vision, accessories.
- ❑ Time to issue and zero individual weapons.
- ❑ Provisions—Suitable uniforms and load bearing equipment, food, water, sleeping gear, tents, maps, compasses, thermal protection, rain gear.
- ❑ Emergencies—Survival skills, field medical skills, helicopter med-evac, signaling devices, hospital coordination, flares, signal smoke, LZ preparation.

Headsets are essential to coordinate tactical operations, but more powerful longer ranged radios will also be needed in the mountains

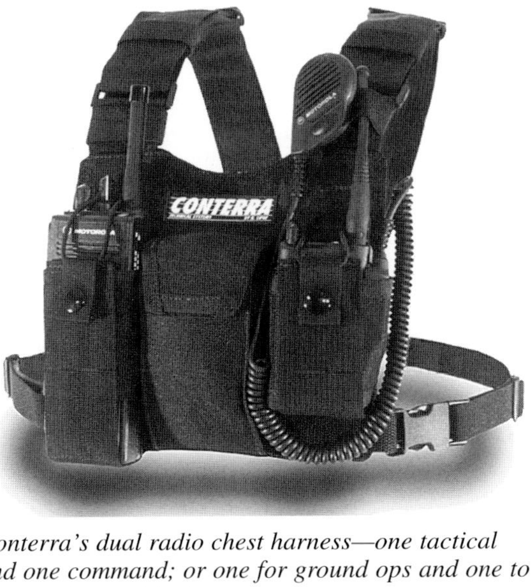

Conterra's dual radio chest harness—one tactical and one command; or one for ground ops and one to communicate with air assets

APPROACH CONSIDERATIONS

Apart from collecting intelligence and maintaining observation and listening posts in the field, the approach is one of the most difficult aspects of wilderness operations.

- ❑ The necessity of a ground insert far enough away from the suspect location so as not to cause alarm
- ❑ Time to insert long rifles and perimeter personnel ahead of the assault team
- ❑ Sufficient time for the insertion and ground approach
- ❑ Use of basic patrolling principles for team security on the approach
- ❑ The necessity to walk considerable distances to reach the suspect location
- ❑ Potential difficulty of finding the location on the ground
- ❑ Potential problems and confusion involved in deploying at night
- ❑ Availability of suitable cover and concealment
- ❑ Possibility of booby traps near the suspect location
- ❑ An Immediate Action (IA) plan if compromised

EXECUTION

In the execution of this type of operation, some of the additional considerations are:

- ❑ On-site observation of the target location up to the final assault
- ❑ Aerial Surveillance / Infra-red scanning (FLIR) up to the last minute
- ❑ The ability to secure the perimeter or avenues of escape
- ❑ Coordination and Timing of multiple ground teams
- ❑ Limited range of radio communications in the mountains
- ❑ A contingency plan in the event of heavy casualties

FINAL ASSAULT

It will be necessary to:

- ❑ Ensure the perimeter is secure
- ❑ Have long rifles cover final movement over open terrain
- ❑ Move quickly to dominate the location and seal exits
- ❑ Be conscious of the potential for friendly cross-fire
- ❑ Use the buddy system or small team concept
- ❑ Search for and secure all occupants
- ❑ Isolate suspects from weapons
- ❑ Touch nothing unnecessarily (booby-traps)
- ❑ Stabilize casualties (yours first)
- ❑ Track fleeing suspects (with caution)
- ❑ Call in helicopter support for casualty evacuation
- ❑ Identify evidence of illegal activity
- ❑ Call in specialists - EOD, ATF, Photographers, Field Interrogators, etc

Mountain operations are complex but not overly difficult provided a considerable effort is put into the planning, preparation and logistical aspects of the action. Neither the suspects nor the conditions should be underestimated. Suspects may be accomplished outdoorsmen, hunters or survivalists so tactical teams may require additional specialized training and equipment. Individual tasks should also be eliminated and maximum use made of small, flexible teams as part of the bigger effort.

Communications are always a problem in the mountains so tactical coms systems will need to be re-evaluated. Land navigation skills are more important and additional time must be allowed for over-land movement, location surveillance and the final approach.

Lastly, there must be some realistic expectations. Without a full military capability, it is impossible to totally seal off a large area, especially if there is dense vegetation. Suspects who escape on foot may be more skilled than the pursuers and lead them into areas where they may have set booby-traps or cached weapons. There is also the danger of ambush, physical injury or getting separated and lost, making the team concept that much more important in the wilderness.

Note: See the STTU training manual "RAIDS" for additional information of the planning and execution of wilderness operations.

Whether in an urban or wilderness environment, sound tactical principles must be observed and adhered to at all times. One of those principles is that it is invariably better to shoot around cover rather than over it.

(Conterra)

32

CONCLUSION

Mountain warfare and alpine rescue each have their own inherent elements of danger that are compounded by the conditions encountered in the high mountain zone. Recent history has shown us that more soldiers fall victim to the elements in the mountains than to enemy bullets; and that even internationally renowned alpinists can lose their lives with a single lapse in judgment.

Extreme cold, steep rugged terrain, thin air, strong winds, deep snow and massive avalanches all work together to create an environment that shows no mercy to the ill prepared or poorly equipped individual or team. Alpine operations require physically fit, highly motivated individuals who display a natural affinity for the mountains; selected for their ability to not only survive but to thrive in conditions that cause others to give-up and need to be rescued.

A teamís first line of defense in the mountains is their overall fitness backed by professional training and sound judgment. Their second line of defense is good clothing systems and top-of-the-line support equipment. The third line of defense is hard earned and time-honored experience.

As stated earlier in this text, the professionals whose duties take them into the mountains must train to a high standard and be equipped with the best equipment since they may often be deployed in less than ideal conditions. Unlike the recreational climber or backpacker who can and should choose the best time to venture into the mountains, the professional may be required to attempt a rescue or launch an operation at any time and in almost any conditions.

The importance of competent leadership in the mountains can also never be overlooked. The team leaders and mountain guides, who must shoulder the responsibility for training and safety, must be held to the highest standards since the lives of others will be dependent upon their decisions and actions. It is all too easy for anyone to print up a business card proclaiming themselves "instructor" or "mountain guide," without the necessary experience or accreditation that goes with these titles. There are no shortcuts to gaining the experience necessary to lead in the mountains and it is only thought standardized training, careful evaluation and nationally recognized certification (such as AMGA) that the standards can be raised and credentials will become verifiable.

However, all this does not negate the myriad of dangers inherent in alpine operations. Not unlike going into combat, extreme activities such as mountaineering carry an unpredictable element of risk that must be understood and accepted by all the players. No matter how well prepared a team or individual is when they venture into the mountains, there is always the possibility that they could fall victim to conditions beyond human control. This has been highlighted by the deaths of so many of the world's top mountaineers.

There will always be the known risk-takers who play on the ragged edge of safety and common sense and who are not destined to be long on this earth; but there are also those who have earned a reputation for common sense the professionalism that lose their lives in the mountains each year. This was never more graphically illustrated than in Jon Krakauer's book "Into Thin Air." To this day, many are still wondering what drove two strong climbers and highly respected expedition guides like Rob Hall and Scott Fischer to ignore their own safety rules on Everest in 1996. By pushing on to the summit after the agreed upon turn-around times they fell victim to the elements and lost their lives in the process.

Then no sooner had we finished reading Anatoli Boukreev's account of the disaster, "The Climb—Tragic Ambitions on Everest," than we learn he has been killed by an avalanche in the Annapurna region. Anatoli was a world-class Russian high-altitude mountaineer and expedition guide who was instrumental in the rescue of several climbers on the ill-fated Everest expedition.

Then in October 1999, one of Americaís strongest and most respected climbers, Alex Lowe, and high altitude cameraman Dave Bridges were killed by a massive avalanche while climbing on the slopes of Shishapangma, a 26,398-foot peak in southern Tibet.

Both climbers along with Conrad Anker were exploring a route up the side of a glacier at about 19,000 feet when the slide started 6,000 feet above them. The climbers were hit by a wall of snow and debris, estimated to be traveling at over 100 miles per hour, that buried all three.

Only Conrad survived by lying down, securing himself with his ice axe and covering his head as the avalanche swept over. Even though sustaining injuries to his ribs, shoulder and head, he was able to dig himself out and immediately began searching for Alex and Dave. The rest of the team also began a search for the victims buried under 20 feet of debris, but after 20 hours and due to hazardous conditions, gave up any hope of recovering the bodies.

The lesson to be learned from these tragic events is that even with the best training and equipment, and a world of experience, the inherent risks of working in the high alpine zone can never be taken for granted. Anyone accepting the challenge of the mountains must first accept the dangers concealed in this awe inspiring but all too often unforgiving environment. Then, only by having a realistic appreciation of one's limitations, will the mountaineer be able to conserve the energy and resources that may be needed for a hasty retreat. To paraphrase the sage advice of high-altitude mountaineer Ed Viesturs, "Getting to the summit is optional—getting down is mandatory!"

The End

This photo hangs on the wall of the Army National Guard Mountain Warfare School house. An appropriate reminder of the hazards of mountaineering.

APPENDIX A

ABBREVIATIONS

AAC—American Alpine Club
AAI—Alpine Ascents International
AGL—Above Ground Level (for aircraft)
AMGA—American Mountain Guides Association
ANG—Army National Guard
AT—All Terrain ski gear, also known as Randonnee
ATC—Air Traffic Controller belay device made by
BD—Black Diamond
BLM—Bureau of Land Management
CCT—Combat Control Team
CG—Center of Gravity
CM—Charlet Moser
CO—Commanding Officer
CP—Command Post
CPR—Cardio Pulmonary Resuscitation
CRTC—Cold Regions Test Center
CSAR—Combat Search & Rescue
ECWCS—Extended Cold-Weather Clothing System
ESD—Emergency Services Detail
EMS—Emergency Medical System
EMT—Emergency Medical Technician
FFP—Final Firing Position
FM—Field Manual
GPMG—General Purpose Machine Gun
GPS—Global Positioning System
HIGE—Hover in ground effect
HOGE—Hover out of ground effect
IC—Incident Commander
ICS—Incident Command System
IG—Instructor Guide
LASD—Los Angeles Sheriff's Department
MEU—Marine Expeditionary Unit
MLC—Mountain Leader's Course
MM—Military Mountaineer
MRA—Mountain Rescue Association
MS—Military Skier
MSI—Military Ski Instructor
MSL—Mean Sea Level (altitude for aircraft)
MSR—Mountain Safety Research
MTT—Mobile Training Team
MWS—Mountain Warfare School
MWTC—Mountain Warfare Training Center
NPS—National Park Service
NSWG—Naval Special Warfare Group
NVD—Night Vision Device
NVG—Night Vision Goggles
NWTC—Northern Warfare Training Center
OIC—Officer in Charge
OP—Observation Post
OR—Outdoor Research
PJ—Air Force Para-rescue Jumper
SAR—Search and Rescue
SAS—Special Air Service—British Special Forces
SBS—Special Boat Service—Royal Marine Commandos
SD—Sierra Designs
SEAL—Sea, Air and Land naval commandos
SEB—Special Enforcement Bureau
SF—Special Forces
SME—Subject Matter Expert
SMG—Sierra Mountain Guides
SOC—Special Operations Capable (MEU-SOC)
SOC—Special Operations Command
SOF—Special Operations Force
SOTG—Special Operations Training Group
STTU—Specialized Tactical Training Unit
TC—Training Circular
TCS—Task, Conditions, Standards
TNF—The North Face
USAF—US Air Force
USANG—US Army National Guard
USFWS—US Fish & Wildlife Service
USMC—US Marine Corps
XC—Cross-country skiing

APPENDIX B

MOUNTAINEERING GLOSSARY

Adze—The flat pick on an ice axe used to cut steps
Ahkio—Sled used by the US Army
Arrete—A knife-edge ridge formed by the ice-shattering of the rock
Belay—To safety the climber by moving the rope in and out as needed through a belay device
Bergschrund—A large lateral crevasse or crack at the base of a mountain marking the change in gradient or start of a glacier. Often difficult for climbers to cross
Camelot—Brand of camming device made by Black Diamond
Chimney—a wide crack or fissure in the rock that can be climbed by stemming the gap
Crevasse—A large crack in a glacier, often hundreds of feet deep
C-rig—A 2:1 hauling system usually made with a single pulley
Crux—Most difficult part of a climb
Cwm—A Welsh term for a heavily iced or glaciated valley
Daisy Chain—length of webbing with multiple loops sewn into it
Dihedral—Meeting of two faces or surfaces; a corner
Dynamic—Rope that stretches under load to cushion a fall
Dyno—To make a dynamic up-ward climbing move
8 Plate—A rappelling and belay device shaped like a figure 8
Figure 8—A knot used a lot in climbing and rescue work, especially to tie into a harness
Fixed Rope—A static rope secured with pitons, pickets or ice screws used to cross an exposed area
Friend—A brand of camming device used for protection
Gibbs—Type of rescue ascender
Glacier—A very slow moving mass or river of solid ice
Gore—The company that makes Gore-Tex. Named after Bill Gore
Jamming—Using a body part to jam into a crack while climbing
Jugging—Jumaring up a rope with ascenders or the use of prusik hitches
Jumar—Mechanical ascender like the Petzl Ascension
Kernmantel—most common rope construction used in climbing ropes where a braided sheath (the mantel) covers a core (kern)
Mantel—A climbing technique where the climber pushes down with the palms of his hands to get up over a ledge
Moraine—Rock debris carried by a glacier and deposited when the ice melts
Pick—The sharp spike on an ice axe
Piton—A spike with a ring in it, usually steel, that can be driven into a crack in the rock
Prusik—Type of hitch used to ascend a rope. Also seen spelt "prussik" and "prusick"
Pulk—Norwegian sled used by the British Royal Marines
Quick-Draw—Two non-locking carabiners connected by a short strap or sling to secure the climbing rope to the protection
Randonnee—All terrain ski gear that can be used for both telemark skiing and downhill
Rap Off—Another term to mean rappel down
Rappel—Descending a rope in a controlled manner using some form of braking device like an 8 plate, ATC, wrapping a rope through a carabiner, or by passing the rope around the back for a body rappel
Runner—A sewn sling used to connect the climbing rope to protection with carabiners

ALPINE OPERATIONS

Serac—Unstable ice formation found in icefalls that can topple at any time and crush a climber
Sketchy—Difficult part of a climb with minimal protection
Skins—Nylon or Mohair strips that are affixed to tele or AT skies to allow up hill movement
Sling—Sewn or tied loop of nylon webbing, usually 12″, 24″ or 48″ loops
Spectra—Synthetic material used in the manufacture of slings and ropes. Stronger and lighter than nylon.
Static—Essentially non-stretching rope, no more than 2% at 200 pounds, used for rappelling and fixed lines.
Stemming—straddling a gap to climb between two faces such as in a chimney
Stitch Plate—Form of belay device
Tele—Telemark skis or skiing
Tubular—Soft nylon webbing constructed in a flat tube and used in make slings
Via Ferrata—Fixed steel cables found on more accessible climbing routes in the European Alps
Whipper—A significant fall while rock climbing resulting in being whipped around on the end of a dynamic rope
Z-rig—A 3:1 mechanical advantage raising system used for hauling and rescue work

(Sgt. Clearman)

APPENDIX C

METRIC / U.S. CONVERSIONS

Quick Conversions:

Feet to meters - multiply by 0.3048
Meters to feet - multiply by 3.2808
Miles to kilometers - multiply by 1.609
Fahrenheit to Centigrade - subtract 32, multiply by 5 and divide by 9.

Commonly Used Elevations:

3,048 meters	10,000 feet
4,267 meters	14,000 feet
6000 meters	19,685 feet
8000 meters	26,246 feet

Distance:

1 mile	1.6093 kilometers or 1,760 yards
1 kilometer	0.62137 mile

Rope Lengths:

45.7 meters = 150´
50 meters = 165´ (approx)
60 meters = 197´
70 meters = 230´

Force: 1kN = 225 pounds

Length:

1 centimeter	0.3937 inches
1 inch	2.54 centimeters
1 yard	0.9144 meters
1 meter	39.37 inches or 1.0936 yards
1 meter	3.2808 feet

Weight:

1 ounce	28.35 grams or 437.5 grains
1 kilogram	2.2046 pounds
1 gram	15.432 grains or 0.03527 ounces
1 kilogram	35.274 ounces or 2.2046 pounds
1 pound	453.6 grams or 0.4536 kilograms
100 pounds	45.359 kilograms
1 metric ton	0.98421 English ton
1 English ton	1.016 metric ton

METRIC / U.S. CONVERSIONS — Continued

Volume:

1 pint	473.17 milliliters
1 fluid ounce	29.573 mls
1 liter	33.814 fluid ounces
1 liter	0.908 quarts dry or 1.0567 quarts liquid
1 quart dry	1.101 liters
1 quart liq.	0.9463 liters
1 cubic yard	0.7646 cubic meter

Temperature conversion:
F = C x 9/5 + 32
C = F - 32 x 5/9

Degrees F.	Degrees C.
212 Boiling	100
104	40
98.6 Human	38
80	27
70	21
50	10
41	5
32 Freezing	0
20	- 6
10	- 12
05	- 15
0	- 17.77
- 10	- 23
- 20	- 29
- 40	- 40

IMPACT CONVERSION FORMULAS

Pounds/Force x 4.448 = Newtons (N)
Pounds/Force x 0.4536 = Kilograms (Kg)
Newtons x 0.2248 = Pounds/Force (lb.)
Newtons + 1000 = Kilonewtons (kN)
Kilograms + 0.4536 = Pounds/Force (lb.)

APPENDIX D

SOURCES of EQUIPMENT & TRAINING

MOUNTAINEERING EQUIPMENT:

ATLAS SNOW-SHOE - 415/703-0414; 800/645-SHOE; Fax 415/252-0354
1830 Harrison St, San Francisco, CA 94103 - www.atlassnowshoe.com
BLACK DIAMOND EQUIPMENT - 801/278-5533; Fax 800/775-7625
www.BlackDiamondEquipment.com
CLIMB HIGH - 802/985-5056; Fax 802/985-9141; www.climbhigh.com
135 Northside Dr., Shelburne, VT 05482
CMC RESCUE - 800/235-5741; 805/562-9120; Fax 805/562-9870
PO Drawer 6870, Santa Barbara, CA 93160-6870 - www.cmcrescue.com
GIBBS PRODUCTS - 801/272-8354; Fax 801/277-6564
MISTY MOUNTAIN - 828/963-6688; Fax 828/963-6810
718 Burma Rd, Banner Elk, NC 28604 - www.mistymountain.com
MOUNTAIN GEAR - 800/829-2009; Fax 509/325-3030
710 N. Hamilton, Spokane, WA 99202 - www.mgear.com
MSR - MOUNTAIN SAFETY RESEARCH - 800/877-9677; 206/624-7084; Fax 206/224-6492
PO Box 24547, Seattle, WA 98124 - www.msrcorp.com
OMEGA PACIFIC (Carabiners) - 800/369-3990; www.omegapac.com
OUTDOOR RESEARCH (OR) - 800/421-2421; Fax 206/467-0374
2203 1st Avenue South, Seattle, WA 98134 - www.orgear.com
PETZL - 801/327-3805; Fax 801/327-3806; www.petzl.com
PMI - 800/282-7673; 706/764-1437; Fax 706/764-1531
PO Box 803, 4466 Highway 27 North, LaFayette, GA 30728 - www.pmi.com
SMC - 800/426-6251; 425/883-0334; Fax 425/869-7152
12880 NE 21st Place, Bellevue, WA 98005-1986 - smc@accessone.com
WILD COUNTRY - 603/356-5590; Fax 603/356-5394; www.charlet-moser.com
230 East Conway Rd, Center Conway, NH 03813
WILD THINGS - 603/356-6907; www.wildthingsgear.com

BOOTS, SOCKS & FOOT BEDS:

BLACK DIAMOND - www.BlackDiamondEquipment.com
DANNER - 800/345-0430; 503/251-1100; Fax 503/251-1119
PO Box 30148, Portland, OR 97294-0148 - www.danner.com
GARMONT USA (Boots) - www.garmontusa.com
170 Boyer Circle, Suite 20, Williston, VT 05495
KOLFLACH - 800/258-5020; 603/880-6143; Fax 603/880-6099
9 Columbia Dr, Amherst, NH 03031
LOWA - 203/353-0116; Fax 203/353-0311
PO Box 407, Old Greenwich, CT 06870
SEALSKINZ - 615/847-6882; Fax 615/847-6556
1002 Industrial Rd, Old Hickory, TN 37138
SMARTWOOL - 800/594-5148; Fax 423/756-4206
Duke Designs Inc, 1401-N Williams St, Chattanooga, TN 37408
LA SPORTIVA - 303/443-8710; Fax 442-7541
3280 Pearl St, Boulder, CO 80301 - www.sportiva.com
SUPERFEET - 800/634-6618; Fax 800/320-2724
1419 Whitehorn St, Ferndale, WA 98248
SUREFOOT - 888/655-8110; www.surefoot.com
TECNICA - 800/832-6422; Fax 603/298-5790; www.tecnicausa.com
19 Technology Dr, West Lebanon, NH 03784

SOURCES—Boots, Socks & Foot Beds (Continued)

THORLO - 800/438-0286; Fax 704/838-7005
PO Box 5399, Statesville, NC 28687-5399 - www.thorlo.com

CLIMBING GEAR & HARNESSES:

BLACK DIAMOND - www.BlackDiamondEquipment.com
2084 East 3900 South, Salt Lake City, UT 84124
CLIMB HIGH - 802/985-5056; Fax 985-9141; www.climbhigh.com
135 Northside Dr, Shelburne, VT 05482
MISTY MOUNTAIN HARNESSES - 828/963-6688; Fax 828/963-6810
718 Burma Rd, Banner Elk, NC 28604 - www.mistymountain.com
OMEGA PACIFIC (Carabiners) - 800/369-3990; Fax 509/244-0954
11919 W. Sprague Ave. Bldg. H-1 Airway Heights, WA 99001 - www.omegapac.com
PETZL - 801/327-3805; Fax 801/327-3806
Freeport Center, M-7, PO Box 160447, Clearfield, UT 84016
TRANGO - 800/860-3653; www.trango.com

CLOTHING & GLOVES:

FIRST LIGHT - Tel/Fax 970/879-3557
GLACIER GLOVES - 800/728-8235; Fax 702/825-6544
WL GORE & ASSOCIATES - 206/325-2929; Fax 206/325-9234
www.wlgore.com or www.gorefabrics.com
JAGGED EDGE - 800/926-0159; Fax 970/728-2056.
129 W. Colorado Ave, Telluride, CO 81435 - www.jagged-edge.com
MARMOT - www.Marmot.com
MARMOT MOUNTAIN WORKS - www.marmotmountain.com
MOONSTONE - 415/550-4891; Fax 415/550-3586
833 Indiana Street, San Francisco, CA 94107 - www.moonstone.com
MOUNTAIN HARDWEAR - 510/559-6700; Fax 510/559-6709; www.mountainhardwear.com
THE NORTH FACE - 510/618-3500; 800/447-2333; 510/347-2520; Fax 510/618-3532 - www.thenorthface.com
PATAGONIA - www.patagonia.com
SIERRA DESIGNS - 800/736-8592; Fax 510/601-5518
1255 Powell Street, Emeryville, CA 94608 - www.sierradesigns.com

COMMUNICATIONS & RADIOS:

TEA Inc. 310/457-7401; 914/763-8893; Fax 914/763-9158 - www.swatheadsets.com
MOTOROLA - www.motorola.com

DRILLS:

PIKA MOUNTAINEERING - 801/485-1686; Fax 485-0578; www.pikamtn.com

FOOD:

ALPINE AIRE FOODS - 800/322-6325; Fax 530/272-2624
PO Box 926, Nevada City, CA 95959 - www.alpineairefoods.com

INSTRUMENTATION - WATCHES, ALTIMETERS, GPS, COMPASSES, ETC:

BRUNTON - 307/856-6559; Fax 307/857-8282; www.brunton.com
620 East Monroe, Riverton, WY 82501
CASIO - 973/442-5707; Info: 800/YO-CASIO
GARMIN - 913/397-8200; Fax 913/397-8282; www.garmin.com
MAGELLAN - 909/394-5000; Fax 909/394-7050; www.magellangps.com

NIELSEN-KELLERMAN (Kestrel) - 800/784-4221; 610-447-1555; Fax 610/447-1577
104 West 15th St, Chester, PA 19013 - www.kestrel-instruments.com or www.nkelectronics.com
SUUNTO - 800/543-9124; 760/931-6788; Fax 760/931-9875
2151 Las Palmas Dr, Suite G, Carlsbad, CA 92009
THOMMEN (Altimeters) - www.thommenag.ch

KNIVES & MULTI-TOOLS:

BENCHMADE KNIFE COMPANY - 503/655-6004; Fax 503/655-6223; www.benchmade.com
BUCK KNIVES - 800/735-2825; Fax 800/733-2825; www.buckknives.com
COLD STEEL - 800/255-4716; www.coldsteel.com
GERBER LEGENDARY BLADES - www.gerberblades.com
LEATHERMAN TOOL GROUP - 503/253-7826; Fax 503/253-7830; www.leatherman.com
SPYDERCO - 800/525-7770; 303/279-8383; www.spyderco.com

LIGHTS & HEADLAMPS:

PRINCETON TEC - 609/298-9331; Fax 609/298-9601
PETZL - www.petzl.com
PELICAN PRODUCTS (Lights & Cases) - 310/326-4700, Fax 310/326-3311
23215 Early Ave, Torrance, CA 90505 - www.pelican.com

PACKS:

GREGORY - 800/477-3420; Fax 909/676-6777; 909/676-5621
100 Calle Cortez, Temecula, CA 92590 - www.gregorypacks.com
ARC'TERYX - www.arcteryx.com
BLACKHAWK - 800/694-5263; 757/436-3101; Fax 757/436-3088
1133 Executive Blvd, Chesapeake, VA 23320 - www.blackhawkindustries.com
DANA DESIGN - www.danadesign.com
EAGLE INDUSTRIES - 314/343-7547; Fax 314/349-0321
400 Biltmore Dr, Suite 530, Fenton, MO 63026 - www.eagleindustries.com
SHOMER-TEC - 360/733-6214; Fax 360/676-5248
Box 28070, Bellingham, WA 98228

RESCUE & MEDICAL:

ADVENTURE MEDICAL KITS - 800/324-3517; Fax 510/261-7419
CHINOOK MEDICAL GEAR - 970/926-9277; Fax 970/926-9660; www.chinookmed.com
CMC RESCUE INC - 800/235-5741; 805/562-9120; Fax 805/562-9870
PO Drawer 6870, Santa Barbara, CA 93160-6870 - www.cmcrescue.com
CMI - www.cmi-gear.com
CONTERRA Inc - 360/734-2311; Fax 360/738-2241; www.conterra-inc.com
1600 Kentucky St, Unit B1, Bellingham, WA 98226
DERMATONE - 860/292-1311; Fax 860/292-1459
80 King Spring Rd, PO Box 3536, Windsor Lock CT 06096
DYNA-MED - 800/854-2706; www.dyna-med.com
GIBBS PRODUCTS (Rescue Ascenders) - 801/272-8354; Fax 801/277-6564
OUTDOOR RESEARCH - 800/421-2421; Fax 206/467-0374; www.orgear.com
PRO-MED - 800/776-5310; www.promedkits.com
SOS - Survival On Snow - 780/973-5412; Fax 780/973-3318
Box 1 - Site 218, RR#2, St Albert, Alberta, Canada T8N 1M9 - www.sos-find.com
SPORT-STAT - Nonin Medical - 800/356-8874; 612/553-9968; Fax 612/553-7807
www.sportstat.nonin.com

SOURCES OF EQUIPMENT & TRAINING (Continued)

ROPES:

CMC RESCUE, Inc. - 800/235-5741; 805/562-9860; Fax 805/562-8260; www.cmcrescue.com
BLUEWATER - www.bluewater-climbing.com
METOLIUS / MAMMUT ROPES - 541/382-7585; Fax 541/382-8531
 www.metoliusclimbing.com
NEW ENGLAND ROPES / MAXIM - 508/678-8200; Fax 508/679-2363
 848 Airport Rd, Fall River, MA 02720 - www.neropes.com
PMI - 800/282-7673; 706/764-1437; Fax 706/764-1531; www.pmirope.com
STERLING ROPE - 800/788-ROPE; 207/885-0330; Fax 885-0033; Fax 800/755-ROPE
 31 Washington Ave, Scarborough, ME 04074-9782 - www.sterlingrope.com

SKIS, SKINS, POLES & SLEDS:

ASCENSION ENTERPRISES (Skins) - Acquired by Black Diamond - www.bdel.com
ATLAS SNOW-SHOE - 415/703-0414; 800/645-SHOE; Fax 415/252-0354
 1830 Harrison St, San Francisco, CA 94103 - www.atlassnowshoe.com
ATOMIC SKI USA - 800/258-5020; 603/880-6143; Fax 603/880-6099
 9 Columbia Dr, Amherst, NH 03031 - www.atomicski.com
BLACK DIAMOND - 801/278-5533 - www.BlackDiamondEquipment.com
CLIMB HIGH - 802/985-5056; Fax 802/985-9141; www.climbhigh.com
K2 - 206/463-3631; Fax 206/463-2861; www.k2skis.com
 19215 Vashon Highway SW, Vashon Island, WA 98070
LEKI - 800/255-9982; Fax 800/291-2161
 356 Sonwil Dr, Buffalo, NY 14225 - www.leki.com
LIFE-LINK INTERNATIONAL - 307/733-8469; 800/443-8620; Fax 307/733-8469; www.life-link.com
TUA-Cima Inc - 303/417-0301; Fax 303/417-0145
 1945 33rd St, Boulder, CO 80301
WILDERNESS ENGINEERING (Sleds) - 888/553-5207

SLEEPING BAGS & BIVI SACKS:

INTEGRAL DESIGNS - 403/640-1445; www.integraldesigns.com
MARMOT - www.Marmot.com
THE NORTH FACE - 510/618-3500; 800/447-2333; 510/347-2520; Fax 510/618-3532
OUTDOOR RESEARCH (OR) - 800/421-2421; Fax 206/467-0374; www.orgear.com
SIERRA DESIGNS - 800/736-8592; Fax 510/601-5518; www.sierradesigns.com

TACTICAL EQUIPMENT:

BIANCHI INTERNATIONAL - 800/477-8545; Fax 909/676-6777; www.bianchiinternational.com
BLACKHAWK INDUSTRIES - 800/694-5263; 757/436-3101; Fax 757/436-3088
 www.blackhawkindustries.com
CMC RESCUE - 800/235-5741; 805/562-9120; Fax 805/562-9870; www.cmcrescue.com
EAGLE INDUSTRIES - 314/343-7547; Fax 314/349-0321
 400 Biltmore Dr, Suite 530, Fenton, MO 63026 - www.eagleindustries.com
PAUL EVERS HOLSTERS - 011/44(151)494-1001; Fax 494-2002
 Unit 6, The Match Factory, 140 Speke Road, Garston, Liverpool L19-2RF, UK
SAFARILAND - 909/923-7300, Fax 909/923-7400
 3120 E. Mission Blvd, Ontario, CA 91761- www.safariland.com
SHOMER-TEC - 360/733-6214; Fax 360/676-5248
TAC-ORD - Tactical Ordnance and Equipment Corp - www.tac-ord.com
 PO Box 428, Meridian, ID 83680 - 208/288-1450; Fax 208/288-1451
TACTICAL & SURVIVAL SPECIALTIES - 540/434-8974; Fax 540/434-7796

WEAPONS, SCOPES & ACCESSORIES:

ROBAR COMPANIES - 623/581-2648; Fax 623/582-0059; www.robarguns.com
21438 N. 7th Avenue, Suite B, Phoenix, AZ 85027
WILSON COMBAT - 800/955-4856; www.wilsoncombat.com
2234 CR 719, Berryville, AR 72616
D&L SPORTS - P.O. Box 651, Gillette, WY 82716 - 307/686-4008
LEUPOLD & STEVENS - 503/646-9171; www.leupold.com
PO Box 688, Beaverton, OR 97075
METRO TACTICAL - 800/735-7030; www.metrotac.com
SHOOTER'S CHOICE / VENTCO - 440/834-8888; Fax 440/834-3388; www.shooter-choice.com
15050 Berkshire Industrial Parkway, Middlefield, OH 44062.
HSM AMMUNITION - 406/777-2106; 406/728-3263; Fax 406/777-3908
PO Box 7465, Missoula, Montana 59807 - www.montana.com/hsm
VULPES VENTURES (custom ammo) - 630/759-1229
PO Box 1363, Bolingbrook, IL 60440-7363
FEDERAL CARTRIDGE CO - 612/323-2300, Fax 612/323-3890
SIERRA BULLET CO - 800/223-8799; www.sierrabullets.com

TRAINING & MOUNTAIN GUIDING:

STTU - Specialized Tactical Training Unit / Alpine Operations - Fax 310/829-0868; www.sttu.com
CMC RESCUE SCHOOL - 800/227-9281; Fax 800/235-8951; www.cmcrescue.com
ALPINE ASCENTS INTERNATIONAL - 206/378-1927; Fax 206/378-1937
121 Mercer St, Seattle, WA 98109 www.alpineascents.com
SIERRA MOUNTAIN CENTER - 760/873-8526; www.sierramountincenter.com
Robert "SP" Parker - PO Box 95, Bishop, CA 93515
KATHY COSLEY/MARK HOUSTON ALPINE GUIDES - www.cosleyhouston.com
NWTC - National Wilderness Training Center - www.theriver.com/Public/nwtc
SAN JUAN MOUNTAIN GUIDES - 970/325-4925; Fax 970/325-7225
PO Box 895, Ouray, CO 81427 - www.ourayclimbing.com
ALASKA MOUNTAIN SAFETY CENTER - Wk/Fax: 907/345-3566 - djamsc@alaska.net
ALPINE SKILLS INTERNATIONAL - www.alpineskills.com
ARCTIC WOLF MASTERS - 303/258-7916; Fax 303/258-0610
ADVENTURE CONSULTANTS - www.adventure.co.nz
ALPINE GUIDES LTD - www.alpineguides.co.nz
AMERICAN ALPINE INSTITUTE - 800/424-2249; 360/671-1570; Fax 360/734-8890
EMS - Eastern Mountain Sports - 800/310-4504; www.emsclimb.com
EXUM Mountain Guides - 307/733-2297; Fax: 307/733-9613; www.exumguides.com
NOLS - National Outdoor Leadership School - www.nols.edu
VERTICAL ADVENTURES - 714/854-6250; Fax 714/854-5249; BGVertical@aol.com
WILDERNESS MEDICINE INSTITUTE - NOLS - 970/641-3572; Fax 970/641-0882
WILDERNESS MEDICAL SOCIETY - www.wms.org/wms

CATALOG SALES:

CCI - 800/704-3891; www.climberschoice.com
CMC RESCUE, INC - 800/235-5741; 805/562-9860; www.cmcrescue.com
CONTERRA - 360/734-2311; Fax 360/738-2241; www.conterra-inc.com
MOUNTAIN GEAR - 800/829-2009; www.mgear.com
REI - www.rei.com; www.rei-outlet.com
SHOMER-TEC - 360/733-6214; Fax 360/676-5248

APPENDIX E

ASSOCIATIONS & USEFUL WEB SITES:

AAC—AMERICAN ALPINE CLUB—303/384-0110; Fax 303/384-0111
710 10th Street, Suite 100, Golden, CO 80401 www.amercianalpineclub.org
ACC—ALPINE CLUB OF CANADA—www.culturenet.ca/acc

AAAP—AMERICAN ASSOCIATION OF AVALANCHE PROFESSIONALS—www.avalanche.org/~aaap

AMGA—AMERCIAN MOUNTAIN GUIDES ASSOCIATION
303/271-0984; Fax 303/271-1377 - www.amga.com; info@amga.com

MRA—MOUNTAIN RESCUE ASSOCIATION—www.mra.org

NASAR NATIONAL ASSOCIATION FOR SEARCH & RESCUE—www.nasar.org

NSP—NATIONAL SKI PATROL—www.nsp.org

NTOA—NATIONAL TACTICAL OFFICERS ASSOCIATION—www.NTOA.org

NWTC—NATIONAL WILDERNESS TRAINING CENTER www.theriver.com/Public/nwtc

PSIA—PROFESSIONAL SKI INSRUCTORS OF AMERICA—www.psia.org

UIAA—UNION INTERNATIONALE DES ASSOCIATIONS D'ALPINISME
www.mountaineering.org

WILDERNESS MEDICAL SOCIETY—www.wms.org/wms

HEALTH, TRAVEL & MEDICAL:

HIGH ALTITUDE MEDICINE GUIDE - www.gorge.net/hamg
CENTERS FOR DISEASE CONTROL AND PREVENTION—CDC
Fax info service: 888/CDC-FAXX - www.cdc.gov
TRAVEL HEATLTH INFORMATION SERVICE: www.travelhealth.com
INTERNATIONAL SOCIETY OF TRAVEL MEDICINE: www.istm.org
AMERICAN SOCIETY OF TROPICAL MEDICINE & HYGIENE: www.astmh.org
STATE DEPARTMENT TRAVEL ADVISORIES—http://travel.state.gov

USEFUL WEB SITES:

www.avalanche.org	Avalanche Info
www.csac.org	Avalanche Info
www.climbing.com	Climbing Magazine
www.climbNET.com	Climbing Info
www.mountainzone.com	General Information & Equipment
www.peakware.com	World Mountain Encyclopedia
www.rockandice.com	Rock & Ice Magazine
www.igf.fuw.edu.pl/hill	Mountain Guide of the World

NWTC

The National Wilderness Training Center for Law Enforcement Officers (NWTC) was founded in 1998 to provide advanced training for law enforcement officers operating in various environments such as Mountains, Forests, Deserts, Arctic and Alpine Tundra, Snow and Ice, Coasts, Rivers, Lakes and Wetlands.

The NWTC offers a variety of courses varying in length and purpose from one-day seminars designed to be added to classes and conferences offered by other agencies, to weeklong stand-alone courses. Backcountry skills and disciplines are specifically tailored for a law enforcement application, and include: Expedition planning and execution, Clothing and equipment selection, Land Navigation, Survival, Tactical tracking, Patrolling, Mountaineering, Snow travel, and Helicopter operations.

Information regarding the NWTC, data supporting the need for such training, or a schedule of current courses, may be received by contacting the NWTC at:

(520) 219-4926; E-mail: nwtc@theriver.com;
Web-site: www.theriver.com/Public/nwtc

Test and Evaluation Program:

August, 1999 marked the initiation of a national test and evaluation program to enhance law enforcement operations in both urban and wilderness settings. As part of The Foundation, a charitable organization providing financial assistance to loved ones who lost their spouse in the line of duty, this program was implemented to promote officer safety by testing and evaluating two sources of law enforcement products used in the field. First, current equipment designed specifically for law enforcement. Second, those commercial products designed for recreational purposes that can be used by law enforcement with minimal modification and/or alteration.

Products tested include all forms of apparel such as extreme weather clothing, non-instrument equipment such as ropes and backpacks, non-electronic equipment such as compasses, and high-technology such as GPS receivers and night-vision devices. Testing ranges from durability in the field to ballistics.

In conjunction with Philadelphia University, School of Textile and Material Technology, the testing facility for this project, a certification board identifies products to be tested and evaluated, sets the standards to be met, and then certifies those products meeting and/or exceeding such standards. The certification board is made up of nationally recognized experts in the law enforcement community, Director of the NWTC, Program staff from Philadelphia University, and representatives from the National Institute of Justice and the Defense Logistics Agency. Similar to Consumer Reports, a report is then provided to law enforcement so that they may be assured of purchasing quality products.

Field evaluations of all products are conducted by the NWTC. Members of Team Extreme L.E., a group of law enforcement officers from across the country who are experienced (at various levels) in the art of mountaineering, Instructional staff, and even students, are used to evaluate the products in the field under various settings.

The program is funded through the tax-deductible donations of corporate sponsors and individuals from around the world. For information regarding the test and evaluation program, receipt of the quarterly reports, or to learn how you can make a donation to the program, you may contact either the NWTC (see prior listed information) or the National Tactical Officers Association at (800) 279-9127 or (215) 230-7616.

NWTC founder Mike Goldsmith

RECOMMENDED READING

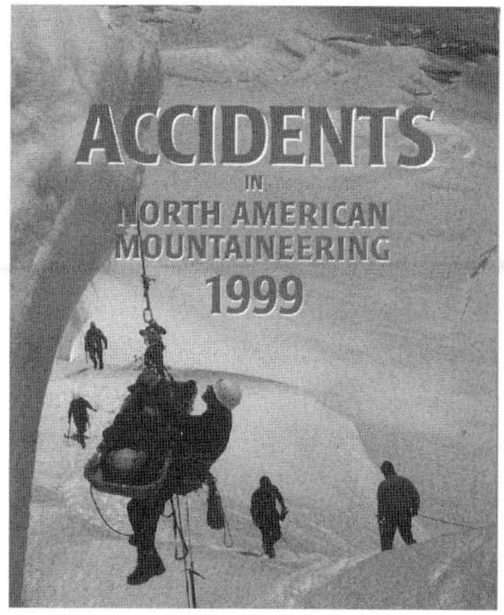

APPENDIX F

RECOMMENDED READING

MOUNTAINEERING - FREEDOM OF THE HILLS - 6TH Edition—The Mountaineers
CMC ROPE RESCUE MANUAL—James A. Frank—CMC
ACCIDENTS IN NORTH AMERICAN MOUNTAINEERING—American Alpine Club
HIGH ANGLE RESCUE TECHNIQUES—2ND Edition—Tom Vines & Steve Hudson - NASAR
THE AVALANCHE HANDBOOK—David McClung & Peter Schaerer
SNOW SENSE—Jill Fredston & Doug Fesler
GLACIER TRAVEL & CREVASS RESCUE—2ND Edition—Andy Selters
HELICOPTERS IN SEARCH & RESCUE OPERATIONS—Charley Shimanski—MRA
SEARCH & RESCUE FUNDAMENTALS—Emergency Response Institute
MORE CLIMBING ANCHORS John Long & Bob Gaines
ICE WORLD—Jeff Lowe
ALTITUDE ILLNESS: Prevention & Treatment—Stephen Bezruchka, MD
GOING HIGHER—4TH Edition—Charles Houston, MD
CMC RAPPELLING MANUAL—James Frank & Donald Patterson—CMC
CMC RESCUE FIELD GUIDE—3rd Edition—CMC
TECHNICAL RESCUE RIGGERS MANUAL—Rick Lipke - Conterra
MILITARY MOUNTAINEERING—TC 90-6-1
TACTICAL TRACKING OPERATIONS—David Scott-Donelan
FREE-HEEL SKIING—Paul Parker
BACKCOUNTRY SKIING By Jean Vives

Related interesting Reading:
INTO THIN AIR—John Krakauer
CLIMB: Tragic Ambitions on Everest—Anatoli Boukreev
EVEREST—Mountain Without Mercy—Broughton Coburn
ALL FOURTEEN 8,000ers—Reinhold Messner
ASCENT OF EVEREST—John Hunt
EPIC—Stories of Survival—Edited by Clint Willis
SOLDIERS ON SKIS—Whitlock & Bishop—Paladin Press

Many of these titles are available from:
ADVENTEROUS TRAVELER BOOKSTORE—800/282-3963; 802/860-6776
www.AdventurousTraveler.com
ALPEN BOOKS—425/290-8587
AMAZON.COM
BARNES & NOBLE.COM
CHESSLER BOOKS—800/654-8502; www.chesslerbooks.com
CLIMBING BOOKS—800/358-6327; Fax 413/662-2408
THE MOUNTAINEERS BOOKS—800/553-4453; www.mountaineersbooks.org
NASAR Bookstore—www.nasar.org—800/893-7788

ADDITIONAL TITLES BY MARK LONSDALE

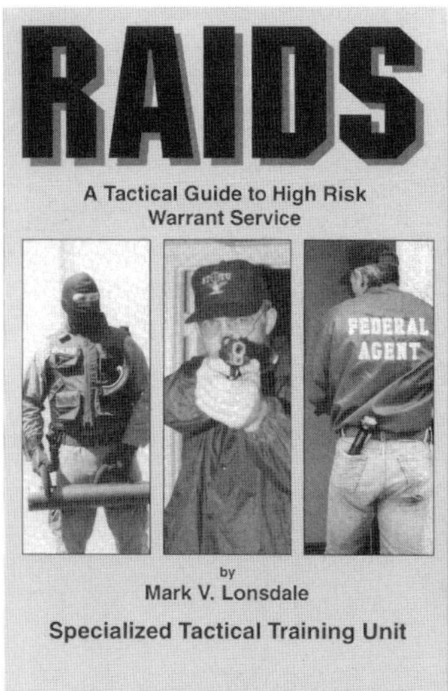

www.sttu.com
amazon.com or barnesandnoble.com

Recommended Magazines:

CLIMBING	www.climbing.com
ROCK & ICE	www.rockandice.com
COULOIR	www.couloir-mag.com
OUTSIDE	http://outside.starwave.com
BACKCOUNTRY	www.backcountrymagazine.com

Additional Titles By MARK LONSDALE Available From STTU:
ADVANCED WEAPONS TRAINING for Hostage Rescue Teams (4th Edition)
C.Q.B.—A Guide to Unarmed Combat and Close Quarter Shooting (6th Edition)
RAIDS—A Tactical Guide to High Risk Warrant Service (3rd Edition)
SNIPER COUNTER SNIPER—A Guide for Special Response Teams (4th Edition)
SNIPER II—An Advanced Guide to Special Operations Sniping (2nd Edition)
SRT DIVER—A Guide for Special Response Teams (Revised 1999 Edition)
BODYGUARD—A Practical Guide to VIP Protection (1st Edition)

www.sttu.com or FAX 310/829-0868
Email: sttubooks@aol.com

Notes: